Reconstruction of Cell Evolution: A Periodic System

Author

Werner Schwemmler
Apl. Professor
Freie Universität of Berlin
West Germany

CRC Press, Inc.
Boca Raton, Florida

Library of Congress Cataloging in Publication Data

Schwemmler, Werner, 1940-
  Reconstruction of cell evolution.

  Translation of: Mechanismen der Zellevolution.
  Bibliography: p.
  Includes index.
  1. Evolution.   2. Life--Origin.   3. Cellular control
mechanisms.   4. Cells--Evolution.   I. Title.
QH371.S3913   1984   575   83-26280
ISBN 0-8493-5532-X

Title of the original edition:

Werner Schwemmler
  Mechanismen der Zellevolution
    Grundriss einer modernen Zelltheorie.
  Walter de Gruyter, Berlin · New York 1979
  Copyright © 1979 by Walter de Gruyter & Co.

  Direct all inquiries to CRC Press, Inc., 2000 Corporate Blvd., N.W., Boca Raton, Florida, 33431.

© 1984 by CRC Press, Inc.

International Standard Book number 0-8493-5532-X

Library of Congress Card Number 83-26280
Printed in the United States

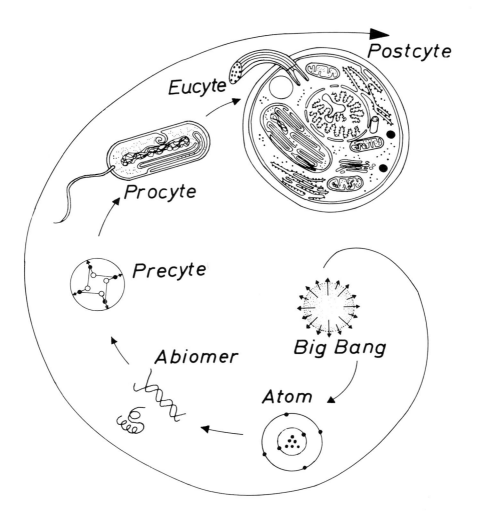

This book attempts for the first time to relate the mechanisms of cell evolution (ur-cells, bacterial cells, higher cells) to cosmic and chemical evolution. An outline of a modern cell theory is developed from the available data. One important aspect of this theory is the derivation of a periodic system of cell types. The analogies between this tentative periodic system of cells and the periodic system of the elements are discussed.

FOREWORD

The goal of this book is to make the whole complex of evolution in its cosmic, chemical, and biological dimensions understandable to the scientifically oriented reader. To this end, the available theoretical and experimental data are incorporated, and, qualitatively at least, their implications are thoroughly discussed. The transition from the quantitative to the qualitative, which is necessary for understanding the whole complex, is thus well defined.

The data are not merely presented in order, but in such a way that basic relationships become evident. This is most clearly expressed in the presentation of comprehensive systems and models. The phenomena of evolution are not only described and explained, but also systemized and arranged into a general concept of evolution. The essential criterion for such a concept is that it provides a least common denominator of logic for the facts and conclusions. The most significant result produced was the derivation of the periodicity of all evolved and evolving systems. Thus, this text not only caters to the need for comprehensive information on the process of evolution, it also offers a new and more complete understanding of evolution. It is the central goal of science to achieve ever more realistic, uniform models of the entire real world. In this sense, this book represents a modest contribution to theoretical biology. In view of the rising flood of data, there is a growing need for such unifying theory. The present inability to solve basic biological problems like cancer, cell differentiation, and endorhythms is not only due to a lack of technical ability or knowledge, but also to the lack of interdisciplinary evaluation of data and to a lack of unifying theory.

This book was developed parallel to a seminar "Evolution of the Cell", with the cooperation of a working committee recruited from the participants in the seminar. I wish to thank all the participants, in particular M. Berthold, M. Herrmann, B. Luuring, H. Mündelein, Ch. Manger E. Schütte-Arnst, S. Vieth, and U. Zabel for their many suggestions and for their constructive criticism. I thank Dr. M. Brewer and P. Bradish for translating the manuscript into English, Ch. Manger and M. A. Biemelt for typing the manuscript, and Ch. Dörgeloh, M. Jupe, G. Kemner, and H. Mattow for the drawings, and P. Holzner for the photos. I am indebted to a number of colleagues, in particular Prof. C. G. Arnold, Prof. C. Bresch, Dr. M. Brewer, Dr. R. Dierstein, Prof. G. Drews, Prof. M. Eigen, Prof. J. Fuchs, Dr. M. Grasshoff, Dr. W. Gutmann, Prof. F. Hinderer, Ass. Prof. A. Karpf, Prof. W. Kaplan, Prof. P. Karlson, Prof. H. Kuhn, Dr. D. Mollenhauer, Dr. D. Peters, Prof. H. Schenk, Prof. P. Sitte and many others for their critical evaluations and helpful discussions of the manuscript. Their criticism made me aware of the difficulty involved when a single individual attempts to present such complex and voluminous material. Along with encouragement, I received such sharp criticism that I was forced to develop the courage to "jump into the gap", for the benefit of the potential readers, and because of the ever more acute need for a comprehensive and unified presentation. I hope that the specialists in those fields where I have no special competence will, in light of the above-mentioned goals of this text, excuse any errors or imprecise formulations which may still remain, in spite of my efforts. I shall also be grateful for any verbal or written criticism of the present work, which may be taken into account in subsequent editions. Finally, thanks are due to the publishers, CRC Press, in particular Sandy Pearlman and Mary Kugler, for their help and their consideration with regard to the author's wishes.

W. Schwemmler

THE AUTHOR

Werner Schwemmler was born in 1940 in Offenbach, Germany (FRG). He studied biology, chemistry and geography from 1961 to 1966 at the Universities of Marburg, Giessen, and Freiburg. From 1968 to 1970, he held stipendia from the "Duisburg-Stiftung" and the "Max-Planck-Gesellschaft" to work with Prof. Vago at the Institute for Comparative Pathology and at the Institute for Invertebrate Pathology of Montpellier University in southern France. The author took his doctorate in 1972 with Prof. Sitte, Freiburg.

From 1972 to 1974 he held a research fellowship for "Habilitation" from the Deutsche Forschungsgemeinschaft (DFG) at the Microbiological Institute of Prof. Drews, Freiburg. In late 1972, he spent a short time at the University of Minnesota (Minneapolis) with Prof. Halberg and Prof. Brooks, with the financial support of the DFG and the "Freiburger Wissenschaftliche Gesellschaft".

From 1974 to 1980 Dr. Schwemmler was an assistant professor at the Freie Universität of Berlin. He took his "Habilitation" in zoology there in 1975 and is the head of an interdisciplinary group working on the physiology, ecology, genetics and evolution of cellular systems. He is author of the book "Mechanismen der Zellevolution. Grundriss einer modernen Zelltheorie", which has been published by De Gruyter, Berlin/New York. Also, he organized with Prof. H. Schenk an International Colloquium on Endosymbiosis and Cell Research, April 11—15, 1980, in Tübingen (FRG). They organized also the Second International Colloquium on Endocytobiology, April 10—15, 1983, in Tübingen. Dr. Schwemmler is co-editor of the proceedings for these colloquiums and co-founder of the new synthetic research area "Endocytobiology". In 1983 he was appointed ausserplanmässinger Professor at the Freie Universitat Berlin.

It is so easy to suggest no theories, with the excuse that the necessary basis of fact is not yet available. As if the theory did not first have to show where one should look for the facts!

These sentences are part of a letter written by the Freiburg zoologist **August Weismann** to the zoologist **Ernst Haeckel**, in Jena, after he had read the latter's treatise ''Die Gastraea-Theorie, die phylogenetische Classifikation des Tierreichs und die Homologie der Keimblätter''.

TABLE OF CONTENTS

Chapter 1

EVOLUTION RESEARCH: GOALS AND PROBLEMS

In the broadest sense, *evolution* is the process which has led from the origin of the universe to the rise of the modern world, including humanity and all other living beings. It can be divided into several phases, *cosmic,* chemical, biological,* and *cultural.* The most important junctions passed in the course of this evolution are the elementary particles, the atom, the cell, and human culture. The essential difference between these phases is the way in which information is stored.

The goal of evolution research is to reconstruct the entire sequence of this historical process, and thus to clarify the underlying mechanisms. In order to understand the nature of this undertaking, we must first explore the extent to which the concept "evolution" can be perceived or expressed in words. There are phenomena in the real world which are less complex than the senses and brain through which we perceive them. This is the world of atoms, and cells and their interactions, which are the main objects of research in the natural sciences. It is the task of *natural sciences* to examine these phenomena by means of *experiment* and to formulate the results precisely, objectively, and logically in terms of reproducible data (Figure 1).

There are other phenomena whose degree of complexity is roughly the same as that of our cognitive apparatus, such as the human being, human consciousness and culture. These are generally the subject of the *social sciences.* The criterium is *experience,* and the result is the assembling of data which are subjectively logical, but only reproducible within certain limits, and thus not amenable to exact formulation. Finally there must be phenomena which are more complex than our cognitive apparatus. Our senses are clearly limited and act as a filter which allows only a part of reality to our consciousness (e.g. we do not hear as well as bats, or see as well as birds of prey, or smell as well as dogs, etc.), quite aside from the limitations of our computer-like brain.[8] This more complex side of the real world is related to the position of humanity in the universe, and is dealt with by *metaphysics.* Metaphysical method is called *synopsis,* or consideration of the whole at once. Its imprecisely formulated statements, e.g. on telepathy, may be statistically only barely defensible, but they are no longer or not yet capable of logical formulation (Figure 2).

Evolution, the most complex and central phenomenon of our real world, belongs to all three areas, the natural sciences, the social sciences, and metaphysics. Therefore, only some aspects of evolution, such as the mechanics of molecular and cellular evolution, can be clearly understood, while other aspects, such as the psychic and intellectual achievements of man, can only be vaguely comprehended and still others, such as the origin and end of cosmic development, lie completely beyond causal explanation, at least for the present. A further problem is that the process of evolution, which has been going on for billions of years, is an historical phenomenon, which we can now only attempt to reconstruct on the basis of fossil and recent evidence. The criterion for the validity of such reconstructions is the lack of contradiction with the previously gathered data. The insight into the entire process of evolution is thus supported by much evidence, but can never be proven.

The lack of cognitive precision is compounded by the limitations of language, which can only describe the often simultaneous in the one-at-a-time of words. For this reason, the combination of language and visual expression in the form of tables and models is the best which can be achieved in representing the cognitively derived complex of evolution.

---

* Since evolution in the Darwinian sense did not occur in this phase, it is better to speak here of cosmic development.

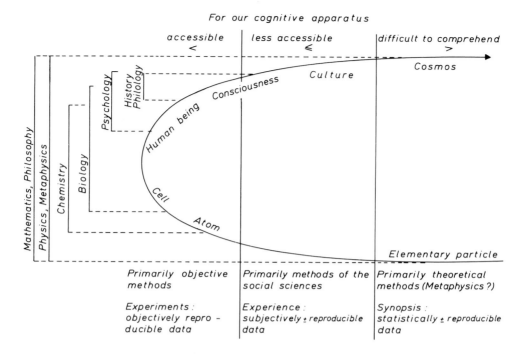

FIGURE 1.   Correlation between the degree of complexity of the human cognitive apparatus, methods of examination, and forms of knowledge about the real world (for details, see text).

The central *theory of biological evolution,* which is now generally accepted, was proposed by the Englishman Charles Darwin ("On the Evolution of Species through Natural Selection", 1859). Darwin deduced the theory for the biological phase of evolution, but its applicability to the chemical phase of evolution has, in the meantime, become apparent. The mechanism of evolution is thus the same in the chemical and biological phases. The different variations (mutations) of those members of a population which are best adapted to their internal and external environment are the most likely to survive (be selected) in the competition for existence:

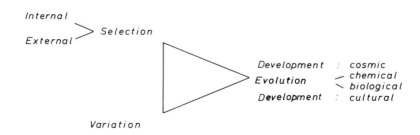

The general mechanism of the whole process of evolution is the modular principle.[20] Each level of evolution is reached through combination of different representatives of the next lower level, each representative being best adapted to a different milieu (Figure 3). How and when such developmental leaps occur is a matter of chance. That they occur, however,

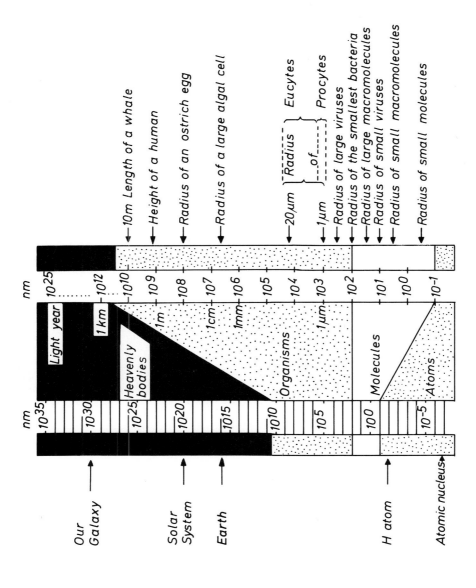

FIGURE 2. Size comparison of the individual evolutionary systems on a logarithmic scale (adapted from Laskowski and Pohlit[12]). 1 nm (nanometer) = 10^{-9} m = 10 Å.

PHASE	MODE OF DEVELOPMENT			COSMIC MILIEU	PERIOD	
Cosmic Development	*?*	*+ ?*	*→Elem.particle*	Big bang	*<-8*	billion years ago
	Elem.particle	*+Elem.particle*	*→Atom*	Galaxy	*<-7*	
	Atom	*+Atom*	*→Molecule*	Solar system	*<-6*	
Chemical Evolution	*Molecule*	*+Molecule*	*→Abiomer*	Earth	*<-5*	
	Abiomer	*+Abiomer*	*→Precyte*	Primal soup	*<-4*	
Biological Evolution	*Precyte*	*+ Precyte*	*→Procyte*	Ocean bottom	*<-3,5*	
	Procyte	*+Procyte*	*→Eucyte*	Ocean surface	*<-2,5*	
	Eucyte	*+Eucyte*	*→ Human*	Land	*<-0,01*	
Cultural Development	*Human*	*+ Human*	*→ Culture*	Space	*- 0,001*	
	Culture	*+Culture*	*→ ?*	Death of the solar system	*< +6*	

FIGURE 3.   Schematic representation of evolutionary processes. Evolution can be divided into cosmic, chemical, biological, and cultural phases. Major milestones along the way are the elementary particles, the atom, the cell, and consciousness of culture. Each higher level of evolution was formed by the combination of representatives of the next lower level (from Schwemmler[19]).

is the result of a causal, highly probable process. To put it another way, this means that the systems obtained through variation and selection are not predetermined with respect to their individual structures, but that they arise in a necessary result of evolution, or a law of nature. This is just as true for the formation of elementary particles as for atoms, cells, humans and cultures.

In the following, this process will be described in general terms, with special attention to the evolution of cells. Where data are totally lacking, we fall back on deductive models, which can only be suggested within the framework of that which is already known. This framework is provided by monophyly, or the historical relatedness of all evolved systems, which demonstrates the universality of the evolutionary process. Thus evolution research always means the reconstruction of the causal chain of systems, based on the deduced mechanisms of evolution (e.g., Darwin[5], Gutmann[9], Peters et al.,[14] Bonik et al.,[2] and Schwemmler.[20] The result is here the foundation of the modern cell theory.

REFERENCES

1. **Ayala, F. J.,** Mechanismen der Evolution, *Spektrum der Wissenschaft,* 5, 9, 1979.
2. **Bonik, K., Grassmann, M., and Gutmann, W. F.,** Funktion bestimmt Evolution, *Umschau,* 77(20), 657, 1977.
3. **Bresch, C.,** *Zwischenstufe Leben: Evolution ohne Ziel?* Piper Verlag, Munich, Zurich, 1977.
4. **Capra, F.,** *Der kosmische Reigen. Physik und östliche Mystik — ein zeitgemässes Weltbild.* 5. Aufl., Otto Wilhelm Barth Verlag, Bern, Munich, Vienna, 1981.
5. **Darwin, C.,** *Die Entstehung der Arten durch natürliche Zuchtwahl.* 6. Aufl. 1972. Reclam jun.; Stuttgart, 1963.
6. **Ditfurth, von H,** *Wir sind nicht nur von dieser Welt. Religion und die Zukunft des Menschen,* Hoffmann und Campe, Hamburg, 1981.
7. **Dobzhansky, T.,** *Evolution,* W. H. Freeman, San Francisco, 1977.
8. **Hubel, D. H.,** Das Gehirn, *Spektrum der Wissenschaft,* 11, 37, 1979.
9. **Gutmann, W. F. and Peters, D. S.,** Konstruktion und Selektion: Argumente gegen einen morphologisch verkürzten Selektionismus, *Acta Biotheoretica,* 22(4), 151, 1973.
10. **Kimura, M.,** Die ''neutrale'' Theorie der molekularen Evolution, *Spektrum der Wissenschaft,* 1, 94, 1980.
11. **Kull, U.,** *Evolution,* J. B. Metzler, Stuttgart, 1977.
12. **Laskowski, W. and Pohlit, W.,** *Biophysik.* Bd. I, II. Georg Thieme Verlag, Stuttgart, 1974.

13. **Osche, G.**, *Evolution: Grundlagen, Erkenntnisse, Entwicklungen der Abstammungslehre.* 3. Aufl. 1974. Herder Verlag, Freiburg, Basel, Vienna, 1972.
14. **Peters, D. S., Franzen, J. L., Gutmann, W. F., and Mollenhauer, D.**, Evolutionstheorie und Rekonstruktion des stammesgeschichtlichen Ablaufes. *Umschau*, 74(16), 501, 1974.
15. **Prigogine, I. and Stengers, I.**, *Dialog mit der Natur. Neue Wege naturwissenschaftlichen Denkens.* 2. Aufl., Piper Verlag, Munich, 1981.
16. **Rechenberg, I.**, *Evolutionsstrategie. Problemata-Reihe.* Fromann-Holzboog Verlag, Stuttgart, 1973.
17. **Riedel, R.**, *Die Ordnung des Lebendigen: Systembedingungen der Evolution,* Paul Parey Verlag, Hamburg, Berlin, 1975.
18. **Riedel, R.**, *Biologie der Erkenntnis. Die stammesgeschichtlichen Grundlagen der Vernunft.* 2. Aufl., Paul Parey Verlag, Hamburg, Berlin, 1980.
19. **Schwemmler, W.**, Allgemeiner Mechanismus der Zellevolution. *Naturwiss. Rdschau*, 28(10), 351, 1975.
20. **Schwemmler, W.**, *Mechanismen der Zellevolution. Grundriss einer modernen Zelltheorie,* De Gruyter Verlag, Berlin, New York, 1978.
21. **Sieving, R.**, *Evolution.* Gustav Fischer Verlag, Stuttgart, 1978.
22. **Timofeev-Ressovskij, N. V., Voroncov, N. N., and Jablokov, A. V.**, *Kurzer Grundriss der Evolutionstheorie,* Gustav Fischer Verlag, Jena, 1975.

Chapter 2

EVOLUTION OF ATOMS

The formation of atoms is a side reaction in the process of cosmic development, the formation of the universe. The discipline concerned with this process is called cosmology, and it comprises three main areas of investigation:

- Structure and form of space
- Formation and development of the objects present in the universe
- Age of the cosmos and the course of its expansion.

These are examined with the current physical and mathematical methods, such as telescopy, spectral analysis, gravitational and magnetic field measurements, geometry, kinematics, etc. Cosmology can only bear fruit through an interplay of theory and observation. A fundamental difficulty of cosmology is that observation is subject not only to technical limitations, but also to those imposed by the very nature of the universe, while the theory should ideally encompass the universe as a whole.

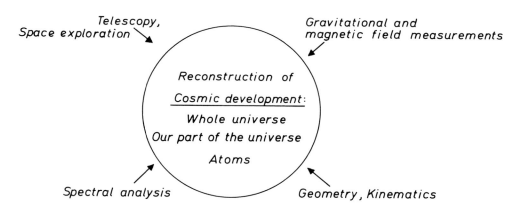

I. THE MILKY WAY

It is difficult, given the unimaginable size of the universe, to generalize about the entire complex. Our knowledge is limited to those areas of the cosmos which we can observe directly, which have a radius of about 5 billion light years.* This we can simply call *our observable part of the universe* (Table 1, Figure 1). At the limits of the observable universe we can just make out the quasars (*quasi*-stell*ar* objects), which have a diameter of 2×10^{18} km and radiate extremely powerful radio waves. They are thought to be young, very distant galaxies. Quasars move through the cosmos with a velocity of 250,000 km/sec, or 84% of the speed of light. Within the limits of the observable universe we can see several thousand spiral nebulae, or *galaxies,* which are collected together in galactic clusters. The galactic cluster nearest to us contains about 20 smaller galaxies, including the two Magellanic Clouds, the Andromeda galaxy, and the Milky Way, which is about 1.8×10^{19} km from the Andromeda galaxy.

---

* A *light year* is the distance traveled by light in one year at a velocity of 300,000 km/sec; this is equal to 9.5 $\times 10^{12}$ km.

Table 1
ASTRONOMICAL DISTANCES

Astronomical numbers	Diameter (Ø) Distance (↔)	Kilometer (km)	Time required for light to travel the distance (velocity of light about 300 000 km/sec)
Interplanetary space	Ø Moon	3 500	<0.01 sec
	Ø Earth (at the equator)	12 756	0.04 sec
	Earth ↔ Moon (average)	384 400	1.28 sec
	Ø Sun	1 500 000	5 sec
	Earth ↔ Sun (average)	149 565 800	8 min
	Sun ↔ Pluto	7 350 000 000	6 h
Interstellar space	Sun ↔ Nearest star	40 680 000 000 000	4.3 years
	Ø Globular cluster	1 000 000 000 000 000	105 years
	Ø Solar system	130 000 000 000 000 000	about 14 000 years
	Sun ↔ Center of the galaxy	236 500 000 000 000 000	about 25 000 years
	Ø Disc of the galaxy	946 500 000 000 000 000	100 000 years
	Ø Halo of the galaxy	1 600 000 000 000 000 000	170 000 years
Intergalactic space	↔ Quasar	2 000 000 000 000 000 000	210 000 years
	Milky way ↔ Andromeda galaxy	21 285 000 000 000 000 000	2 250 000 years
	Milky way ↔ Quasars (Limit of the observable universe)	50 000 000 000 000 000 000 000	5 250 000 000 years
	Distance traveled by the cosmic background radiation since the Big bang	about 100 000 000 000 000 000 000 000	about 10 500 000 000 years (Age of the universe)

The *Milky Way* is a galaxy consisting of a nucleus and spiral arms, the disk (which is about 10^{18} km in diameter), and the halo (which is a spherical collection of globular clusters, single stars, and interstellar gas and dust), with a diameter of 1.6×10^{18} km. The galaxy as a whole contains thousands of gas and dust nebulae, about 400 globular clusters, and 100 billion single stars. One of these single stars, which lies toward the edge of the disk, is the sun, with a diameter of 1.5×10^6 km. It is about 2.4×10^{17} km from the center of the galaxy and revolves around it with a velocity of 250 km/sec and a period of about 250 million years. The sun, in turn, is encircled by nine planets. One of these is the *earth*, with a diameter of 1.3×10^4 km, a velocity of 30 km/sec, and a period of one year. Satellites revolve around the planets with a velocity of about 1 km/sec and a period on the order of months. The earth has only one satellite, the moon, with a diameter of 3.5×10^3 km. The moon and earth together make up less than 1% of the total matter in the solar system. In comparison to the universe, our planet is infinitesimally small.

It happens that we can draw upon direct observation of conditions as they must have been in the course of our own planet's development. This is due to the fortunate circumstance that light emitted by various celestial objects is always under way for a certain length of time before we can observe it with our telescopes. If we observe stars which are 1 billion light years away, then the light reaching us now was radiated 1 billion years ago, and represents conditions as they were at that time. Using these directly observed data and extrapolating from them, cosmologists attempt to reconstruct the formation of our world.

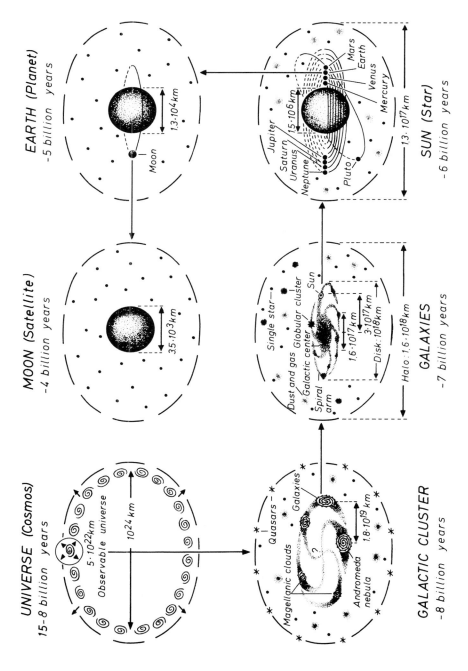

FIGURE 1. Hierarchy and developmental phases of the *cosmic systems* in the observable portion of the universe, from the galactic clusters to galaxies, globular clusters, stars, planets, and their satellites. This representation of the universe should not be taken as a three-dimensional model (details, see text). The schemes are not drawn to scale with respect to each other.

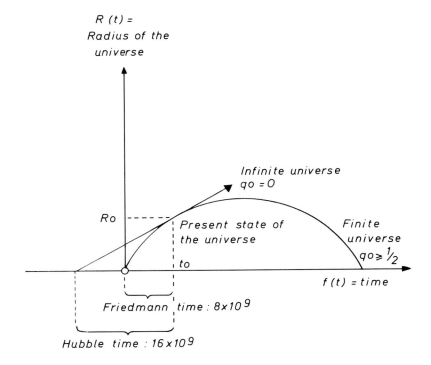

FIGURE 2.   Relativistic model of a *closed, finite universe* with an age of about 8×10^9 years (Friedmann time): t signifies time, t_0 the present, R(t) the radius as a function of time, which describes the expansion of the universe, R_0 is the present radius of the universe, q the expansive acceleration of the celestial bodies, and q_0 their present acceleration. The tangent to the R(t) curve represents the model of an *open, infinite universe*, with an age of 16×10^9 years (Hubble time). It is obtained by linear extrapolation of the expansion equation, according to which the velocity of the celestial bodies increases in proportion to their distance. Neither model can be proven, since the physics of the very large, like that of the very small, must rest upon unprovable basic assumptions (adapted from Dehnen[6]).

It is believed that over 8 billion years ago, all the matter of the cosmos was concentrated in a very small space. Its estimated density, 10^{14} g/cm³, was many orders of magnitude greater than the present cosmic density of 10^{-30} g/cm³, which corresponds to only three to four hydrogen atoms per cm³. An event which is called the *big bang* by cosmophysicists must have heated the matter to trillions of degrees (2×10^{12}K*), giving it an expansive velocity close to that of light (Figure 2). The matter expanded spherically and homogeneously in a shock wave, so that its density decreased rapidly. After about 1/2 billion years, as the temperature dropped to about 4000K (3700°C), and the radiation pressure sank correspondingly, a critical point was reached. As the expansion slowed down, the attraction force of gravity began to outweigh it locally. Local inhomogeneities in the matter appeared, similar to turbulence in fluids, and initiated the formation of primitive nebulae. These began to rotate as they condensed, so the gaseous matter was now moving in two modes, the original expansive mode and the local rotational mode. An unknown number of local turbulences arose in the cloud of gas, evenly distributed in space. These served as condensation nuclei for galactic clusters,** and later for about 10 billion galaxies with approximately 2×10^{11}

---

* Zero on the Kelvin scale lies at the absolute or thermodynamic zero of temperature (-273.15°C). The freezing point of water under atmospheric pressure is 273.15K (0°C).

** The theory of the development of the galactic cluster is still being discussed and not yet conclusively proven to be correct.

stellar masses each. These rotating islands of gas were held together by the gravitational forces of their total material content. However, the gravitational force, with its tendency to contract the gas masses, was opposed by the centrifugal forces of rotation. The result was that the gas spheres could only contract in one direction, namely along the axis of rotation. In this way the rotating disks of the spiral galaxies, with greater densities of matter in the center, arose about 7 billion years ago.

Within each galaxy, the continued concentration of matter produced still smaller centers of turbulence: the globular clusters (about 100,000 per galaxy) of about 10^{12} km in diameter and containing 10^5 to 10^7 stellar masses each. The rotational velocity increased, due to conservation of angular momentum, as the radius of the gas nebula decreased. However, radiated electromagnetic fields (background radiation) allowed the rotational angular momentum to be dissipated, so that the process of material concentration could continue without being outweighed by centripetal force. Ever smaller centers of turbulence continued to condense out of the gas and dust cloud to form celestial objects with ever increasing concentrations of matter. Starting about 6 billion years ago, the stars began to condense out of the globular clusters or directly out of the gas and dust clouds. Starting 5 billion years ago, the planets of our solar system with their atmospheric shells began to form (estimated age of the earth: 4.7 billion years) out of the sun, which was a star with only a little remaining gas and dust surrounding it in elliptical clouds. Starting about 4 billion years ago, the satellites began to form out of the planets.* The system of celestial bodies is thus a hierarchy consisting of at least five systems, formed in an analogous manner: galaxies → globular clusters → stars → planets → satellites.

The satellites revolve about the planets, the planets move more rapidly around the stars, the stars about the galactic centers, the galaxies still more rapidly about the possible centers of the galactic clusters, and all these systems are moving apart linearly at an ever increasing velocity. The combination of these many motions, linear and circular, centrifugal and centripetal, faster and slower, produces the apparent confusion with which the celestial bodies move through the universe.

Up to this point, our description has been based only on the observed conditions at present, namely the radial expansion of the celestial bodies in the observable universe, the matter contained in them, and the velocities of the objects, and on the laws of physics and the strict extrapolation of these laws. This is, however, the limit of the observable. Beyond this limit there are, for the present, only mathematical postulates on the nature and development of the universe.

II. THE UNIVERSE

Space, time, and matter are summarized in the concept of the cosmos. The cosmos is described in a system of at least four coordinates: the three coordinates of three-dimensional space or matter, and one coordinate for time. As far as we can "see", space is evenly filled with matter, and the expansion of matter with time is also identical everywhere. If we generalize this finding for the whole of cosmic space, then it is homogeneous and isotropic.** To be sure, this is only statistically true, because on a smaller scale, space is neither homogeneous nor isotropic, due to the existence of rotating galaxies. Based on the assumption of statistical homogeneity and isotropy, and also on the basis of other criteria which will not be discussed here further (e.g. the *luminosity-red-shift relationship**** for galaxies), it

---

* According to another theory, at least some of the satellites are small planets which were "captured" by larger ones.

** This proposal as to the isotropic nature of cosmic space has come under fire in recent times.

***All terms whose meaning is not obvious from the context are defined in the glossary (Appendix II).

follows that the universe must be finite, since otherwise the celestial bodies would have to attract each other with infinitely great force, which is not the case. According to Albert Einstein's theory of general relativity (1916), one can presume that space is *positively curved*, and thus closed or finite (Figure 2). However, for the sake of completeness, it must be mentioned that there is also evidence for a flat or even *negatively curved*, and thus an open, infinitely large universe. It has not yet been definitely decided whether the universe is positively or negatively curved, even though a positive curvature would be more attractive from an evolutionary point of view, since only this kind of space would be capable of periodic expansion and contraction, and periodicity is a phenomenon which is observed in all areas of our world (Chapter 9, III). The concept of a periodically expanding and contracting universe was formulated by Hughes in 1863 *(pulsation theory)*. In the following we will examine more closely the properties of positively curved, periodically variable spaces.

In such spaces, parallel lines intersect each other at a certain point, and lines are closed upon themselves. To make this idea clearer, one can imagine walking on the (two-dimensional) surface of a sphere in a straight line. One would eventually return to the starting point. The universe can be compared with the surface (not the interior) of an expanding three-dimensional balloon in a four-dimensional space* (the present diameter of space may be more than 10^{23} km).

The universe is thus not bounded and ends nowhere, but its space is not infinite and has a finite volume, which, however, is constantly increasing by expansion. In the model of a finite universe, the radius of the universe at the time of the cosmic origin, or the big bang $(t = 0)$, and also at the end of its contraction phase, approaches zero. The universe is thought to have expanded from a point of minimal radius and will, according to the pulsation theory, eventually collapse into a point of minimal radius. Space, time, and matter came into existence from nothing and will disappear again into nothing. Many cosmic physicists see in this concept an elegant solution to the ancient debate over whether the universe is finite or infinite. However, it is nearly as difficult for us today to imagine a positively curved, periodically variable universe as it once was for people to imagine a round, rotating earth with its inhabitants "down under".

How can one imagine the development of such a pulsating universe? Let us begin with the phase of minimal diameter, and thus maximal concentration of matter (Figure 3). This phase begins when the gravitational attraction of matter (in the language of Newtonian mechanics, the potential energy, E_{pot}) outweighs the expansive inertia (in Newtonian mechanics, the kinetic energy, E_{kin}). This can only occur if the material density of the universe is greater than the critical average density of 5.7×10^{-30} g/cm³, a condition which is still being examined by the astronomers. If this condition is fulfilled, and our universe is closed, then in the course of further expansion, the previously described instability must eventually occur. Then the matter would contract, slowly at first, and then ever more rapidly, until it finally collapsed into a minimal space. This collapse "heats" the matter enormously (all energy converted to E_{kin}), leading to a gigantic explosion, the so-called big bang. The result of the explosion is a linear, expansive momentum of the matter *(big bang phase)*. The matter now spreads out evenly on all sides as a homogeneous gas. In this phase, gravitational forces play no role in the structure of the gas cloud. However, as the expansion gradually diminishes, gravitation becomes important. As described in Chapter 2, I, local areas of greater density, or turbulences arise, which then give rise to the galaxies *(phase of galactic formation)*. The galaxies continue to recede from each other in all directions. This is the phase in which the universe seems to be at present. As the universe ages, the attractive forces gradually begin to outweigh the repulsive forces. Eventually the negative contribution of the gravitational

---

* The fourth dimension should be thought of as a fourth spatial dimension (not time) which we could only experience indirectly.

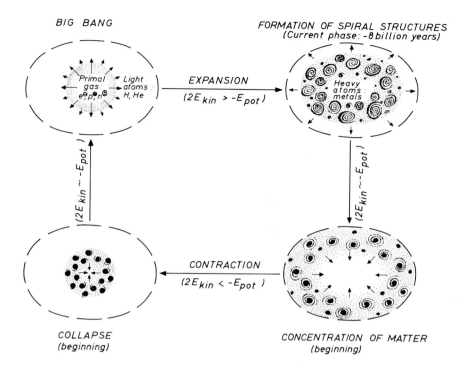

FIGURE 3. Hypothetical scheme of a *developmental cycle for the entire universe*, including the formation of light and heavy atoms, based on the pulsation theory. The representation of the univers? is not intended to be a three-dimensional model. It is based rather on the concept of a three-dimensional balloon expanding in a four-dimensional space (details, see text).

force, or the potential energy E_{pot} exceeds twice the contribution of the expansive inertia or the kinetic energy E_{kin} ($2 E_{kin} < - E_{pot}$). The point of maximum diameter and minimal matter density is reached. The matter begins gradually to become concentrated again, and the radius of the universe to shrink *(phase of material concentration)*. This then leads at last to another rapid accumulation of matter *(collapse phase)* followed by another explosion, which ushers in a new developmental cycle.

The total amount of matter and energy in such a cycle remains constant and can be calculated. One can also calculate the length of such a cycle from the Einstein equations (about 80×10^9 years; the present cycle is about 7.5×10^9 years old).

In view of the enormous dimensions of time and space, one must doubt that our planet has any special place in the universe. The question of whether there is life, or even intelligent life, on other planets belonging to other stars immediately rises. The answer is practically certain, in light of what has been said above about the structure and development of the universe. According to the hypothesis presented above, the process of planet formation must have occurred for all stars with up to about two solar masses. There are about 50 billion such stars per galaxy, of which about 10 billion are similar enough to the sun with respect to mass, radius, and temperature to possess earth-like planets. In the total of 10 billion galaxies in the universe, there must therefore be about 10^{20} earth-like planets. It also must follow that organic life, and in some cases, even intelligent life, would have developed on these earth-like planets.

Up to this point, it seems justifiable to invent hypotheses or to speculate about the structure and development of the universe. We must realize, however, that there will always be ultimate questions about the origin of matter which cannot be answered by scientific methods, and that there are paradoxical questions, such as "what was there before the beginning" or "what is there outside the universe" which it is not sensible to pose. In our exploration of the largest dimensions of space, and also of the smallest dimensions of atoms, we come upon limits to our understanding which cannot, at least at present, be overcome (see Figure 1).

III. ATOMS

Having first followed the large-scale changes in the form of matter in the course of cosmic development, we must now consider the development of its fine structure, which is now reasonably well understood. Let us begin again with the big bang. Maximal temperatures of 2×10^{12} K were reached as a result of the big bang. Atoms are not stable at such high temperatures; they dissociate into their components, i.e. into elementary particles. At this stage of cosmic development, there existed a primitive gas consisting of all possible free elementary particles* in thermodynamic equilibrium, and at extremely high temperatures and densities, as is still postulated today in the theory of elementary particles. Thus, in this primitive gas, all elementary particles were in thermodynamic equilibrium, that is, they collided often enough with each other to produce or destroy each other continually. However, as the gas expanded after the explosion, the density and temperature fell rapidly. Within the first 30 seconds, the temperature fell to about 4×10^8 K, within a year to 2×10^6 K, and at the end of the rapid expansion phase after 1 million years, it had fallen to 2×10^3 K. For some of the particles (neutrinos or photons), the collisions were not frequent enough to allow them to reach equilibrium with the other particles. They became "uncoupled" from the rest of matter and cooled independently (to about 3 K). They form the background radiation which can now be measured at 2.7 K.

In the first 30 minutes, the temperature was above 10^{10} K, which is high enough for nuclear fusions to take place.[15] Only during this short span did the energy and density conditions allow for sufficiently frequent collisions of elementary particles to yield atomic nuclei. The first step was the formation of Helium (He) from protons (hydrogen nucleus: H^+) by way of the fusion of two nuclei. The primitive gas was thus transformed into a gas of light atoms, consisting of 70 mass per cent hydrogen, 30 mass per cent helium, and traces of other elements. The atomic gas was cooled in this condition.

Further development of the *heavy atoms* proceeded only during the formation of the galaxies between 8 and 7 billion years ago, when matter was again brought to extremely high temperatures and densities. As the galactic clouds collapsed toward their centers, the mass became concentrated into the variously sized celestial bodies. The gravitational energy released in this process heated the bodies until the hydrogen gradually began to fuse into helium in the nucleus of the newly formed stars. This process continued until about 40% of the hydrogen in the star had been consumed. Then the star contracted still more and thus produced temperatures greater than 10^8 K. At these high temperatures, the helium nuclei began to fuse. Three helium nuclei fused to form carbon (C), four helium nuclei produced oxygen (O), and five helium nuclei formed neon (Ne). The energy released by the nuclear fusion led in turn to such enormous increases in temperature that the stars sometimes exploded, producing new and heavier atoms in the process. At 8×10^8 K, magnesium atoms (Mg) were formed; at 1.5×10^9 K, aluminum (Al), silicon (Si), sulfur (S), and phosphorus (P) were formed; and at 2.4×10^9, titanium (Ti), chromium (Cr), manganese (Mn), iron (Fe), cobalt (Co), nickel (Ni), copper (Cu), zinc (Zn), and the other heavy atoms were formed. At present, 80% of the matter in the universe is hydrogen, and hydrogen and helium together make up 99.9% of the matter, so that the other elements account only for 0.1% (Table 2).

---

* The particles which are most important, because they are the most stable, are the electron e^-, the proton p^+, the neutron $n^\pm$, the neutrino ν^0 and the photon γ^0.

Table 2
RELATIVE FREQUENCIES OF THE ELEMENTS IN THE UNIVERSE[4]

For	*1 000 000 atoms*	*hydrogen*
there are	*160 000 atoms*	*helium*
	700 atoms	*oxygen*
	600 atoms	*neon*
	300 atoms	*carbon*
	100 atoms	*nitrogen*
	30 atoms	*silicon*
	20 atoms	*magnesium*
	10 atoms	*iron*
	5 atoms	*argon*
	2 atoms	*sodium*
	2 atoms	*aluminium*
	2 atoms	*calcium*
and less than	*1 atom*	*all other elements*

From Bresch.[4]

The fate of celestial bodies depends and depended on their size. The smallest ones, like the moon, cooled further and further in the course of billions of years until they finally reached the temperature of the surrounding space, 3 K. In the larger celestial objects, two forces are balanced against one another: gravitational forces, which would cause the material to collapse, and the pressures produced by nuclear fusion and the associated extreme temperatures. When the lighter nuclei have been consumed by fusion reactions so that the expansive force decreases, then the gravitational force becomes dominant and causes the star to collapse. If the total mass of the star is less than a certain critical mass (M_2) then it flares up to become a *red giant** before subsiding into an extremely dense *white dwarf** with about the mass of the sun, but a diameter of only a few kilometers. On the other hand, if the original mass (M_1) is greater than M_2, the star collapses and explodes, by a mechanism similar to that producing the big bang, and what remains is either a *black hole,** a *neutron star,** or perhaps a *pulsar.** Our sun, a medium-sized star, can be expected to expand into a red giant so that all life on our planet will be extinguished by heat, before it finally goes out. This, however, should not happen for about 6 billion years (Figure 4).

IV. MECHANISMS OF COSMIC DEVELOPMENT

There are typical mechanisms involved in cosmic development which can be generalized. After the big bang, the system of celestial bodies** (and atomic systems), which was (is) locally highly organized, developed from the relatively great disorder of the primal gas cloud. According to the Second Law of Thermodynamics, however, a closed system tends toward the state of highest entropy or disorder. In the case of cosmic development, the increase in order is achieved at the expense of gravitational potential energy. The decrease in the entropy of the system due to the increase in order is compensated by the increase in

---

* Definitions of terms not explained in the text are included in the glossary.
** It should be mentioned that, according to another concept, the structure of a planetary system with its harmonic orbits is comparable to the shell structure of an atom (compare Figure 1 with Chapter 9, Figure 4). This idea has its roots in the "Harmony of the Spheres" of Johannes Kepler (1571—1630).

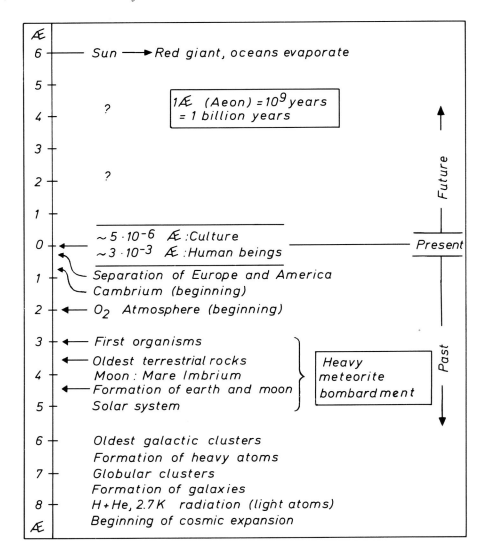

FIGURE 4.   Time scale of the presumed course of development of the universe and the earth (adapted from Dehnen[6]).

entropy associated with the dispersal of gravitational energy. The endogenic properties of atoms, such as electron charge and spin, play essentially no part in cosmic development and can be neglected in comparison to the exogenic force of gravitation. These properties became important later, during chemical evolution, when the capacity of molecules for self-organization came into play (see the next chapter).

If one compares total cosmic development from the exploding primal cloud through the local concentration of matter to the final collapse of matter (big bang) into itself with the development of individual cosmic systems from hot, gaseous spiral nebulae to stars and their gravitational collapse, it becomes apparent that the individual systems more or less completely follow the essential steps of the entire system. In both cases, finely distributed matter is slowly concentrated into a small volume, until the release of rest mass energy causes it to explode (Figures 1 and 3).

REFERENCES

1. **Barrow, J. D. and Silk, J.,** Die Struktur des Universums, *Spektrum der Wissenschaft,* 6, 79, 1980.
2. **Bergamini, D.,** *Das Weltall,* Time-Life Buch, Taschenbuch Rowohlt, Hamburg, 1975.
3. **Brandt, S.,** Elementarteilchen mit Charme, *Naturwissenschaften,* 64, 229, 1977.
4. **Bresch, C.,** *Zwischenstufe Leben; Evolution ohne Ziel?* Piper Verlag, München, Zürich, 1977.
5. **Breuer, R.,** Der Tod der schwarzen Löcher, *Umschau,* 76(15), 495, 1976.
6. **Dehnen, H.,** Die Struktur der Welt, *Umschau,* 76(7), 209, 1976.
7. **Einstein, A.,** Zur Electrodynamik bewegter Körper, *Ann. Phys.,* 17, 891, 1905.
8. **Herrmann, J.,** *Astronomie,* dtv-Atlas, München, 1973.
9. **Hughes, J. L.,** Possible application of the mass-energy cycle on a micro- and macroscopic scale in the universe, *Nature,* 197, 441, 1963.
10. **Karpf, A. D.,** *Struktur der Elementarteilchen,* Universitätsverlag, Konstanz, 1976.
11. **Laskowski, W.,** *Der Weg zum Menschen,* De Gruyter Verlag, Berlin, 1968.
12. **Meier, D. L. and Sunyaev, R. A.,** Galaxien im Frühzustand, *Spektrum der Wissenschaft,* 1, 23, 1980.
13. **Schotte, U.,** Quarks mit Charme und Farbe, und andere neue Ideen in der Elementarteilchenphysik, *Naturwiss. Rdschau.,* 29(6), 204, 1976.
14. **Unsöld, A.,** Kosmische Evolution, *Naturwiss. Rdschau.,* 28(1), 3, 1975.
15. **Weinberg, S.,** *Die ersten drei Minuten,* R. Piper Verlag, München, 1977.

Chapter 3

EVOLUTION OF MOLECULES

The evolution of molecules was a direct continuation of the evolution of atoms. Chemical evolution on the earth was a part of this universal process. Its course is presently being reconstructed, using the methods of cosmology, geology, paleochemistry, and molecular biology. The goal of this research is first to reconstruct the development of the earth and then to simulate the conditions of the primitive earth, in order to repeat some aspects of chemical evolution in vitro. Stanley Miller introduced this line of research with his classical experiments in 1953 (Figure 1).

The efforts of a number of research groups have led to the simulated abiogenic synthesis of nearly all the basic molecules of the living, or biogenic world. Such *micromolecules,* the so-called abiomonomers, include amino acids, nucleobases, sugars, and lipids. Even more complex molecules have been artificially synthesized from these abiomonomers. They can be called abiooligomers or abiopolymers such as proteins, nucleic acids, saccharides, and lipids *(macromolecules).* The general process of chemical evolution can be reconstructed, at least in outline form, from the above data. This reconstruction is now generally accepted (Figure 2), and is in agreement with all the pertinent data.

I. PRIMITIVE EARTH

Like the present-day earth, the primitive earth was divided into three regions, the lithosphere (lithos, Greek = stone; sphaira, Greek = sphere), the hydrosphere (hydor, Greek = water) and the atmosphere (atmos, Greek = air). The *lithosphere* was formed in the same way as other smaller celestial bodies, as described in Chapter 2, III. The gas of light and heavy atoms formed by nuclear fusion began to condense about 5 billion years ago to form the earth. The gravitational energy released in the process heated the planet sufficiently to produce a glowing, fluid ball. As it cooled, the components became stratified. The heavy atoms of nickel (Ni) and iron (Fe) collected in the center to form the hot, fluid *nickel-iron core of the earth* (Figure 3). It still has a temperature of about 4000 K and a radius of about 3300 km. Around the core there is a fluid-to-solid *mantle,* about 3000 km thick, consisting principally of heavy magnesium silicates with admixtures of other atoms. Material from the mantle may still reach the surface as magma from volcanic eruptions. The magma must penetrate a 10 to 40 km thick *crust,* which is cold and solid, and forms the outermost layer of the earth. The composition of the earth's crust is varied. The dominant components are the relatively light, hydrated basalts, granites, and minerals which contain aluminum (Al), iron (Fe), sodium (Na), and potassium (K).

After the crust had solidified and cooled, the *hydrosphere* began to form (Figure 3). After the temperature of the surface had cooled below the boiling point of water (depending on the pressure, <373 K ≈ 100°C), the water vapor in the atmosphere condensed out in mighty downpours of rain. Upon reaching the warm surface of the earth, some of the water re-evaporated and fell again as rain. Slowly, the *primitive seas* formed, with depths up to 3 km. Their volume was only 10% of that of the modern ocean, so the primitive seas could hardly have been connected with each other. Instead, they must have been scattered over the entire earth. The oceans did not reach their present volume until the beginning of the Cambrium, when they began to fill with water vapor released through volcanic outgassing and hot springs (geysers) from the interior of the earth. With the formation of the hydrosphere, the erosion of the surface also began, favored by the processes of water circulation. The rainwater of that time dissolved NH_3 from the atmosphere, forming a basic solution of

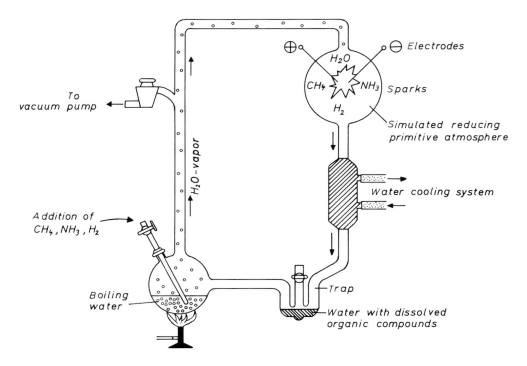

FIGURE 1.   Experimental set-up used by Miller[12] to study the formation of organic compounds in a simulated primitive reducing atmosphere subjected to arc discharges (adapted from Fox and Dose[7]). The gases NH_3, CH_4, H_2, and water vapor were circulated through a system of tubes, and energy was applied in the form of electric discharges. The experiment was ended after several days, and the contents of the trap were analyzed.

As an alternative to Miller's investigations, recent experiments carried out by Freund (e.g. Atomarer Kohlenstoff. Nachrichten aus Chemie, Technik und Laboratorium, *Ges. Dsch. Chem.*, 29, 301, 1981) raise the possibility of the prebiotic origin of carbohydrates by way of purely inorganic reactions, i.e. from reactive carbon atoms in crystals.

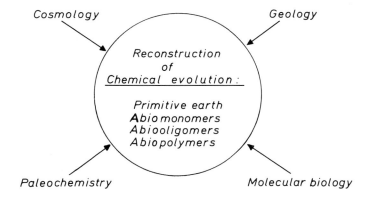

NH_4OH. This solution speeded the weathering of the rocks, which contained the more acidic silicates (see Figure 3), and the resulting mixture produced a mildly basic sea water with a pH between 8 and 9. The partial dissolving of the materials washed into the sea increased the salt concentration (anions, cations) of the water. However, most of the material washed into the sea settled onto the bottom and was gradually transformed into *sedimentary rocks,* which had not previously existed.

Simplified scheme of chemical and prebiotic evolution

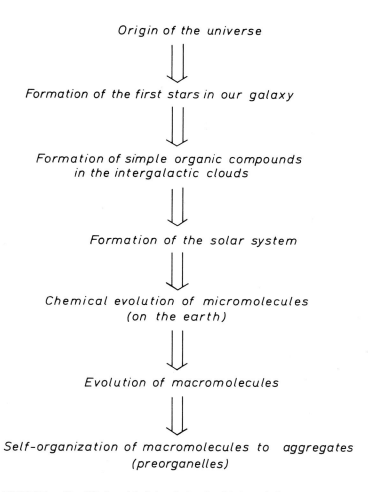

FIGURE 2.   Simplified model of chemical and prebiotic evolution now generally accepted (adapted from Dose and Rauchfuss[6]).

The *atmosphere* developed simultaneously with the lithosphere and hydrosphere. The first or *primary atmosphere* present as the earth formed consisted mostly of hydrogen (Chapter 2, Figure 3). However, this atmosphere was lost during the period of extremely high temperatures produced by the condensation of the earth. Since hydrogen is very light, the gravitational attraction of the earth was not sufficient to prevent it from escaping into space. As the earth cooled, volcanic action released the gases of a *secondary atmosphere* between 5 and 4.5 billion years ago. This atmosphere contained small amounts of free hydrogen, but most of the hydrogen was present in the form of compounds: methane (CH_4), ammonia (NH_3) and water (H_2O). These gases had large enough masses to be retained by the earth's gravitational field. However, between 4.5 and 4 billion years ago, the solar radiation oxidized them partly to nitrogen (N_2), carbon dioxide (CO_2) and carbon monoxide (CO), thus forming the *tertiary atmosphere*. The primary, secondary, and tertiary atmospheres were all reducing

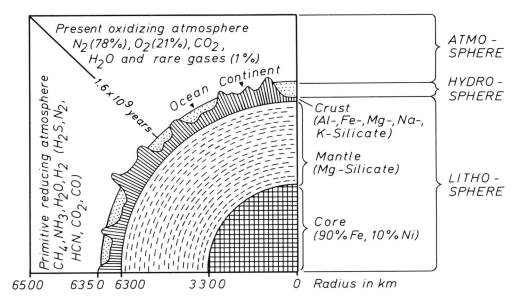

FIGURE 3.   Schematic cross section of the primitive and present-day earth, suggesting the composition of the litho-, hydro- and atmospheres (not drawn to scale).

atmospheres.* The present or *quaternary atmosphere*, on the other hand, is oxidizing.** It is at most 150 km thick and is composed of about 78% nitrogen, 21% oxygen, and only about 1% of other substances, including carbon dioxide, water vapor, and rare gases. It must have developed long after the origin of life, when water-cleaving, oxygen-producing bacteria arose (see Chapter 6, II). This development can still be traced in geologically very old layers. Here we find sandstone containing pyrite grains (FeS_2) whose position and characteristics show that they were long exposed to the atmosphere of that time. Since pyrite is geologically unstable in the presence of oxygen, we can be sure that the primitive atmosphere contained no significant amounts of oxygen. The most recent pyrite-containing layers are 1.8 billion years old. On the other hand, the oldest stones containing fully oxidized iron (Fe_2O_3), which must have been laid down under an oxidizing atmosphere, are 1.4 billion years old. Thus the transition from a primitive reducing atmosphere to an oxidizing atmosphere must have occurred between these two points in time.

The material conditions for chemical evolution on the primitive earth included a reducing atmosphere (composed of methane, ammonia, water vapor, hydrogen, etc.), an alkaline, widely scattered hydrosphere (containing the cations, Al^{3+}, Fe^{3+}, Ca^{2+}, Na^+, K^+, and Mg^{2+} — some present only in trace amounts — and the anions Cl^-, PO_4^{3-}, CO_3^{2-}, NO_3^-, SO_4^{2-}), and finally, a primitive lithosphere with a crust that was cracked and of varying temperature. The energetic conditions are also important for the course of chemical evolution. The most important energy forms on the primitive earth were ultraviolet (UV) solar radiation, radioactivity (α and β particles and γ rays) from the interior of the earth and from space, the electric discharges of lightning, and heat from the interior and from the sun (Table 1). Heat and lightning would have been the quantitatively least important, but the qualitatively

---

*   Some arguments seem to support the theory of an oxidizing primitive atmosphere. An example of reduction is the formation of metallic iron from iron oxide (Fe_2O_3) : $2\ Fe^{3+} + 3\ O^{2-} + 3\ H_2 \rightarrow 2\ Fe + 3\ H_2O$. Here the iron is reduced by the addition of electrons, while the hydrogen is oxidized to H^+, which then combines with the O^- to form water.

**   In the present atmosphere, metallic iron tends to rust, which is just the reverse of the reaction shown above.

Table 1
**AVAILABLE ENERGY SOURCES ON THE PRIMITIVE EARTH
BETWEEN 4.5 AND 4 BILLION YEARS AGO**

Form of energy	Total energy	
	(cal /cm2/ year	J /cm2/year
Total radiation of the sun....... .... 170 000		711 762.8
of this, UV<2 000 Å ········ ····· 30		125.6
Radioactivity from the earth's crust to 30 km depth (α,β,γ)....... 47		196.8
Electric discharges in the atmosphere (lightning)......... 4		16.7
Heat of the earth's crust....... ....>1		> 4.2

Adapted from Dose and Rauchfuss.[6]

Table 2
**APPROXIMATE ENERGY CONSUMPTION FOR THE SYNTHESIS OF 1
MOL GLYCEROL UNDER THE CONDITIONS OF THE PRIMITIVE
ATMOSPHERE**

Form of energy	Energy needed to produce glycerol	
	cal /mol glycerol	J/mol glycerol
Heat.................... ......2 000		8 373.7
Electric discharge (lightning).... ......2 400		10 048.4
Radioactivity (β)................. ....10 000		41 868.4
UV radiation (sun)............... .. 220 000		915 104.8

Note: The effectiveness of the four forms of energy in forming the compound is inversely proportional
to the amount of energy available from them on the primitive earth (see Table 1).

Adapted from Fox and Dose.[7]

most effective forms of energy (Table 2). However, even those forms of energy which were
only occasionally active, such as ultrasonic waves resulting from meteorite collisions, must
have played an important role in chemical evolution. Not only the amount of energy present
is decisive, but, furthermore, its distribution over space and time; every source of energy,
when evenly applied, will finally destroy as much as it created. In the end, chemical evolution
depended on the interplay of all the energetic and material factors and their spatial and
temporal distribution to produce the starting materials for life.

II. ABIOMONOMERS

The synthesis of abiomers on the earth may have begun about 4.5 billion years ago.
Ultraviolet radiation, radioactivity or electric discharges broke the covalent (electron-pair)
bonds of gaseous methane, ammonia, water and hydrogen, either symmetrically to form
radicals, or asymmetrically to form ions:

$$A \cdot \xrightarrow{\quad\quad} \cdot B \xrightarrow{\quad\quad} A^{\cdot} + \cdot B$$
radicals

$$A \quad \cdot \longrightarrow \cdot B \xrightarrow{\quad\quad} A^{+} + B^{-}$$
ions

(UV radiation with wavelengths greater than 2000 Å was not effective, since it was not absorbed by the gases). The radicals and ions now reacted in the atmosphere to form intermediate products such as hydrogen cyanide (HC≡N), formaldehyde (HC$\overset{O}{\underset{H}{\diagdown}}$), formic acid (HC$\overset{O}{\underset{OH}{\diagdown}}$), acetylene (HC≡CH) and ethylene (H$_2$C=CH$_2$) etc. These intermediate products could also be formed directly and much more gently by heat at a temperature of about 200°C (473 K). However they were formed, the products were washed out of the atmosphere by the rain and landed in the hydrosphere and on the lithosphere. Some of them may also have been brought to the earth by meteorites, since we still find comparable intermediate products in interstellar space.

The abiomonomers then formed in various ways from the intermediate products in the atmosphere, the hydrosphere, and the lithosphere (Table 3, Figure 4). The simple carbo-hydrates were joined together by carbon, which is particularly suited to the formation of larger molecules, due to its quadruple valency. The rates and kinds of synthesis were presumably controlled by periodic changes in the environment, such as those connected with day and night, the tides, summer and winter, etc. The results were abiomonomers such as amino acids (peptides), nucleobases (nucleotides), sugars (saccharides), fatty acids (lipids), etc. These were qualitatively and quantitatively identical to the biomonomers, except that the abiomonomers were present as mixtures of optical D-, L-isomers,* while biomonomers are always present as pure optical isomers: L-amino acids and D-sugars (see Chapter 5, III).

The abiomonomers accumulated in ponds, lakes, and oceans. However, this cannot have led to a uniformly thick ''primal soup'', because mixtures of abiomonomers are not indef-initely stable. They eventually disintegrate in a reversal of their synthesis. Instead, the abiomonomers must have been locally isolated and concentrated by selective, periodic proc-esses, such as adsorption to active surfaces (foam, mud, clay, and minerals) and desorption, or filtration, sedimentation and recrystallization, or dehydration and redissolving in the hydrosphere. This variety of reaction conditions separated in time and space was necessary for the further synthesis of abiopolymers from abiomonomers.

III. ABIOOLIGOMERS AND ABIOPOLYMERS

The formation of abiopolymers must have begun about 4 billion years ago, partly in the atmosphere, but mostly in the hydrosphere and lithosphere. First abiomonomers with two reactive ends polymerized to form bifunctional *abiodimers* (Table 4, Figure 4). These in turn grew on both ends to form longer or shorter *abiooligomers,* such as oligopeptides, oligo-nucleotides, oligosaccharides, lipids, and porphyrins. For all these compounds polymeri-zation took place, both in aqueous and dry environments, by *condensation,* whereby a molecule of water is excluded for each new bond formed between the reactants. When the

---

* Optically active compounds include at least one carbon atom with four different substituents which can be arranged in two different ways. The two compounds are mirror images of each other and are called optical isomers (antipodes). Pure solutions of the isomers rotate polarized light passing through them in opposite directions, to the right or to the left. The optical rotation of a compound is designated with the symbol (+) for those which turn to the right (dextrorotatory) and with (−) for those which turn to the left (levorotatory): e.g. D(−) lactate, but D(+) fructose. The prefix D or L indicates the relative configuration, related on glyceraldehyde.

Table 3
SUMMARY OF POSSIBLE SYNTHESIS CONDITIONS OR SYNTHETIC PATHWAYS FOR THE MOST IMPORTANT PREBIOTIC MONOMERS.

CLASS of COMPOUND	STARTING MATERIALS	MEDIUM, ENERGY	ABIOMONOMER
Amino acids AA	H_2O + CH_4 + NH_3 Water Methane Ammonia	Primitive atmosphere / all energy forms	$R-CH(NH_2)-COOH$ + H_2 ↗ (at least 14 AAs)
Carboxylic acids: Fatty acids (Lipid**s**)	CO_2 H_2O $P-CH_3$ + $COO^{\ominus}$ + $\cdot OH$ + $H^{\oplus}$ Alkanes Carbon dioxide Water	Primitive atmosphere / ionizing radiation	$R-CH_2-COOH$ $-H_2O$ (R = max. 10 C atoms)
Nucleobases: Purines Pyrimidines	CH_4 N_2 CH_2O, NH_3 $H-C{\equiv}C-C{\equiv}N$ + $(NH_2)_2CO$ Cyanoacetylene urea HCN Hydrogen cyanide	Primitive atmosphere and hydrosphere / several energy forms	Cytosine Uracil }pyrimidines (as yet, no thymidine) Adenine Guanine }purines
Sugars: Pentoses Hexoses	CH_2O Formaldehyde	Primitive atmosphere and hydrosphere / all energy forms	Ribose (pentose)
Pyrroles	CH_4 $CH{\equiv}CH$ + NH_3 Acetylene Ammonia	Primitive atmosphere and hydrosphere / all energy forms	Pyrrole + H_2 ↗

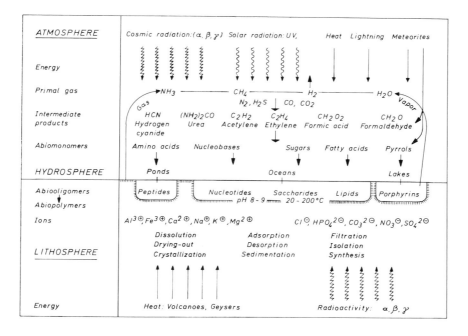

FIGURE 4. Schematic representation of the primitive earth as a "chemical laboratory" for the synthesis of abiomonomers, abiooligomers, and abiopolymers.

condensation occurred on hot lava as a surface catalyst, the condensed water would have evaporated immediately, so that the reaction was essentially irreversible. In an aqueous medium, however, the condensation would only have taken place if the water released was

Table 4

COUPLING OF THE ABIOMONOMERS BY CONDENSATION TO FORM THE MOST IMPORTANT PREBIOTIC ABIOOLIGOMERS AND ABIOPOLYMERS

immediately bound by some active substance.* On the primitive earth, this function appears to have been performed by the abundant hydrogen cyanide and its derivatives, cyanamide and dicyanymide. Cyanamide, for example, can bind the condensed water released by the dimerization of two monomers by being hydrolyzed to urea.

$$R^1 \boxed{OH + H} R^2 + H_2N-CN \longrightarrow R^1-R^2 + H_2N-CO-NH_2$$

monomers   cyanamide   dimer   urea

*   In the cell, this process is catalyzed by dehydrogenases which link the reaction to the hydrolysis of ATP (Adenosine TriPhosphate). This nucleotide has two functions: it provides energy for the reaction, and absorbs water from the environment of condensing molecules:

$$ATP + H_2O \longrightarrow ADP + H_3PO_4 \; (P_i)$$

Table 5

THE INTRA- AND INTERMOLECULAR BONDS OF ABIOMONOMERS OR BIOMONOMERS AND THEIR AGGREGATES RESPONSIBLE FOR CHAIN OR SPATIAL STRUCTURE

	Covalent bonds		Ionic bonds	
Primary bonds	$R-S:S-R$	Disulfide bonds (proteins)	$R-COO^{\ominus}H^{\oplus}$	Carboxyl group
	$R-C \overset{O}{\underset{\approx}{:}} NH-R$	Peptide bonds (proteins)	$R-^{\oplus}NH_3$	Amino group (proteins)
	$R^{5'}-O: \overset{O}{\underset{OH}{P}}-R^{3'}$	Phosphate esters (nucleotides, phospholipids)	$R-O-\overset{O}{\underset{\underset{H^{\oplus}}{O^{\ominus}}}{\overset{\|}{P}}}-O-R$	Phosphate group (nucleic acids)
	$R-O:R$	Acetal bonds (saccharides)		
	$R-C \overset{O}{\underset{}{:}} O-R$	Ester bonds (lipids)		
	Hydrogen bonds		**Hydrophobic bonds** [a]	
Secondary bonds	$-\overset{H}{\underset{\|}{N}}H\ldots\ldots\bar{I}\bar{O}=C$	Hydrogen-oxygen bridges	$R-C\overset{CH_3}{\underset{CH_2-CH_3}{\overset{\diagup}{\underset{\diagdown}{-}}}CH_3}$	Carbohydrate chains (proteins)
	$-OH\ldots\ldots IN \diagdown$ $\underbrace{\quad\quad}_{<2,8\ \mathring{A}}$	Hydrogen-nitrogen bridges	$CH_2-O-C\overset{O}{\overset{\diagup}{=}}(CH_2)_x-CH_3$ $CH-O-C\underset{O}{\diagdown}(CH_2)_x-CH_3$ CH_2-O-R	Carbohydrate chains (lipids)

[a] Hydrophobic bonds form when hydrocarbon groups are pressed together by the surrounding aqueous medium. The mechanism is the same as that which causes two small oil drops in water to unite — it reduces the area of contact between hydrocarbon oil and water molecules.

Montmorillonite $Al_2(Si_4O_{10})(OH)_2 \cdot n\ H_2O$, a component of the clays on the primitive earth, could have favored condensation in the same way. The abiooligomers were in principle quantitatively and qualitatively the same as the biooligomers, except that they included unphysiological bonds between their monomers and were more highly branched (e.g. 3′-5′ ester bond of nucleotides[5]).

The development of the abiooligomers proceeded directly to that of long-chained abiopolymers such as proteins, nucleic acids, carbohydrates, and fats. The sequence (in proteins, the primary structure) of monomers in these polymers has not been purely random, because the charge distribution and structure of the end monomer influenced the choice of the next monomer to bind to the chain (Table 5). Also, the coupling of the same kind of monomers (homomers) into a chain was energetically more favorable, and thus more frequent, than the formation of heterogeneous chains (heteromers) (see Chapter 5, III). The spatial structure of a polymer depended in turn on its sequence (in proteins, the secondary and tertiary structure); macromolecules do not occur as long, straight chains. Instead they are folded, wound in spirals, and/or coiled, in such a way that the water-soluble or hydrophilic groups tend to be on the outside, while the water-insoluble or hydrophobic groups tend to be on the inside. The chain structure in turn determines the specificity of a molecule, i.e. its ability to interact with other polymers via primary or secondary bonding (see Table 5). These other polymers may be the same or different from the first. The formation of specific complexes by several chains is called, in proteins, quaternary structure and results in their highly specific spatial structure (enzymes, see Chapter 5, III). The ability of micromolecules to organize themselves into macromolecules led to ever more complex abiogenic aggregates. These abioaggregates were not, in contrast to the biogenic aggregates, controlled by a genetic system. Their organization depended only on the information for self-organization contained within them.

The formation of stable abiopolymers and their self-aggregates would only have been possible in the depths of the hydrosphere. Here the polymers would not have been immediately depolymerized (photolysis) by ultraviolet radiation which, for lack of an ozone layer, reached the earth's surface with its full intensity. Indeed, the polymers must have become selectively enriched in small, closed reaction areas, where they could react undisturbed to precellular or pre-organelle-like aggregates. Hollows in clay particles or "primal soup droplets" (small, water-containing spheres bounded by lipid films, or protein spheres) would have been particularly favorable sites for the formation of aggregates.

IV. MECHANISMS OF CHEMICAL EVOLUTION

The prerequisites for chemical evolution on the earth were a spatially varied and temporally periodic environment, which were provided by the diversity of the primitive atmosphere, hydrosphere, and lithosphere, and by the earth's rotation. These in turn were the results of dissipation processes set in motion by the gravitational contraction of matter in the universe. Thus chemical evolution is a continuation of cosmic development on the molecular level. The same mechanisms are at work, supplemented by others which apply only at the molecular level.

In both systems, energy is "invested" to achieve an increase in order or complexity. On the cosmic scale, gravitational potential energy is converted to heat, as a diffuse cloud of gas contracts to form stars and planets. On the chemical scale, the heat derived from the parent body (both earth and sun) promotes the formation of chemical bonds in molecules of ever increasing complexity. The tendency to form such complexes *(self-organization of matter)* is, in the final analysis, a property of matter itself which is expressed under the appropriate circumstances. Since the complexes contain large amounts of chemical potential energy, they are rather unstable and do not accumulate in a constant environment. Herein lies the importance of a periodically changing environment, which provides changes in temperature and thus changes in reactivity, drying out and redissolving (caused by the tides and seasonal changes), and exposure to energy, followed by its removal from the reaction mixture by rain and tides. In such a variable environment, many interrelated processes can be constantly repeated, so that the chemical systems are forced back and forth from one extreme state to the other (energy dissipation). This results in the variety of systems necessary for selection to occur. Natural selection itself is constantly pendling between periods of *divergence* and *convergence** (Figure 5).

Kuhn[9] suggests that during the divergent phases, many variants with similar chances for survival arise, and no decisive changes occur. Then, suddenly, a mechanism begins to fulfill a new purpose, and this brings a swing into a convergent phase. This phase is strongly selective, and the selection is directed: those variants of the molecule which best serve the new purpose survive. The transition from the divergent to the convergent phase occurs by chance, but the event must occur sooner or later. Those events which determine evolutionary progress are, to be sure, very unlikely in themselves, but with sufficiently large populations of molecules and/or long enough times, they will eventually take place. Thus it is not the evolutionary step in itself, but the time at which it occurs which is a matter of chance. The

---

*   The term convergence is used here in a different sense than the biological term for a parallel, nonhomologous evolution.

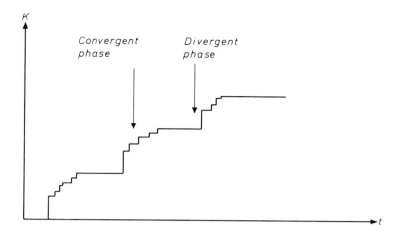

FIGURE 5.   General mechanism of chemical evolution. Knowledge K is a measure for the adaptation of the evolving system to the environment as a function of
the evolutionary time t (from Kuhn[9]).

interplay between selection and self-organization resulted, in the course of chemical evolution, in the unbroken chain of ever more complex chemical systems, leading at last to the
precellular aggregates (see Figure 4).

However, before we can follow the development of abiomers to precellular and then living
systems, we must first define *life*. Only then do we know which structures and processes
must have developed in the evolution of life. The representation of the characteristics of
life is only intended to be a short outline. Details can be found in the literature (e.g. Kaplan[8]).
The basic concepts have been taken from this body of literature, especially from Barckhausen
et al.[1]

REFERENCES

1. **Barckhausen, R., Heger, W., Hollinn, K.-U., Lootz, J., Lüdcke, J., Pulvermacher, C., Schwemmler, W., Seipel, S., and Timner, K.,** *Kompendium Biologie für Mediziner,* 2. Auflage, Gustav Fischer Verlag, New York, 1981.
2. **Berckhemer, H.,** Die Entwicklung der Erdrinde, *Naturwissenschaften,* 68, 323, 1981.
3. **Buvet, R. and Ponnamperuma, C.,** Chemical evolution and the origin of life, *Proc. 3rd Int. Conf. on origin of life,* North Holland, Amsterdam, 1971.
4. **Calvin, M.,** *Chemical evolution: molecular evolution towards the origin of living systems on the earth and elsewhere,* Oxford University Press, New York, 1969.
5. **Calvin, M.,** Chemische Evolution, *Naturwiss. Rdschau.,* 29(4), 109, 1976.
6. **Dose, K. and Rauchfuss, H.,** *Chemische Evolution und der Ursprung lebender Systeme,* Wissenschaftl. Verlagsgesellschaft; Stuttgart, 1975.
7. **Fox, S. W. and Dose, K.,** *Molecular evolution and the origin of life,* Freeman, San Francisco, 1972.
8. **Kaplan, R. W.,** *Der Ursprung des Lebens. Biogenetik, ein Forschungsgebiet heutiger Naturwissenschaft.* 2. Aufl. Georg Thieme Verlag, Stuttgart, 1978.
9. **Kuhn, H.,** Model consideration for the origin of life, *Naturwissenschaften,* 63, 68, 1976.
10. **Küppers, B.,** The general principles of selection and evolution at the molecular level, *Progr. Biophy. Molec. Biol.,* 30(1), 1, 1975.
11. **Laskowski, W.,** *Elemente des Lebens,* de Gruyter Verlag, Berlin, 1966.
12. **Miller, S. L.,** A production of amino acids under possible primitive earth conditions, *Science,* 117, 528, 1953.
13. **Schidlowski, M.,** Die Geschichte der Erdatmosphäre. *Spektrum der Wissenschaft,* 4, 17, 1981.
14. **Wagenführ, W.,** *Biogenese — theoretische und experimentelle Aspekte,* Staatsexamensarbeit, Freie Universität Berlin, 1973.

Chapter 4

DEFINITION OF LIFE

As we know today, all the organisms on earth are composed of cells. This applies both to single-celled organisms and to plants, animals, and humans. The development of the latter group can even be traced back to a single cell — the egg in the case of sexual reproduction, and the initiator cell in vegetative reproduction. Phylogenetically speaking, the development of each group of organisms can be traced ultimately to an original cell. The cell is thus the smallest independent unit of structure, function, information, and evolution which is capable of survival and reproduction. This view is based partly on the cell theory developed by Matthias Jakob Schleiden (1804—1881) and Theodor Schwann (1810—1892), and partly on the theory of evolution deduced by Charles Darwin (1809—1882). Rudolf Virchow (1821—1902) summarized this fundamental principle of biology in the sentence

"Omnis cellula e cellula" (every cell comes from a cell).

The phenomenon, life, is therefore inseparable from the structure, function, information, and evolution of the cell. All the properties which have developed in the course of the evolution of the more highly organized multicellular species, such as feeling, speech, and thought, are, strictly speaking, unimportant for the definition of life. This can be concisely formulated by changing the above sentence to

"Omne vivum e cellula" (all life comes from the cell).

But what is the specific feature of a cell that determines its quality of life? The definitive criterion of life is, according to Jacques Monod,[7] the capacity of the cell for *autonomous morphogenesis*, that is, the ability to direct its own structural development. This ability is based specifically on a self-regulating anabolism and energy metabolism, constant and identical reproduction, and an inheritable mutability or changeableness. These individual properties must be regarded as the elemental criteria of life, and they are only realized in the cell.

I. METABOLISM

Every cell is surrounded by a membrane which separates it from its environment. This is called the *plasma membrane*. In electron micrographs, membranes appear to be fairly uniform. They consist of a central, lightly stained area bordered on both sides by a dark layer. According to the generally accepted membrane model (fluid mosaic model), the lightly stained area represents two layers of lipid molecules (Figure 1). The hydrophobic ends of the lipid molecules are directed towards the middle of the film, and the hydrophilic ends are directed outwards. Proteins are embedded into or on both sides of the hydrophilic outer layer of the film. The lipid film is also interrupted in many places by proteins which extend all the way through it to form tunnels. This produces "hydrophilic holes" in the film, through which substances can be actively transported into or out of the cell.

Due to its compact construction, the membrane serves as a boundary between the cell and its surroundings (physiological barrier). On the other hand, the membrane's permeability enables it to actively regulate and control the exchange of substances between the internal and external media of the cell. Substances are selectively transported into the cell by the membrane. Inside the cell they are metabolized and some are again transported as metabolic

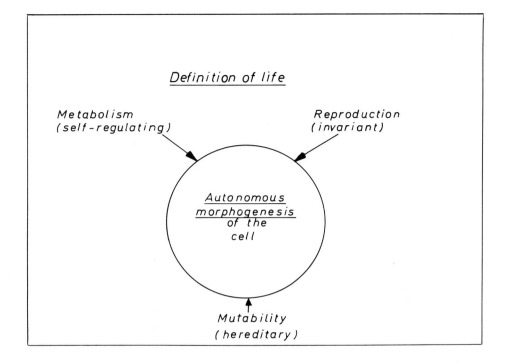

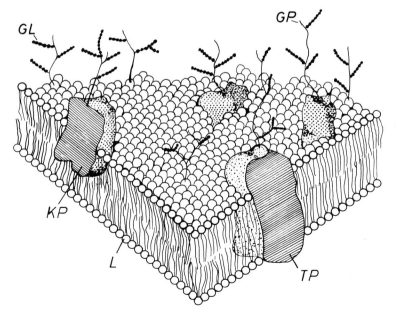

FIGURE 1. "Fluid mosaic model" of the cell membrane as proposed by Singer and Nicolson.[9] L = bimolecular lipid layer; P = tunnel proteins (TP) or spherical proteins (KP); G = polysaccharide chains of the glycocalix attached either to lipids as glycolipids (GL) or to proteins as glycoproteins (GP).

products out of the cell. The cell is thus constantly exchanging material with its surroundings. It is a *dynamic system.* Such open systems have properties differing from those of the closed systems of physics and chemistry. Given a constant chemical composition and environment,

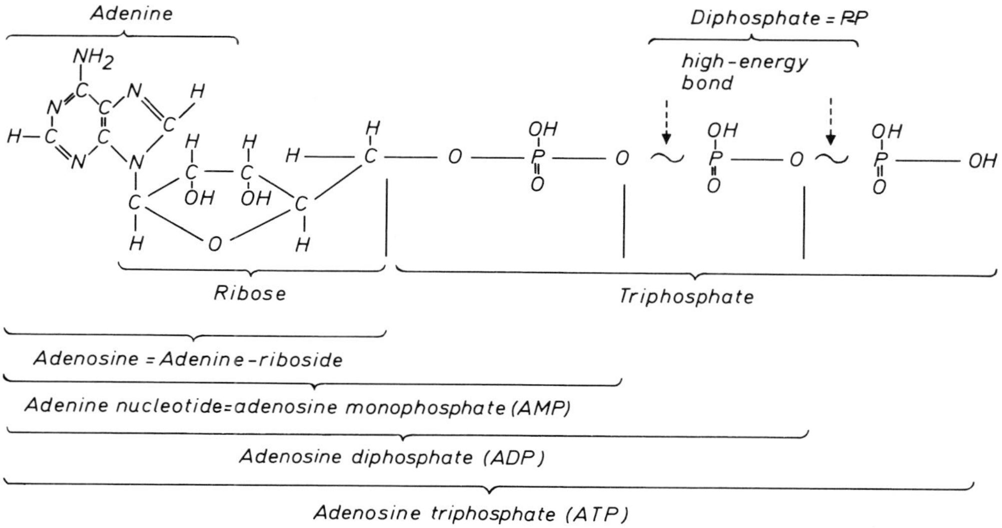

FIGURE 2.   Molecular structure of ATP and its derivatives (from Kaplan[4]).

dynamic systems tend toward a *dynamic steady state* ($\Delta G \neq 0$). They achieve this through internal regulatory mechanisms, a property which is an important characteristic of life.

The turnover of substances within the cell can be divided into *anabolism*, or synthesis of building materials, and *catabolism*, or energy-producing reactions. The turnover in general is called *metabolism*. The building materials, or anabolites, comprise in particular the proteins, sugars (carbohydrates), lipids (fats), and nucleic acids (DNA, RNA). Catabolites are formed by the breakdown of other substances, including energy-rich substrates. One example is the production of pyruvic acid by hydrolysis of sugar phosphates in glycolysis. The energy released is stored in molecules of the phosphate-group carrier *adenosine triphosphate* (ATP). The stored energy is made available for synthesis by splitting off phosphate groups (ATP → ADP → AMP). ATP is the most important energy carrier of the cell and serves as a general "currency" for cell energy (Figure 2).

However, ATP is not only an energy carrier. It is also one of the four building blocks of cellular nucleic acid. The nucleic acids are the carriers of genetic information in the cell (see next chapter). They control the synthesis of proteins, including that of *enzymes*. The enzymes in turn regulate, by means of their catalytic activity, the entire metabolism of the cell. Synthesis and energy metabolism are consequently tightly coupled; their regulation is autonomous. This is another criterion of living systems.

II. REPRODUCTION

The most important characteristic of living systems, however, is their ability to reproduce themselves. The *reproduction* of a cell is achieved by division. The products of division, the daughter cells, have the same properties as the parent cell. It is apparent that information is passed on from generation to generation, which determines the characteristics of the descendants. These instructions are termed the genetic information of the cell. In 1944, Avery and his co-workers demonstrated that this information is contained in the cell's nucleic acids. But how is the genetic information organized within the nucleic acids, and how is it distributed within the cell and passed on to the offspring?

A. Structure of Nucleic Acids
There are two kinds of nucleic acids: deoxyribonucleic acid *(DNA)* and ribonucleic acid

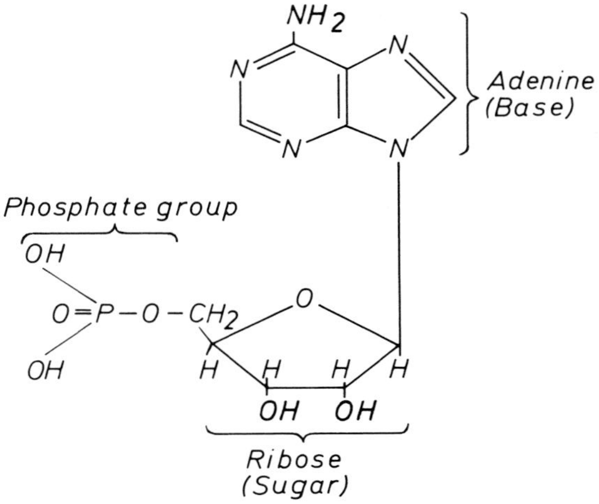

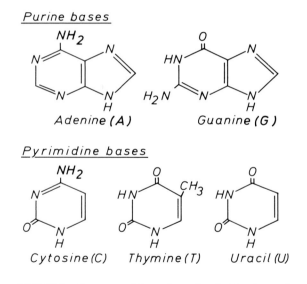

FIGURE 3.   Structure of a nucleotide: AMP (adenosine 5'-monophosphate).

FIGURE 4.   Structures of the purine and pyrimidine bases of nucleic acids.

(RNA). DNA is the actual carrier of genetic information. RNA is used only to realize this information. DNA and RNA are composed of individual building blocks, called *nucleotides*. Each nucleotide consists of a basic ring compound, a sugar molecule and a phosphate residue (Figure 3). The nucleotides of DNA contain the sugar deoxyribose, and as basic ring compounds the purine bases adenine and guanine and the pyrimidine bases thymine and cytosine (Figure 4). RNA nucleotides are identical, except that they contain ribose instead of deoxyribose, and uracil instead of thymine. The nucleotides of DNA or RNA are linked together to form unbranched nucleic acid strands (Figure 5). RNA is present in cells in the form of single-stranded molecules, some of which are folded together, whereas DNA is always found as a double strand.

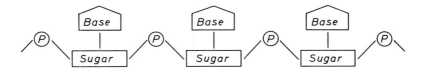

FIGURE 5. Nucleotide linkages in nucleic acid chains (adapted from Barckhausen et al.[2]).

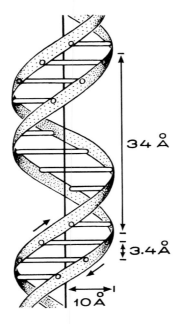

FIGURE 6. Spatial structure (configuration) of a DNA molecule according to Watson and Crick.[11] The "rungs" of the DNA "ladder" symbolize the base pairs.

In 1953, working from x-ray structure analyses and other physical-chemical data, Watson and Crick proposed a double spiral or *double helix model* for DNA. According to this model, the DNA double strand consists of two nucleotide chains which are twisted around each other (Figure 6). The bases form hydrogen bonds with their opposite numbers on the other strand (see Chapter 3, Table 5). The molecular structures of the bases are such that hydrogen bonds can only be formed between adenine and thymine (uracil) or guanine and cytosine (Figure 7). This principle of *base pairing* has the consequence that the sequence of bases on one strand dictates the sequence on the other in complementary form. While RNA is formed as a synthesis product of a given DNA segment, DNA is always generated by self duplication, or *replication*.

B. DNA Replication and the Genetic Code

Before every cell division, the cell's genetic information, and thus the DNA, is exactly duplicated or copied (Figure 8). In this way it is insured that the daughter cells receive the same information. To replicate, the two strands of the DNA double helix separate in zipper fashion. Free nucleotides which have been produced by the cell's metabolism arrange themselves along each single strand, each free nucleotide pairing with a complementary base in

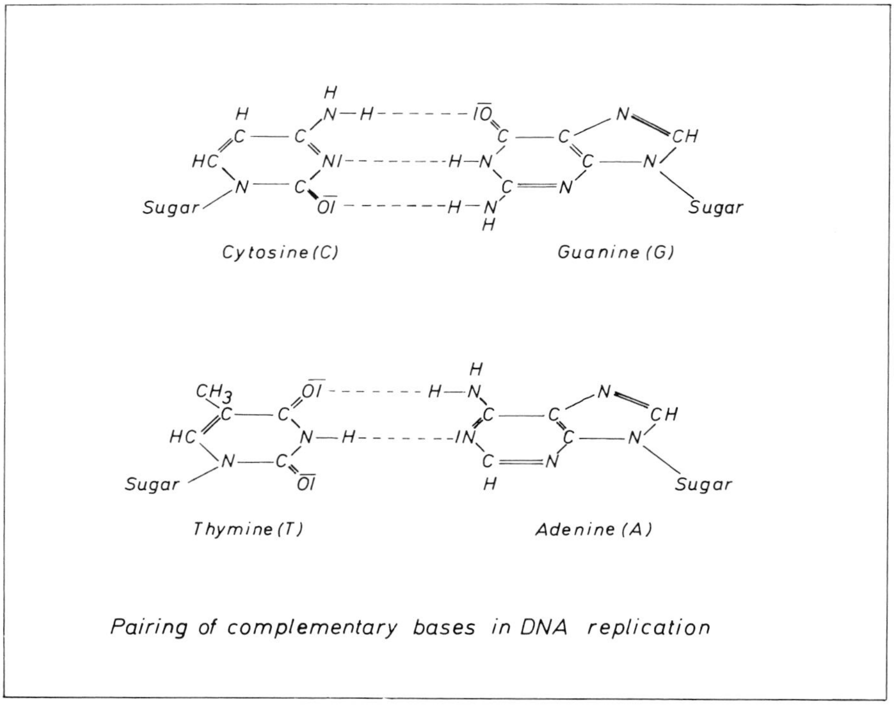

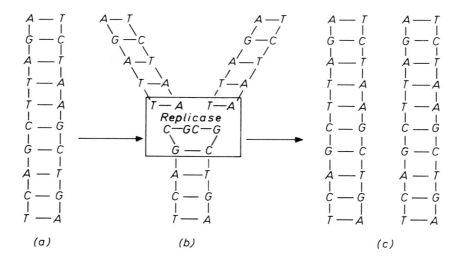

FIGURE 7.   The hydrogen bridges formed between guanine and cytosine and between adenine and thymine in DNA base pairing (see secondary bonds, Chapter 3, Table 5).

FIGURE 8.   Schematic representation of replication (adapted from Barckhausen et al.[2]) (a) = original double strand; (b) = zipper-like separation of the double strand and lining up of free nucleotides to form base pairs; (c) = two identical double strands after replication.

the strand. In this way, a new partner strand is formed for each of the original strands. This process is called *semiconservative replication.* Each new double helix contains one old strand and one new one. The synthesis of the new strand is catalyzed by a protein, the enzyme *DNA replicase* (DNA polymerase).

Table 1
PROTEIN FUNCTIONS IN THE CELL

Functional type	Examples	Occurrence (function)
Structural protein	Keratin	Skin, hair, nails: protective outer layer
Motile protein	Actin, myosin	Muscle myofibrils: motion
Transport protein	Hemoglobin	Erythrocytes: oxygen transport
Storage protein	Ovalbumin	Egg white: embryonic nutrition
Hormones	Insulin, glucagon	Pancreas: regulation of the blood sugar level
Enzymes	Cytochrome c	Mitochondria: electron transport ⎫ energy
	Hexokinase	Cytoplasm: glucose phosphorylation ⎬ production

Note: adapted from Barckhausen et al.[2]

The genetically determined functions of the cell, such as synthesis and energy metabolism, are controlled for the most part by proteins. Proteins (as structural, motile, transport, and storage elements), hormones, and above all enzymes, are involved in a wide variety of cell functions (Table 1). As we have seen (Chapter 3, III), they are macromolecular chains of amino acids (AA). About 20 different amino acids are found as subunits of cellular proteins (see Chapter 5, Table 2). Therefore, there must be a mechanism in the cell which translates the information in the DNA nucleotide alphabet into the amino acid alphabet of the proteins. There are only four letters in the DNA alphabet, namely the bases adenine, guanine, cytosine and thymine (uracil). Groups of three consecutive bases encode the information for each of the 20 amino acids. This basic unit of genetic information is called a *codon* or *triplet*. There are 64 different triplets in all: 60 codons for amino acids, one codon as a start signal, and the other three as stop signs for protein synthesis. Thus several different triplets are available for most of the amino acids, and the code is therefore referred to as degenerate. Figure 9 represents the so-called "*code-pie*", in which the individual triplets are related to the corresponding amino acids or stop and start signals (the figure gives the code for RNA, so U is shown in place of T).

The expression of genetic information, i.e. the synthesis of cell-specific proteins, is achieved in two steps: first the reading or *transcription* of the information in the form of a template, and then the synthesis or *translation* of a protein from this template using free amino acids in the cell. The DNA segment responsible for the synthesis of an individual template or protein is called a *gene*. The genes are linearly arranged in the *chromosomes* of the cell. The *genome* of a cell is the sum of its chromosomes.

C. Transcription and Translation

Proteins are not synthesized directly on the DNA, but on special structures called *ribosomes*. The ribosomes are composed of three special kinds of RNA molecules of different lengths (molecular masses of prokaryote ribosomal RNA: 2900, 1540, and greater than 120 daltons) held together by proteins, some of which have enzymatic properties (Figure 10). A particular kind of RNA, *messenger RNA* (mRNA), carries information from the DNA to the ribosomes. Like the other kinds of RNA, mRNA is a single strand synthesized by complementary base pairing with one strand of the DNA (the codogenic strand). Each mRNA is thus the template or complement of the nucleotide sequence of a gene. The process of transcribing the DNA nucleotide sequence into a mRNA nucleotide sequence is called transcription (Figure 11).

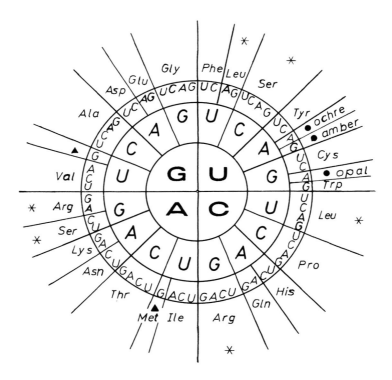

FIGURE 9.   "Code pie" (from Bresch and Hausmann[3]). The condons read from the middle outward; they give the base sequences of the RNA codons for the amino acids shown on the periphery (abbreviations, see IUPAC-IUB Rules, Appendix I). * = amino acid which occurs twice; ▲ = start codons; ● = stop codons.

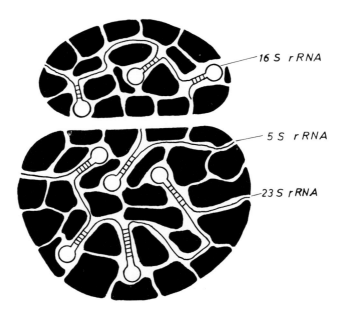

FIGURE 10.   Conceptual model of prokaryote ribosome structure. The position and shapes of the molecules are arbitrary (from Bresch and Hausmann[3]). S = Svedberg unit, a sedimentation constant.

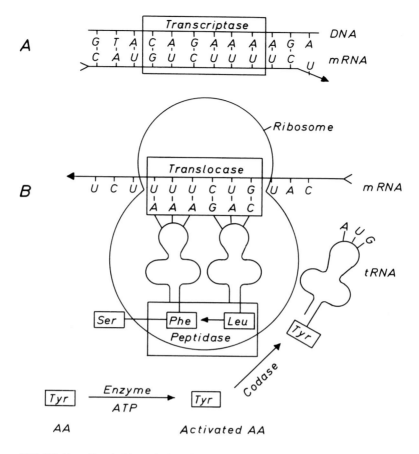

FIGURE 11. Protein biosynthesis (adapted from Barckhausen[2]). A = transcription (in the nucleus); B = translation (in the cytoplasm); AA = amino acid.

The mRNA attaches itself to the ribosomes, where catalytic enzymes for reading the mRNA and coupling the amino acids to proteins accomplish the *translation* of the code *(protein biosynthesis)*. In preparation for protein synthesis, free amino acids in the cytoplasm are first activated by the energy-rich compound ATP. Then the activated amino acids are enzymatically attached to special coupling RNA molecules, the *transfer RNAs* (tRNA). There is at least one specific enzyme (aminoacyl-tRNA synthetase = codase) and one specific tRNA for each of the 20 different amino acids. The tRNAs are composed of about 80 nucleotides. They contain some areas where the bases pair intramolecularly, so that the molecules have clover-leaf structures. Four regions of these molecules are essential for their function (Figure 12):

- a region for recognition of the amino acid coupling enzyme
- a binding site for the amino acid
- a region which recognizes the appropriate nucleotide sequence, the *anticodon*
- a region for binding to the ribosome.

The tRNA, charged with the appropriate amino acid, positions itself with its anticodon adjacent to the complementary codon of the mRNA (Figure 11). This brings its amino acid into position. Now the mRNA proceeds to the next codon. The next tRNA finds its place

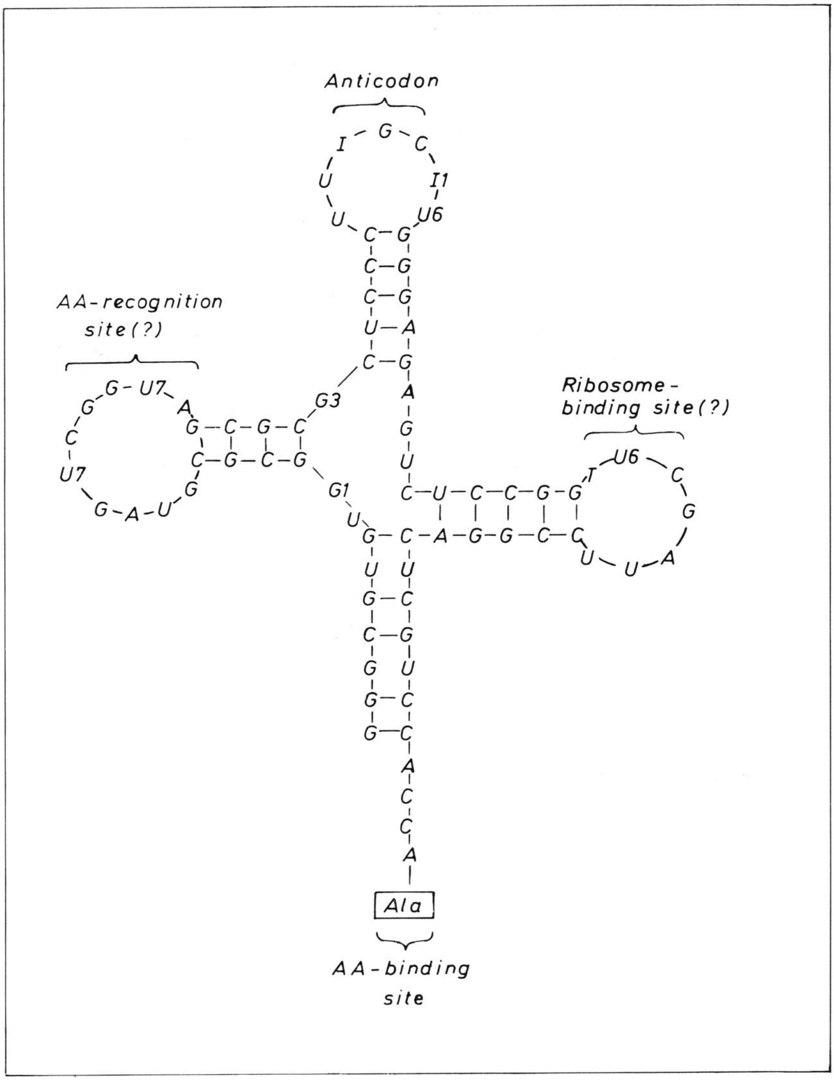

Anticodon

```
              I ⌐ G ⌐ C
             /            ⟍
            U             I1
             ⟍           /
              U ⌐ C — G ⌐ U6
                 C — G
                 C — G
                 U — A
                 C — G
                      A
                      |
                      G
```

AA-recognition site (?)

Ribosome-binding site (?)

```
        G - U7
      G        A
     /          ⟍
    C    G — C — G — C  G3
     \   C — G — C — G
    U7           G
     ⟍          /
       G - A - U  G1
                  U
              G — C — A — G — G — C — C
              U  U
              G — C
              C — G
              G   U
              G — C
              G — C
                  A
                  C
                  C
                  A
```

```
    C — U — C — C — G — G  T ⌐ U6 ⌐ C
    |   |   |   |   |   |  /           ⟍
    G — C — A — G — G — C — C           G
                              U ⌐ U ⌐ A
```

┌─────┐
│ *Ala* │
└─────┘

A A - binding site

FIGURE 12.   Structural model of a transfer RNA: the alanine tRNA from yeast.

with the help of its anticodon, and brings its amino acid into position. A peptide bond is next formed enzymatically between the two amino acids. The first tRNA is then released from the ribosome, and the ribosome advances one place on the mRNA. Now the open site can be occupied by another charged tRNA, and so on. The process is repeated until the mRNA is completely translated, or until a stop codon ends the synthesis. The polypeptide chain is then released from the ribosome, which can be used for another synthesis. Since only two successive positions on the mRNA can be occupied at any one time, the codon-anticodon reading is strictly regulated. The two genetic processes of nucleic acid synthesis and protein synthesis are the basic elements of the phenomenon life.

III. MUTABILITY

The descendants of an organism cannot be absolutely invariant or identical to each other.

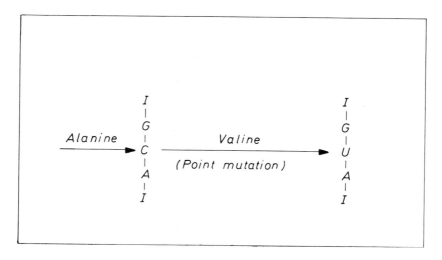

FIGURE 13.   A point mutation.

Otherwise, there would be no evolution, for evolution can only occur when better adapted individuals are favored in reproduction, and thus gradually supplant less well adapted forms. This process depends on the regular occurrence of slight changes or *mutations* in the course of transmission of genetic information. These changes are passed on to the offspring. *Point mutations* occur when the wrong nucleotide is incorporated during DNA replication. This has the effect, via transcription and translation, that a protein with a defective amino acid sequence is synthesized. For example, if the triplet GCA, which codes for alanine, is changed by a point mutation to GUA, which codes for valine (see Figure 13), then valine, instead of alanine, will be incorporated at the appropriate position in the protein. Other types of mutation are due to the incorporation of extra nucleotides or to the failure of all the nucleotides to be included. Even large segments of chromosomes or entire chromosomes can be affected by mutational restructuring, loss, or addition. In the latter case one speaks of *chromosome* or *genome mutations*.

Mutation is the "raw material" of evolution which, together with selection, constantly drives the development of organisms. Mutability is therefore another essential characteristic of living systems. It is mutability which makes life a progressive, not just a conservative phenomenon. We can agree upon a definition of life as that total phenomenon in which the properties of metabolism, reproduction, and mutability are simultaneously present. It thus seems logical to define the beginning of life with the first cell (precyte) which combined these three properties.

IV. HISTORY OF BIOGENETICS*

The question of the genesis of the first cells (precytes) is closely connected to the question of the origin of life. This question has occupied humanity through all the ages. In antiquity and the Middle Ages, the teaching of Aristotle (384—322 B.C.) that life could arise spontaneously from the non-living at any time was widely accepted. Not only plants, but also worms and insects, etc. were supposed to arise from dew, mud and manure by *spontaneous generation* (generatio spontanea). Even in the 17th century, famous scholars claimed that rats arose from rags, ducks from leaves, and sheep from fruits (Figure 14).

---

* Biogenesis used in the sense of origin of life and not used in the sense of genetic continuity of organelles.

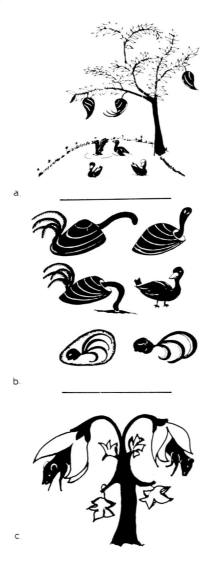

FIGURE 14.   A sketch of medieval concepts (from Rahmann,[8] according to Oparin 1957): a. "Geese tree" b. Development of geese from fruits or leaves of the "geese tree" which have fallen into the water. The transition stage is the barnacle. c. Development of sheep from the flowers or fruits of the "sheep tree".

Exact experiments, including those of Louis Pasteur (1822—1895), were needed to prove scientifically that life always comes from life. William Preyer later generalized this rule to *"omne vivum e vivo"*. In the further course of scientific development, however, the paradox had to be solved, that at the beginning of biological evolution, life must indeed have developed from non-living material. One logically possible solution was the *panspermy hypothesis* of Svante Arrhenius (1908). He supposed that life might have come from outer space and colonized the earth. However, this only shifted the problem of the origin of life to the universe at large. It was the trail-blazing *simulation experiments* of Stanley Miller (1953), who synthesized organic molecules as abiomers under conditions simulating those of the primitive earth (Chapter 3, Figure 1), which led to a scientifically sound concept of the origin of life (biogenesis). This field is known as *biogenetics*. It should not be confused with genetics, the study of inheritance. According to biogenetics, non-living matter organized itself to living matter under certain physical-chemical conditions on the primitive earth. In the first phase, during *chemoevolution* (Chapter 3), abiomonomers and abiopolymers formed out of the gases of the reducing primitive atmosphere (Chapter 3, Figure 4). During the second phase of biogenesis, *biological evolution,* the abiomers reacted with each other and, through their inherent capacity for self-organization and the action of selective external influences, developed further to precytes.

REFERENCES

1. **Avery, O. T., MacLeod, C. M., and McCarty, M.,** Studies on chemical nature of substance inducing transformation of pneumococcal types; induction of transformation by desoxyribonucleic acid fraction isolated from pneumococcus type III, *J. Exp. Med.,* 79, 137, 1944.
2. **Barckhausen, R., Heger, W., Hollihn, K.-U., Lootz, J., Lüdcke, J., Pulvermacher, C., Schwemmler, W., Seipel, S., and Timner, K.,** *Kompendium Biologie für Mediziner.* 2. Auflage. Gustav Fischer Verlag; Stuttgart, New York, 1981.
3. **Bresch, C. and Hausmann, R.,** *Klassische und molekulare Genetik.* 3. Auflage. Springer Verlag; Berlin, Heidelberg, New York, 1972.
4. **Kaplan, W.,** *Der Ursprung des Lebens: Biogenetik, ein Forschungs gebiet heutiger Naturwissenschaft.* 2. Auflage. Georg Thieme Verlag; Stuttgart, 1978.
5. **Karlson, P.,** *Biochemie.* 7. neubearbeitete Auflage, Georg Thieme Verlag; Stuttgart, 1970.
6. **Miller, S. L.,** A production of amino acids under possible primitive earth conditions, *Science,* 117, 528, 1953.
7. **Monod, J.,** *Zufall und Notwendigkeit,* Piper Verlag, Munich, 1971.
8. **Rahmann, H.,** *Die Entstehung des Lebendigen,* Gustav Fischer Verlag, Stuttgart, 1972.
9. **Singer, S. J. and Nicoloson, G. L.,** The fluid mosaic model of the structure of cell membranes, *Science,* 175, 720, 1972.
10. **Strassburger, E.,** *Lehrbuch der Botanik.* 30. Neubearbeitete Auflage. Gustav Fischer Verlag, Stuttgart, 1971.
11. **Watson, J. D. and Crick, F. H. C.,** Molecular structure of nucleic acids, *Nature (London),* 171, 737, 1953.

Chapter 5

EVOLUTION OF THE PRECYTES

The evolution of procytes (prokaryotes) from precytes (prekaryotes) represents the second and last phase of biogenesis. It began about 4 billion years ago at the latest, after the formation of the abiomer aggregates (abioids). In any case, it must have been in full swing 3.4 billion years ago, since the oldest cell-like microfossils date from that time. These were found in the Onverwacht Formation in South Africa, which is one of the oldest dated layers on earth (Greenland sediments dated older). The fossils represent spherical structures 10 to 30 μm in diameter (e.g. Dose and Rauchfuss;[5] see also Figure 2). However, they cannot be definitely identified as precytes. Even in more recent geological layers, there are no unequivocal, direct proofs for the existence of precytes. Even deductive evidence is rare. Research on precyte evolution thus has yet to advance beyond the stage of plausible hypotheses. At the moment, work in this field consists of creating possible models or theories. In the future, it should be possible to decide between alternative hypotheses by experiment or other studies. However, it must always be remembered that, no matter how plausible our theories may be, the actual prehistoric development might have taken a completely different course.

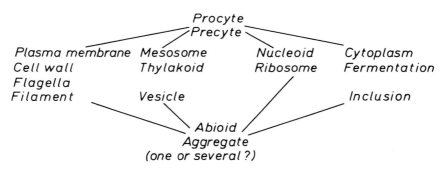

Basically, two alternative possibilities for precyte evolution are currently being discussed (Figure 1). According to the *multistep hypothesis,* the precytes developed directly and successively from one abiomer aggregate. In the opposing hypothesis, the *multilist hypothesis,* it is assumed in the extreme case that several different kinds of aggregates or abioids united to form the original precyte. In the following the two alternative paths of precyte development will be summarized, in order to clarify the principal differences.

I. WORKING HYPOTHESES

The *multistep hypothesis* is comparable in mechanism to the compartment hypothesis for eucyte evolution (reviewed in Schwemmler[54,55,56]). In both cases, a development via direct, continuous differentiation of a system is postulated. In the case of precytes, this process is thought to have been started by the formation of nucleotides from the gases of the reducing primitive atmosphere (Figure 1,A). According to Kuhn,[32] the nucleotides polymerized in aqueous solution to the various types of nucleic acids, the *information molecules.* The self-reproducing system of protein synthesis gradually developed from the nucleic acid system as abiogenic amino acids were included. The proteins serve as *working molecules.* On one hand they form enzymes through their particular spatial folding; on the other, they cooperate with abiogenic lipids to form cell membranes (biomembranes). The differentiation of the

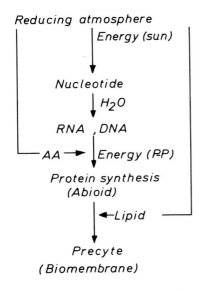

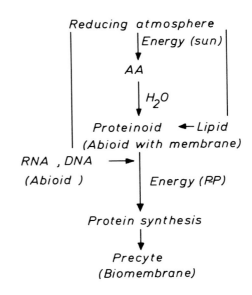

A. MULTISTEP HYPOTHESIS   B. MULTILIST HYPOTHESIS

FIGURE 1.   Highly simplified representation of the alternative working hypotheses on precyte evolution. (A) Multi-step hypothesis, formulated as Kuhn's ''nucleic-acids-first'' hypothesis.[32,33] (B) Multi-list hypothesis, in the combined form of Fox and Kaplan's ''proteins-first'' hypothesis[18,19,25,26] AA = amino acid; P–P = Diphosphate (energy-rich).

nucleic acid components before or without the protein components is called *genobiosis.* It stands in contrast to *holobiosis,* in which nucleic acids and proteins developed together.

Eigen,[6] for example, argues in favor of such a holobiotic cooperation of nucleotides and peptides from the very beginning. He assumes that various proteins and nucleic acids at first formed self-regulating cycles independently of each other. In such cycles, the synthesis of each member is catalyzed by its predecessor. ''Parasitic'' branching of the cycle cannot, however, be eliminated. This is only possible by coupling of protein and nucleic acid cycles to form autocatalytic *hypercycles.* In these hypercycles, each information molecule (nucleic acid) controls the synthesis of an action molecule (protein). This is achieved in the sense that a nucleic acid promotes the formation of a protein, which in turn accelerates the formation of another nucleic acid. The latter promotes another protein, and so on, until a final nucleic acid promotes the formation of a protein which accelerates the formation of the first nucleic acid, thus closing the circle. According to Eigen, only such hypercycles (''self-instructive catalytic hypercycles'') can lead to evolving systems. Evolution follows from self-organization of the macromolecules involved. The impulse for self-organization comes exclusively from the macromolecules themselves. The environment is ideally structureless, both physically and temporally.

Kuhn,[33] in contrast, sees the decisive drive toward self-reproducing systems in a physically variable environment and in an externally dictated periodicity. He describes evolution as a series of *divergent* and *convergent* phases. In the divergent phases, many types of molecules have similar chances of surviving, resulting in a diverse population. Then a chance mutation begins to fulfill a new function, and gives its possessors a selective advantage over the rest of the population. This is the turning point to a convergent phase, which is highly selective. Those mutants which best carry out the new function survive. The change from the divergent to the convergent phase is accidental, and thus very improbable. However, it must have occurred, given a large enough molecule population and long periods of time.

The combined multilist hypothesis[19,25,26] has, in its mechanisms, much in common with the endosymbiont theory of eucyte evolution.[36,54,55,56] In both theories, the new system arises by integration of several different individual elements (Figure 1,B). In the case of precytes, the process is thought by Fox to have begun with the formation of amino acids from the gases of the primitive reducing atmosphere. The amino acids condensed on hot lava to polymers, so-called *proteinoids*. As the lava cooled, these precipitated in water as cell-like microspheres. The spheres formed boundary membranes, which were later stabilized by incorporation of lipids to form abiomembranes. Then, according to Fox, the nucleic acids were integrated synchronously and successively into these proteinoid microsystems. In this way, a holobiotic precyte which was capable of protein synthesis developed.

According to Kaplan, the various types of nucleic acids could have been preformed in genobiotic systems, before they were combined to a holobiotic system. Coupled protein and nucleic acid synthesis would have been catalyzed by primitive enzymes (ur-enzymes) which had already been formed as abiogenic polymers.

The main disagreement in the hypotheses described above is the difference in the starting molecules (Figure 1). According to Fox's theory, the process of evolution was initiated by proteins. This version appears in the literature as the *"proteins-first"* hypothesis. According to Kuhn's concept, in contrast, the process began with nucleic acids. One could designate this as the *"nucleic-acids-first"* hypothesis. Another basic difference is the derivation of the biomembrane. In Kuhn's concept, membranes were developed after the protein biosynthesis machinery. Fox postulates that abiogenic membrane precursors were developed before protein biosynthesis. It remains to be seen what bearing the (to date very sparse) findings on the origin and nature of the interactions between proteins and nucleic acids on the one hand, and between proteins and lipids on the other, may have on this problem.

II. EXPERIMENTAL CLUES

The reconstruction of precyte evolution has already been approached in various ways. Geological studies have yielded chemofossils and have revealed the conditions under which prebiotic evolution must have occurred. These conditions were applied in simulation experiments, in which the biologically most important kinds of molecules (i.e. proteins, nucleic acids, and lipids) were abiogenically synthesized and analyzed by biochemical methods. The abiomers were allowed to react with each other, and their many-sided interactions were studied. Models of the coupled abiogenic nucleic-acid-protein synthesis and of the prebiotic formation of membranes were proposed. Some of these simulation models were checked by analysis of biogenic, cell-free systems. In a few cases, they were even subjected to statistical analysis by computer, or the probability of their occurrence was mathematically determined.

A. Geological Analyses

At present, we have no exact knowledge of the physicochemical conditions on the earth 4 billion years ago (see Chapter 3, I). We can be sure, however, that the surface conditions of the prebiotic earth were subject to great spatial and temporal variation, which must have had a decisive influence on the evolution of prebiotic systems.

No direct evidence can be obtained from chemical fossils, due to the intervening geochemical changes, but the results suggest that peptides, nucleotides, and lipids must already have existed side by side 4 billion years ago (Figure 2). These three kinds of molecules could thus have interacted with each other from the beginning of their existence.

B. Simulation Experiments
1. Protein-Nucleic Acid Interactions

The oldest models of possible precytes are the *coacervate systems,* with which Jong[23] and

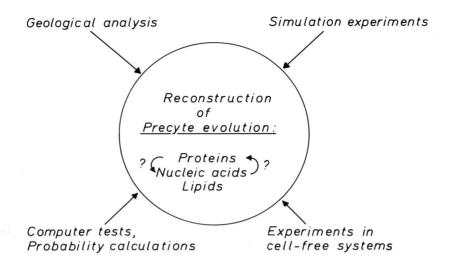

Oparin[42,43] and their co-workers conducted the first simulation experiments. Coacervates are obtained as liquid precipitates or droplets of various kinds of macromolecules separating out of solution. They form in certain concentration ranges when the pH is suddenly changed, or due to differences in the charges, and thus differences in the attraction for water molecules, of the macromolecules involved. The concentrations of macromolecules in the droplets may be as much as one hundred-fold higher than in the surrounding solution. Coacervates may form from many different kinds of molecules and have widely varying properties. For example, they have been produced from acid polysaccharide (gum arabic) and basic protein (histone). Phosphate-splitting enzymes, such as hexokinase and polynucleotide phosphorylase (PPase) were included in these coacervates. Glucose and the nucleoside triphosphate ATP were then added to the solution. These diffused into the coacervates, where the ATP phosphorylated the glucose to glucose-6-phosphate in a reaction catalyzed by the hexokinase. Adenosine diphosphate (ADP) was released and degraded by the PPase to adenosine monophosphate (AMP) and phosphate (P_i). The resulting AMP residues immediately polymerized to polynucleotides. These accumulated within the system, while the glucose-6-P and P_i diffused into the medium.

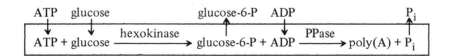

Coacervates thus represent systems which can, in principle, take up "food" and excrete "wastes". They are also capable of building polynucleotides from monomers. Their primitive metabolism attains a synthesizing and degrading steady state with the environment ($\Delta G \neq 0$). They have these properties in common with cellular systems.

Complex coacervates are also comparable to protoplasm in terms of colloid chemistry. Such cell-isomorphic systems, composed of biogenic materials, can be called *bioids*. They have the characteristics of preorganelles. Nonetheless, they do not represent precytic, much less living systems. The highly specific enzymes added to the coacervates are of biological origin. Furthermore, the coacervates can respond only to very limited problems. Even if one assumed a coincidental assembly of all the elements necessary for life in coacervates, their further development would be impossible. The droplets dissociate with small changes in pH or under mechanical stress. There is no boundary membrane. This makes reproduction

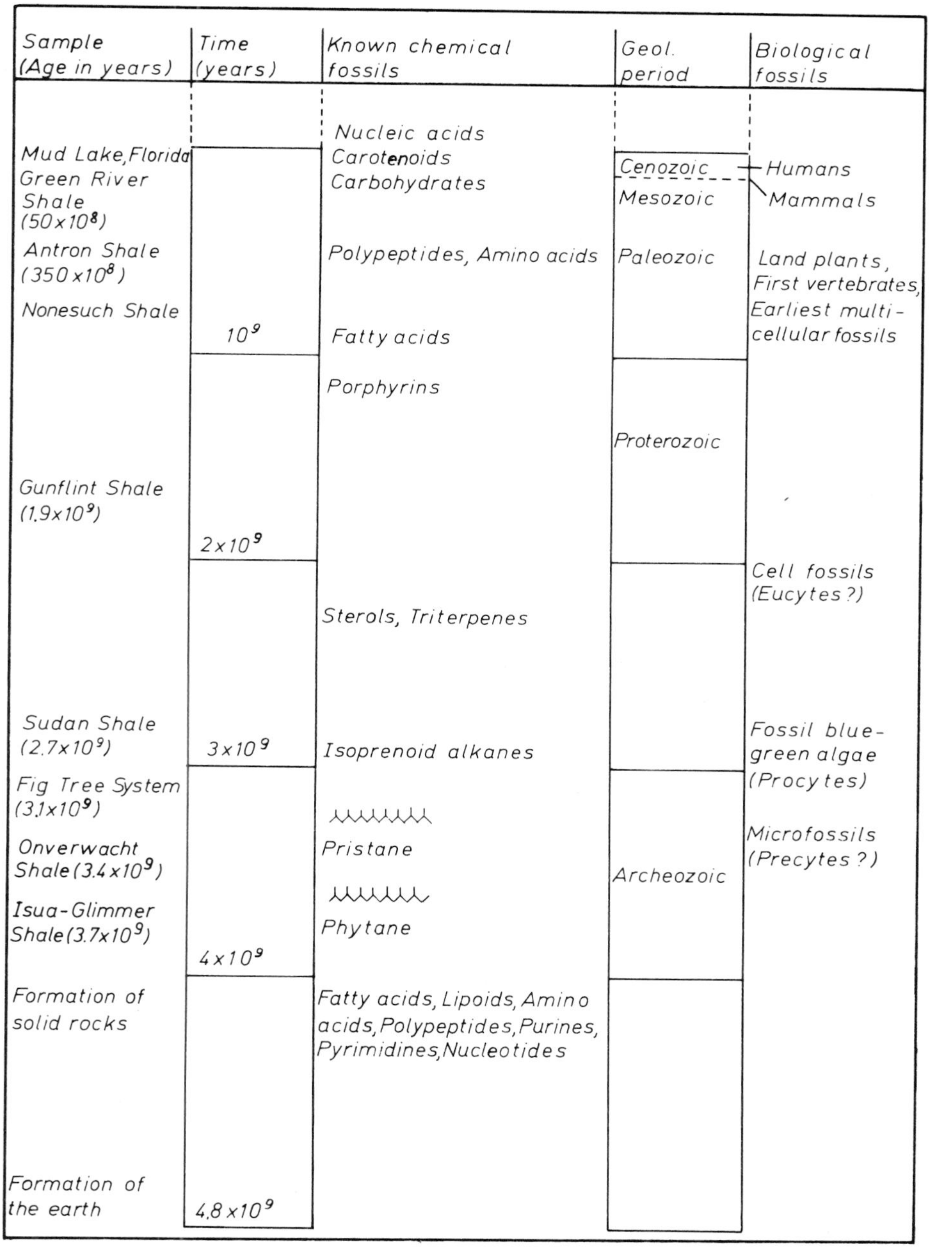

Sample (Age in years)	Time (years)	Known chemical fossils	Geol. period	Biological fossils
Mud Lake, Florida Green River Shale (50×10^8)		Nucleic acids Carotenoids Carbohydrates	Cenozoic Mesozoic	Humans Mammals
Antron Shale (350×10^8)		Polypeptides, Amino acids	Paleozoic	Land plants, First vertebrates,
Nonesuch Shale	10^9	Fatty acids		Earliest multi-cellular fossils
		Porphyrins	Proterozoic	
Gunflint Shale (1.9×10^9)	2×10^9			Cell fossils (Eucytes?)
		Sterols, Triterpenes		
Sudan Shale (2.7×10^9)	3×10^9	Isoprenoid alkanes		Fossil blue-green algae (Procytes)
Fig Tree System (3.1×10^9)		Pristane		Microfossils (Precytes?)
Onverwacht Shale (3.4×10^9)		Phytane	Archeozoic	
Isua-Glimmer Shale (3.7×10^9)	4×10^9			
Formation of solid rocks		Fatty acids, Lipoids, Amino acids, Polypeptides, Purines, Pyrimidines, Nucleotides		
Formation of the earth	4.8×10^9			

FIGURE 2. Geological time scale of chemical and biological evolution (adapted from Dose and Rauchfuss,[5] in turn from Calvin[2]).

impossible. Reproduction, however, is a prerequisite for any further development. Therefore, *Oparin's coacervate hypothesis* cannot serve as a model for precyte evolution. However, his basic approach of explaining the origin of life as an automatic physico-chemical process served as a basis for subsequent work.

Fox et al.[17,18,19] built upon the coacervate hypothesis. Their contribution, however, is more inclusive. The *proteinoid microspheres* which they introduced are stable systems. For

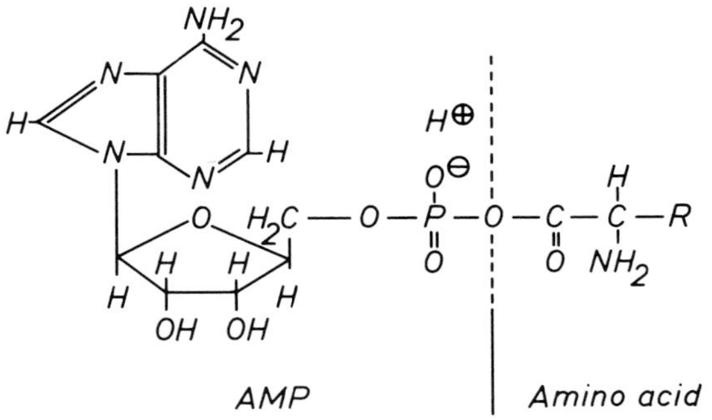

FIGURE 3.   Activation of amino acids by esterification with adenosine monophosphate (AMP).

this reason alone, they are more suitable for precyte simulation experiments than the coacervates. In addition, Fox generated the starting materials for his microsystems under simulated abiogenic conditions. He heated dry mixtures of predominantly acid amino acids in a nitrogen stream to 200°C. The amino acids condensed under these conditions to polyamino acids (polypeptides) with molecular weights of up to 300,000. Fox then dissolved this protein-like (proteinoid) material in warm sea water. A large number of small globules of the same size and shape, which he called proteinoid microspheres, were formed. The size and shape of the globules depended greatly on the kind, concentration, and proportions of amino acids in the starting mixture, as well as on the solvent and temperature. The sizes ranged from 0.5 to 80 μm. The spheres are bounded by a proteinoid membrane, which has a double-lamellar structure at higher pH values. Like biomembranes, the boundary membrane is selectively permeable. It also gives the microspheres their high degree of mechanical stability. The spheres can be centrifuged, embedded, and sectioned, so that they can be examined in an electron microscope. The fine structure of the microspheres resembles that of both the earliest cell fossils and of bacteria. Like yeasts and bacteria, the microspheres are able to form buds. The buds can be separated by thermal, electric, or mechanical shocks from the "parent particles". In saturated protein solution, the buds grow back to the original size, which is called "heterotrophic growth" by Fox. As another similarity to bacteria, the microspheres can divide across the middle by binary fission. However, the division is a purely physical result of increasing surface tension as the pH value is raised.

There are also chemical parallels. For example, there are both Gram-positive and Gram-negative staining forms of both bacteria and microspheres. Furthermore, proteinoid-zinc microspheres to which ATP has been added are capable of active motion. Microspheres react to being placed in solutions of lower or higher osmotic pressure by swelling or shrinking, respectively. The addition of lipids improves this osmotic-like property. The microspheres also display enzymatic properties, due to their proteinoid constituents. They are capable, among other things, of hydrolysis, decarboxylation, amination, deamination, and oxidation-reduction. This primitive enzyme effect is, to be sure, several orders of magnitude less active than that of bioenzymes.

Polynucleotides can also be built into the microspheres. They couple best with basic proteinoid which is composed of more than 50% lysine. In this way the so-called *nucleo-proteinoid microspheres* are formed. These are well suited as model systems for the study of interactions between nucleotides and peptides. Such interactions can be especially well analyzed when homomeric polynucleotides are used. For these studies, the nucleoproteinoids are incubated with amino acids which have been activated by esterification with adenosine monophosphate (AMP; Figure 3). Certain types of amino acids are preferentially coupled,

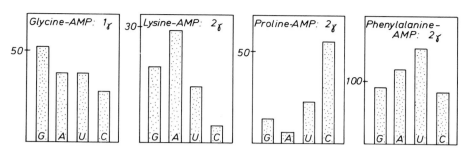

FIGURE 4. The relative polymerization effects of microparticles composed of lysine-rich proteinoids and various homologous polynucleotides with each of the four activated amino acids. The correlations have a code-like quality (from Fox and Dose[19]).

depending on which of the homomeric polynucleotides is used. The coupling is even codon-consistent. That is, poly (G) favors the selective incorporation of glycine-AMP, poly (A) the incorporation of lysine-AMP, poly (C) that of proline-AMP, and poly (U) that of phenylalanine-AMP (Figure 4). This, however, corresponds to the present genetic code for homogeneous triplets (glycine, GGG; lysine, AAA; proline, CCC; phenylalanine, UUU; — see Table 2). These results are so clear that one may infer a stereochemical basis for the genetic code. Fox and co-workers interpret the nucleoproteinoid microspheres as the precursors of ribosomes.

Microsystems thus possess many properties which they share with cellular systems (summary, Table 1). In spite of the indisputable wealth of cell-like properties, however, they are not precytes. They possess neither a self-regulating metabolism nor constant reproduction, nor true mutability, nor a truly autonomous morphogenesis. They lack an information system which could coordinate structure and function. Thus they represent, at most, preliminary stages of precytic organelles (preorganelles or abioids*). The *proteinoid hypothesis,* strictly speaking, can only serve as a partial model for precyte evolution. Nevertheless, Fox's research has served as a point of orientation for all subsequent studies.

The studies of Krampitz[28] and Paecht-Horowitz[47] and their co-workers built on the foundation laid by Fox's experiments. Their experiments were concerned with further clarification of the coupling of peptides on nucleotides under abiogenic conditions. For this purpose, they carried out simultaneous, enzyme-free (autocatalytic) polycondensation of amino acids on nucleotides. As starting material, they again used activated amino acids in the form of AMP esters (see Figure 3). Amino acid AMP esters are supposed to have been plentiful in prebiotic times. They are also the first step in the normal biological amino acid activation, although they are here bound to enzyme complexes. Mixtures of such activated amino acids were heated together with nucleotides for 24 hr at 250°C. Among the reaction products were found not only polypeptides, but also, in all probability, polynucleotides with molecular masses up to 100,000 daltons. The yield, polymer length and reaction velocity of the *simultaneous synthesis* were considerably increased by the use of natural clays, capable of swelling, such as montmorillonite,** as catalysts. Figure 5 represents a possible mechanism of the reactions.

Under certain conditions, nucleoprotein microspheres can form from the polycondensates. These are more stable than the purely proteinoid microsystems. It can be concluded from this that copolymers of protein and nucleic acid apparently form more stable structures than

* Abioids are cell-isomorphic structures which are composed of abiogenic material.
** According to Paecht-Horowitz, certain clays and minerals exhibit selective polymerization characteristics.

Table 1
COMPARABLE PROPERTIES OF PROTEINOID MICROSPHERES AND CELLS

- Size between 0.5 and 80 μm
- Structural stability
- Fine structure of the interior
- Membrane-like outer layer, sometimes two-layered
- Stainable with gram stains
- Selective diffusion
- Osmotic properties
- Enzymatic catalysis
- Selective binding of nucleotides
- Ability to change structure
- Tendency to unite
- Mobility
- Growth
- Budding and division
- Reproduction

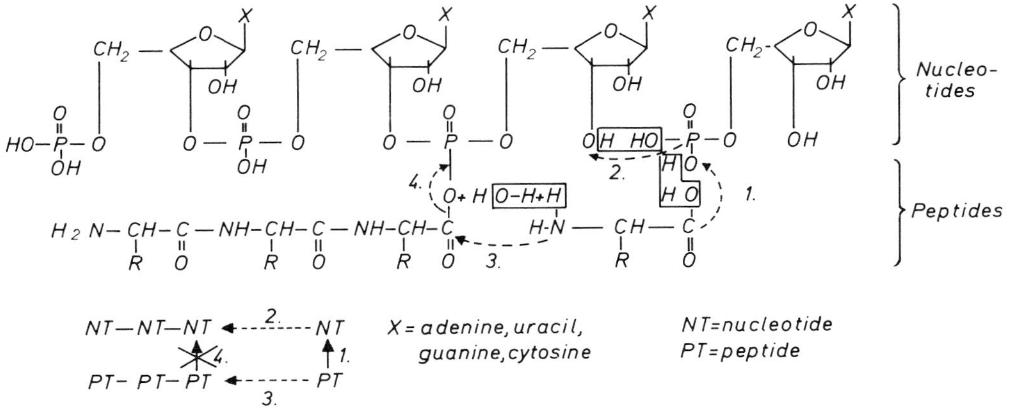

FIGURE 5. Hypothetical reaction mechanism for the simultaneous polycondensation of peptides on nucleotides (adapted from Calvin[3]; originally from Paecht-Horowitz[47]). In the first step, the coupling of the nucleotide (NT) to the amino acid (AA) with the elimination of a water molecule is assumed. Next the dimer couples its nucleotide to an existing polynucleotide chain. In the third step the peptide part of the dimer binds to the polynucleotide, simultaneously breaking the amino-acid-nucleotide bond in a fourth step (see the diagram below).

polymers of only one kind of macromolecule. Also, the results of the simultaneous synthesis indicate that proteins and nucleic acids could have been formed simultaneously and in close interaction with each other on the primitive earth. The interactions apparently did not take

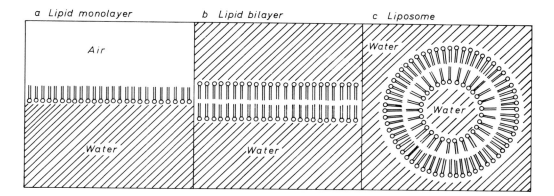

FIGURE 6. Model membranes composed of amphipathic molecules such as lipids (adapted from Nöll;[41] see also Figure 7). (a) One-layered (monolayer); (b) two-layered (bilayer); (c) liposome surrounded by a bilayer (bioid).

the form of simple addition reactions; rather they were similar to stereochemical transformations. "Nucleo-templates" must have affected the formation of proteins, which in turn stimulated the synthesis of nucleic acids. In this way not only proteins, but also nucleic acids of not entirely random sequence could have been formed through mutual interaction.

Other experiments were designed to determine the exact nature of the stereochemical interactions between proteins and nucleic acids. Nucleotides like adenine and cytosine were bound to the synthetic polymer polystyrol,[22] and the velocity with which the amino acids glycine and phenylalanine coupled to the two bases was measured. The following relative binding frequencies were found for the four possible combinations: Phe-A: 6.7%; Phe-C: 2.9%; Gly-A: 10%; Gly-C: 6.5%. The greatest difference in the reactivities was thus a factor of three. This experiment showed once more (compare with Fox's nucleoproteinoid microspheres) that even when only one base and one amino acid are considered, there is a kind of selectivity, i.e. specific affinity of certain amino acids to certain bases. The converse approach has also been taken. The specificity of binding of nucleotides to polypeptides was determined.[35] An atomic model based on this experiment was built and showed that three nucleotides corresponded to each amino acid residue. Selective coupling was observed, for example, between lysine and the base triplets AAA or AAG. That, however, is exactly the modern triplet code for this amino acid. The model also shows that the third base does not interact directly with the lysine residue. This indicates a degeneracy of the code words, as is actually found in the modern code. The experimental arrangement thus represents, up to a point, a model of a primitive transfer RNA. These experiments support the hypothesis that the present genetic triplet code had a prebiotic precursor based on stereochemical, kinetic, and/or thermodynamic factors.

2. Protein-Lipid Interactions

The interactions between proteins and lipids are another subject of intensive simulation research. Protein-lipid complexes are the basis for stable functional membranes. Membranes are the most important structural elements of cells, and thus also of precytes. In the past few years, therefore, interest in *membrane models* has grown rapidly (e.g. Sitte,[59] Calvin,[3] and Nöll[41]). The simplest membrane model is the *lipid monolayer* (Figure 6a). It can be produced by carefully placing a lipid/Dekan®* solution on the surface of water. A monomolecular layer of lipids can then form on the boundary surface between water and air. The lipid molecules are oriented so that their polar, hydrophilic "heads" point toward the

* Dekan is a nonpolar solvent for hydrophobic substances.

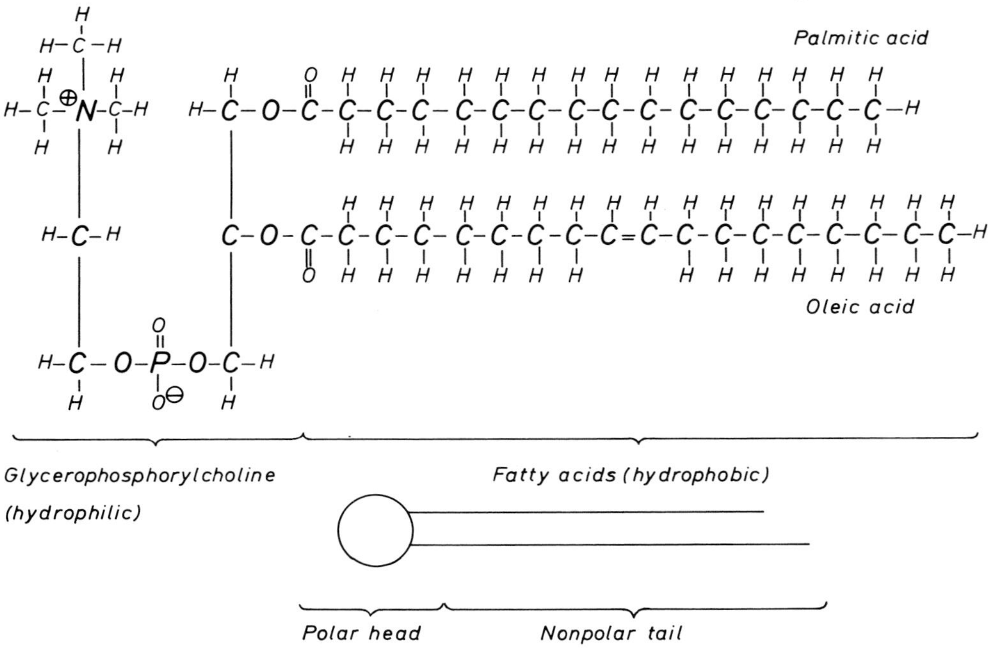

FIGURE 7.   Basic structure of a lipid molecule (lecithin) with a polar and therefore hydrophilic (water-soluble) "head" and nonpolar or hydrophobic (water-insoluble) "tail". Lipids are formed as esters of fatty acids. Phospholipids are fatty acid esters of the triple alcohol glycerol in which one of the OH-groups of the glycerol is esterified with a phosphate group (see also Chapter 3, Table 4). The best known biogenic lipids, the lecithins, are phospholipids in which the phosphate group is esterified to the amino alcohol choline.

water, and their nonpolar, hydrophobic "tails" into the air (Figure 7). On account of this contrasting behavior, such molecules are also called amphipathic.

To obtain two-layer lipid films, so-called *lipid-bilayers* (black films), one applies a lipid-Dekan solution to a hole, several millimeters in diameter, in a thin Teflon® or glass plate which is submerged under water during the operation. The total thickness of the film is 0.006 μm, which is about twice the length of a lipid molecule. The lipid molecules are here arranged so that their hydrophilic heads point outward, and their hydrophobic tails point toward the middle of the film (Figure 6b). The incorporation of proteins (for example cytochrome *c*) stabilizes the bilayer and gives it certain transport properties. Interestingly enough, these lipid bilayers, when stained and viewed in the electron microscope, show the same typical double lines that characterize biomembranes under the same conditions. This has led to the presumption that the amphipathic lipids are responsible for the basic structure of biotic membranes. The proteins probably serve only to stabilize the membrane (structural proteins) and to carry out various catalytic functions (functional proteins; see also Figure 8).

Finally, lipids can be dispersed to form *liposomes* by application of ultrasonic vibrations. The liposomes are spherical, with a diameter of more than 10 μm (Figure 6c). They are usually bounded by a bimolecular lipid membrane. Ions are taken up into the interior of the vesicles at a rate which depends on the nature of the lipids forming the membrane.[4] For example, membranes composed of the lipids phosphatidylserine and phosphatidylcholine are more permeable for Cl^- than for K^+ and Na^+. If the membrane is composed of the lipids phosphatidic acid and phosphatidylinositol, the diffusion of cations is favored. It is assumed that these different diffusion rates are due to the orientation of the dipoles and the

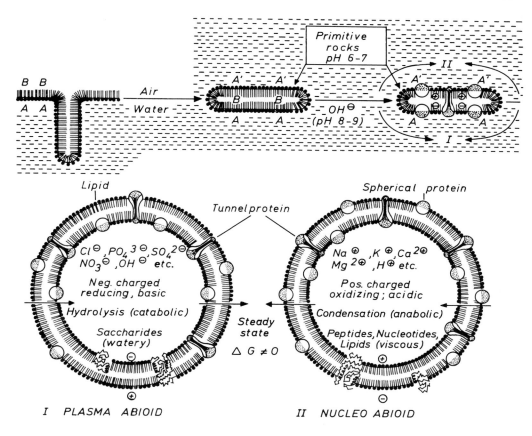

FIGURE 8. Abiotic formation of two-layered, asymmetric abiomembranes through differential incorporation of amphipathic, i.e. hydrophilic (A = dark) and hydrophobic (B = light) lipids or proteins. There are two possibilities for the formation of liposomes (see Figure 6): type I (plasma abioid) with negatively charged, reducing, basic, sugar-containing, catabolic sol medium (sugar cleavage) and type II (nucleo abioid) with positively charged, oxidizing, acidic, lipid-peptide-nucleotide-containing, anabolic gel medium (peptide-nucleotide simultaneous synthesis).

This model is purely hypothetical; as long as experimental data are lacking, it is meant only as a stimulus for further research. A′ = the less negative or more positive (in comparison to A) membrane layer. A tunnel or spherical protein is represented in cross-section in the lower part of each abioid.

charged groups in the membrane. The various kinds of lipids are also present in different amounts in the inner and outer layers of natural membranes. For example, the lipid phosphatidylcholine is preferentially incorporated into the inner layer of the erythrocyte membrane, while seryl-phosphatidylinositol is primarily present in the outer layer. This *structural asymmetry* is then, for example, responsible for the above-mentioned variable diffusion of water with respect to cations and anions.

The structure, osmotic properties, transport properties, etc. of liposome membranes resemble those of biomembranes. Thus the liposomes represent cell-isomorphic structures, which, due to their biogenic components, can be designated as bioids. Bioids are organelle precursors or preorganelles. Strictly speaking, neither the liposome membranes nor the monolayers or bilayers can be taken as models of precyte membranes, because they are produced from contemporary, biogenic substances. Nevertheless, it is assumed that such systems could have been formed under prebiotic conditions. Modern fats or lipids are not necessary. Fatty acids or ionized lipids will also do. They are also amphipathic molecules and can substitute functionally for the lipids in the membrane. They must also have been capable of forming asymmetric membranes.

C. Experiments in Cell-Free Systems

Cell-free, in vitro systems can also serve as models for exploring the origin of life, and thus precyte evolution. In these experiments, the individual components of cells are isolated from the cell complex and then recombined in vitro to determine their capacity for self-organization. The results of these experiments shed some light on the functioning of possible abiogenic systems. For example, it was shown with ribosomes and viruses that simple addition of all the molecular components to the cell-free system was sufficient for these highly organized systems to assemble themselves to functional units (e.g. Garret and Wittmann[20]). A ribosome, it should be remembered, consists of more than 50 different macromolecules. The RNA phage QB is even capable of reproduction in a cell-free system, if all the factors necessary for growth, such as nucleotides and RNA replicase are included.[31,38,61,62] After several replications, mutants even appear. They reproduce significantly faster, but become shorter and thereby lose their infectious capacity. In the given system, which contains an excess of nutrients, selection favors the most rapid growth rate. Here we have an *evolution experiment in vitro.*

In this sense, laboratory experiments have already succeeded in producing an artificial genobiotic system. These first experiments indicate that the evolution theory of Eigen,[6] which holds that the origin of life is coupled to the close mutual development of proteins and nucleic acids (hypercycle), is correct. To be sure, only contemporary biogenic macromolecules have so far been used in cell-free experiments. However, it is to be expected that such experiments will also be conducted in the near future with abiogenic starting materials. Eigen and co-workers[8-10] are attempting in this way to reconstruct the primitive forms of life from the clues which have been preserved in the biomolecules, the witnesses of genesis. By comparing the sequences of various tRNAs, they have deduced what may be the sequence of *ur-gene.* They are now attempting to synthesize the corresponding *ur-protein* and to test its enzymatic properties.

D. Computer Tests and Probability Calculations

Another method of precyte evolution research is *computer simulation,* in which the models developed from the results of simulation experiments, as well as from cell-free systems can be tested. Kuhn and co-workers[14,15,34] studied the selection processes involved in the RNA replication by computer simulation. They discovered, among other things, that the probability of error in the base pairing during RNA replication could not exceed *one* error per 1,000 bases, if selection was to occur. However, there is normally one error per 100 bases in non-enzymatic RNA replication (see also Naylor and Gilham[40]). Therefore, the RNA molecules must have consisted of fewer than 100 nucleotides at the beginning of evolution, in order for selection to have taken place.

Eigen and his co-workers[7] have also run computer simulation tests. They discovered in one such test that the mutation rate of RNA/DNA must lie between 1% and 2%. Above and below that rate there is no selection; above 2%, the collected "positive" genetic information is too quickly "forgotten", and below 1%, the development rate is too slow for the organisms with new genetic information to gain a competitive advantage. Furthermore, Eigen was able to confirm by computer tests the reaction kinetics of the *hypercycle* which he had developed. The results are so clear that they confirm the concept of the hypercycle.[6,8-10,29,53] The hypercycle has thus become a fixed part of the current theory of the origin of life.

Other aspects of precyte evolution have not yet or only partially been tested by computer simulation. For the present, only mathematical models in the form of reaction kinetics and probability calculations exist for these aspects. Eigen has set up quantitative calculations of reaction kinetics for his evolution theory. These are based on the Prigogine-Glanzdorff instability conditions of irreversible thermodynamics. This condition is regarded as the underlying principle of selection and evolution.[6]

Kuhn[32-34] also postulated his evolution theory as a physicochemical process which necessarily occurs under certain conditions. In contrast to Eigen, however, he proceeds from one developmental stage to the next in small, comprehensible steps. At each level, he evaluates the possibilities leading to the next stage in the least possible time.

Kaplan[25,26] has also developed a mathematically based evolutionary theory of the origin of life. It includes, among other things, probability calculations on the minimal complement of information and working molecules needed by a precyte, on the frequency and effectiveness of ur-enzymes, and on the problem of the reproduction rate of precytes. He calculated a doubling time of more than ten years for the first precytes. Unfortunately, the nature and scope of this publication do not permit a detailed mathematical derivation and discussion of the various evolution theories. The interested reader is referred to the original works.

The results of the various simulation analyses clearly indicate a very early common, *autocatalytic interaction between proteins and nucleic acids*. Such interactions appear to be prerequisites for the evolution of the precytes, which would contradict both the proteins-first and the nucleic acids-first hypotheses. To be sure, it is very probable that this common precyte evolution was preceded by a partly separate, parallel development of proteins and nucleic acids, up to the level of preorganelles. However, it is very likely that the precursors of transfer RNAs, ribosomes, sugar hydrolysis, etc. were first developed up to the level of preorganelles in separate areas, and were then integrated into a precyte. It is practically impossible for these basic elements of precyte evolution all to have developed in the same area, because the conditions under which they could have developed are quite different for each group. For example, the polycondensation of proteins and nucleic acids occurs most readily at an acidic pH, whereas hydrolysis of sugars takes place most rapidly in alkaline media. The construction of precytes out of preformed preorganelles would basically conform to the multilist hypothesis, and would contradict the multistep hypothesis.

Furthermore, the beginning of precyte evolution is dependent on the existence of some sort of *semipermeable membrane*. Only such a membrane can provide and maintain a medium in which the evolution of primary life can begin. As the experiments with membrane models show, structures basically comparable to biomembranes can arise spontaneously from lipids and proteins. They can also grow and "reproduce". The necessary, functionally equivalent molecules (lipoids) were already present abiotically. The formation of lipid-protein complexes, like protein-nucleic acid interactions, seems due to the inherent capability of these molecules for self-organization. The presence of membranes at the beginning of precyte evolution is, however, not in accordance with the multistep hypothesis, which holds that the membrane evolved together with protein biosynthesis, but it does agree with the multilist hypothesis, which assumes the abiogenic existence of membranes.

The following hypothetical model of precyte evolution is therefore based primarily on the multilist hypothesis, but it is supplemented by some aspects of the multistep hypothesis which have proved to be better supported by the evidence.

III. HYPOTHETICAL RECONSTRUCTION OF PRECYTE EVOLUTION

A. Abioids

The possibly oldest abiogenically formed aggregates of organic substances that have been reported thus far are 3.7 billion years old.[39] They come from the Isua-Glimmer Metaquarzite layer in Greenland. It is not possible to determine whether they represent preorganelle-like aggregates (abioids), but abioids must have arisen between 4 and 3.5 billion years ago. Their existence can be rather convincingly argued on the basis of the inductive and deductive evidence gathered so far. The basic elements of every cell, i.e. structures and mechanisms for energy supply and processing, as well as for storage and processing of information, may have developed in different abioids. As a structural precursor, the abiotic membrane was formed; the energetic precursor was probably the primitive hydrolysis of sugar with simultaneous formation of energy-rich pyrophosphate (P-P), and the informative precursor was the simultaneous polycondensation of proteins and nucleic acids.

1. Abiomembranes

One can presume that the prebiotic formation of abiomembranes was analogous to the experiments with model membranes (Figures 6 and 7). Monomolecular films of fatty acids or lipids could have first been formed in dark places on the surface of the primitive oceans, such as grottos (Figure 8). The motion of the water surface or wind might have caused these films to fold together to form bilayers, which float on the surface of the water. It is more likely, however, that bilayers arose spontaneously by enrichment with more lipids. The bilayers would have been stable over long periods, if they were surrounded on all sides by the primitive ocean. The double film would have been symmetric at first, but if one of the two membrane layers was in contact with another phase, such as the acidic surface of primal rocks, then the structure eventually became asymmetric through restructuring of the lipid molecules or incorporation of other molecules. Lipids (lipoids) with more negatively charged heads would be repelled by the surface of the rocks and would become enriched in the side of the film facing the alkaline sea water. The heads of the more positively charged or neutral lipids, on the other hand, would be concentrated toward the negative surface of the stone. Later, proteins would also collect in the two membrane layers, with their hydrophilic and hydrophobic parts seeking the corresponding parts of the membranes. Spherical proteins submerged to various depths in the lipid layer *(structural proteins)* are thought to have stabilized the membrane through more or less stable bonds to the lipoids/lipids. Proteins reaching all the way through the membrane and containing tunnel-like hydrophilic channels *(functional proteins)* may already have controlled the transport of ions and abiomers through the membrane. In this way asymmetric abiomembranes of characteristic structure, with negative upper surface and positive lower surface, could have arisen. They may well have been quite similar in structure to biomembranes.

Fine turbulences of the water surface, which might have been caused by supersonic vibrations, for example when meteorites struck, would then have dispersed such abiomembranes, forming closed, membrane-bound abioids (Figure 8). More often, however, the abiomembranes would have been curved due to their asymmetric composition, until they spontaneously closed to form spheres. As semipermeable membranes, which could be penetrated by water molecules but not by all dissolved substances, these abioids, like cells, were in a steady state equilibrium with their environment. Two types of abioids with different diffusion properties would have been formed, depending on whether the positive side of the bilayer was turned to the inside or the outside. Type I, with the positive membrane layer on the outside, must have preferentially absorbed anions, including OH^-, while Type II, with the negative membrane layer facing out, would have taken up more cations, including H^+. The acidic or basic character of the abioid would be determined by these different diffusion processes, and the reducing or oxidizing nature of the internal medium of the abioid would be correlated with the electrochemical potential.

In this way, an anabolic internal medium would soon have formed in the cation-rich, oxidizing, mildly acidic abioid Type II. This medium would favor the polycondensation of peptides, nucleotides, and lipids, which would have been preferentially taken up into these abioids. The polymerization of the peptides and nucleotides would have been partly simultaneous, thus representing the first steps toward the beginnings of a primitive genetic apparatus. For this reason, the Type II abioid has also been called the *nucleo-abioid*.[55]

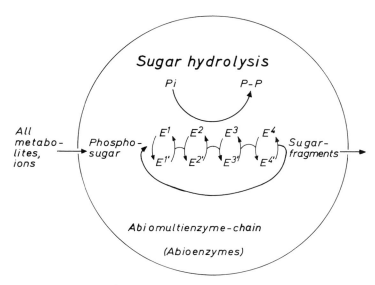

FIGURE 9.   Hypothetical scheme of plasma abioid function. In its hydrolytically active, catabolic, watery interior, sugar biphosphates are hydrolyzed under simultaneous formation of energy-rich pyrophosphate (P–P). Such reactions could already have been regulated by abiogenic multi-enzyme chain.[3]

The anion-rich, reducing, strongly basic internal medium of Type I abioids, on the other hand, would be catabolic, with a tendency to hydrolyze the saccharides selectively enriched by the permeation properties of the Type I membrane. The hydrolysis of sugars thus initiated the beginnings of an energy metabolism through the formation of energy-rich pyrophosphates. Therefore, the Type I abioid is also called the *plasma abioid*.

The changes in the internal milieu of abioids brought about by periodic changes in the environment will be discussed below.[64]

2. Plasma Abioids

Given large enough populations of abioids, and sufficiently long periods of time, eventually some would be formed which were capable of hydrolyzing abiotic sugar phosphates and simultaneously binding the energy released in the form of energy-rich pyrophosphates (P-P; Figure 9). The chemical energy stored in the P-P bonds was then available for other synthesis reactions (Figure 11). These reactions may have been catalyzed by *abioenzymes*, which might already have been present in the form of *abiomultienzyme-chains*.[3]

Fox's experiments with proteinoids produced under simulated primitive earth conditions (see Section II.B) make it appear probable that *abioproteins* exhibited *enzymatic capabilities*. In general, enzymes work by holding the reactants in position on a catalytically active site by means of a characteristic fissure or cleft in the surface of the protein. Their specificity is determined by the precise way in which the peptide chains are folded, so that only the appropriate molecules fit in the correct orientation for the reaction. Abioenzymes probably had no specific clefts, and no catalytic specificity. It is entirely possible, though, that a few amino acids in the appropriate relative positions would have served as naked ''active sites'' with a modest catalytic activity. The hydrolysis of sugar phosphates by such abioenzymes

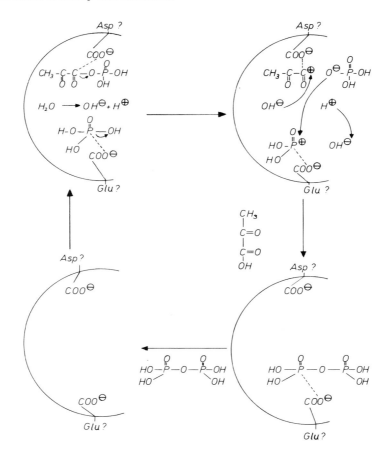

FIGURE 10. Hypothetical mechanism for the abioenzymatic hydrolysis of sugar phosphates to form energy-rich pyrophosphate. The half-circle represents the cleft in the abioenzyme. The two amino acid residues which form the active center are not adjacent to each other in the primary structure of the protein. The hydrolysis of phosphoenolpyruvate serves as an example here. The catalytic action of one of the acid residues causes a redistribution of the electrons in the substrate, resulting in the breaking of the phosphoric anhydride bond and the formation of a negative phosphate residue and a carbonium cation. The carbonium cation is bound by the negatively charged amino acid residue and there hydrolyzed to pyruvate. In the neighboring position, the other acid amino-acid residue simultaneously causes an electronic shift in an inorganic phosphate, resulting in the loss of an OH⁻; the positively charged phosphate residue is bound by the negative charge of the amino acid residue. Finally, the negative phosphate residue from the substrate is bound to the inorganic phosphate to form pyrophosphate. The enzyme can then start the cycle over again.

could have occurred as shown in Figure 10. The phosphate group was hydrolyzed when the sugar was bound to an acidic amino acid; the phosphate group could now be transferred to another phosphate bound in a neighboring position to another acidic side chain, either on the same abioenzyme or on another. In this way a high-energy pyrophosphate bond would be formed.

The hydrolytic activity of the plasma abioids must have been entirely dependent on the *periodic changes* in their environment. When the temperature was below 20°C, as in winter or at night, the abiomembranes were probably permeable to larger hydrophilic substances

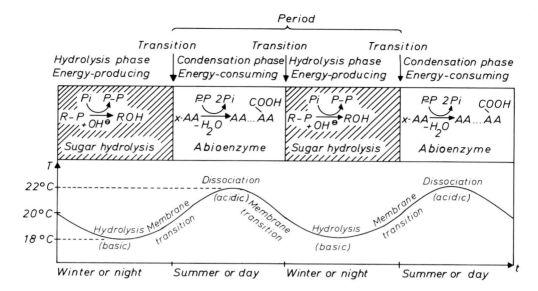

FIGURE 11. Hypothetical periodic changes in the internal milieu of the plasma abioid. Below temperatures of 20°C, as occur in winter or at night, sugar hydrolysis took place. At these temperatures, larger hydrophilic molecules such as sugar phosphates, amino acids and ions, including inorganic phosphate, could have passed through the membrane into the abioid. As the temperature gradually rose above 20°C in summer or in the daytime, the permeability of the membrane changed to exclude all but very small molecules or fat-soluble substances. In addition, the dissociation of H⁺ would have increased at higher temperatures, causing the pH value of the milieu to switch from basic to mildly acidic. This in turn stopped the hydrolysis and allows the condensation of peptides, from which the abioenzymes for the sugar hydrolysis were recruited.

like sugar phosphates and amino acids and to ions, as are modern biomembranes.[50,64] They would have served the abioids as new raw materials. On the other hand, these temperatures and the basicity of the internal milieu would have favored the exergonic hydrolysis of sugars (hydrolysis phase, Figure 11). As the temperature gradually rose above 20°C, as in summer or in the daytime, the permeability of the membranes would have changed significantly (transition phase), so that they only let small or hydrophobic molecules through. Furthermore, as the temperature rose, the dissociation of H⁺ increased, so that the pH of the internal milieu went from alkaline to mildly acidic. This in turn stopped the hydrolysis and encouraged the endergonic condensation of peptides and other molecules (condensation phase). According to the concept of catalytic cycles, the peptides catalyzed their own synthesis.[6] The abioenzymes for sugar phosphate hydrolysis were then recruited from these polypeptides. The energy-rich pyrophosphates accumulated during the hydrolysis phase would have served as an energy source for the formation of the peptide bonds. The yields of such reactions would naturally have been very low. It must have taken years for significant amounts of polypeptides to accumulate in this way (see Section II.D and Kaplan[25]).

3. Nucleo Abioids

Among the condensation-favoring nucleo abioids, some must have arisen, given large enough populations and enough time, which were capable of *simultaneous synthesis* of peptides on nucleotides (Figure 12). A prerequisite for this synthesis was the binding of a nucleotide to each amino acid (codimer). The formation of such bonds prebiotically, must have been possible either catalyzed by abioenzymes or by inorganic catalysts at sufficiently high temperatures (i.e. References 47 and 48). These bonds activated the amino acids energetically and made them reactive. Modern protein biosynthesis takes place in a similar

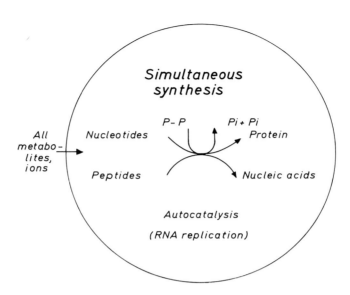

FIGURE 12. Hypothetical scheme of nucleo abioid functions (see also Figure 8). Simultaneous polycondensation of peptides and nucleotides to proteins and nucleic acids takes place in its condensation-favoring, anabolic, viscous milieu. These reactions are strongly endergonic and are autocatalytically driven by the energy-rich nucleotides (e.g. ATP) themselves. In the relatively protected milieu of the nucleo abioids, the RNA molecules should have been able to replicate themselves nonenzymatically by forming double helices (see Figure 13).

way, except that here the nucleotide is part of a chain of nucleotides (tRNA). The codimers then react with each other in such a way that nucleotide-nucleotide and peptide bonds are formed as the amino-acid-nucleotide bonds are broken (for a possible reaction mechanism, see Figure 5). Due to the large amount of energy stored in the amino-acid-nucleotide bond, the simultaneous synthesis could proceed *autocatalytically,* that is, without enzymes or extra sources of energy. This would be particularly true of the energy-rich nucleotide triphosphates, which were primarily involved in such reactions. The formation of bonds between peptides and nucleotides under physiological conditions, but non-enzymatically, is still an unsolved problem.

The simultaneously formed polypeptides and polynucleotides could not have been of purely random sequence, since there are selective affinities between amino acids and nucleotides. Homomers must have occurred with an extremely high frequency. The affinity of identical or homomeric subunits for each other is greater, for steric and energetic reasons, than the affinity of unlike units. The simulated separate polymerization of amino acids and nucleotides also led to such results[2,5,63] (see Table 2). Homomers with about 10 subunits were particularly stable. These 10-membered chains duplicated themselves to 20-membered chains, etc. For the reasons described above, certain combinations appeared relatively frequently (see Figure 18).

The combined synthesis of peptides and nucleotides requires a high degree of cooperation

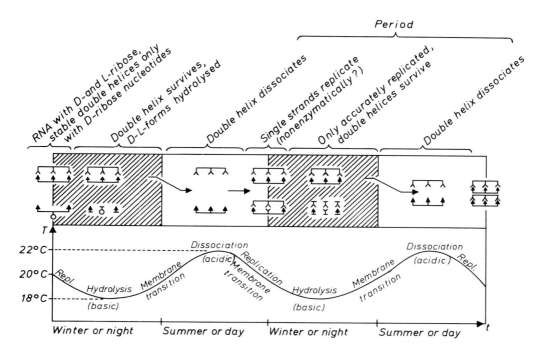

FIGURE 13. Purely hypothetical representation (adapted from Kuhn[32]) of the periodic changes in the internal milieu of nucleo abioids (the schematic structure of the RNA is always shown without helical winding). Below 20°C, at night or in winter, only those double helices composed exclusively of D-ribose would have escaped hydrolysis. Mixed polymers containing both D- and L-ribose could not form stable double helices and were hydrolyzed. When the environment warmed to 20°C by day or in summer, the permeability of the membrane changed to allow only small or hydrophobic molecules to enter. Furthermore, the H⁺ concentration rose from ≥ pH 7 to the acidic range, causing the double helices to dissociate. As the medium gradually cooled, conditions arose which allowed the nonenzymatic replication of the individual strands. As soon as the temperature sank below 20°C, the abiomembrane again became permeable to larger hydrophilic molecules, such as nucleotides, amino acids, and ions, including inorganic phosphate, and the pH value rose into the mildly alkaline range. In this hydrolytically active milieu, all imperfectly replicated double helices were degraded, whereas perfect copies survived to dissociate and replicate in the following phase.

in yet another respect. The D- and L-forms of the optically active amino acids appear not to have copolymerized equally well with either form of ribose or deoxyribose. It may be that, for sterical reasons, L-amino acids fit better with D-sugar nucleotides and that these codimers coupled more rapidly with one another than the other pairing. In any event, the modern biogenic systems contain L-amino acids and D-sugars exclusively.

Apparently, the nucleo abioids first formed their nucleic acids exclusively from ribonucleotides (those with ribose as the sugar moiety: ribonucleic acid, or RNA). Representatives of the DNA type (deoxyribonucleic acid) presumably arose less frequently or later in the course of evolution. D- and L-ribose were originally present in RNA molecules with equal frequency. However, Kuhn[32] has calculated that among $2^{20} = 10^6$ RNA molecules consisting of 20 nucleotides, one containing either all D- or all L-ribose would occur. (If the two forms were not incorporated with equal frequency, as suggested by the copolymerization hypothesis, there would have been more RNA molecules containing only one form of sugar.) The fate of these molecules would have been determined by the periodic changes in the abioid environment (Figure 13[32]). Below about 20°C, at night or in winter, only the uniformly built RNA molecules would have survived, because only they would have been able to form compact and hydrolysis-stable double helices by attracting the complementary bases along

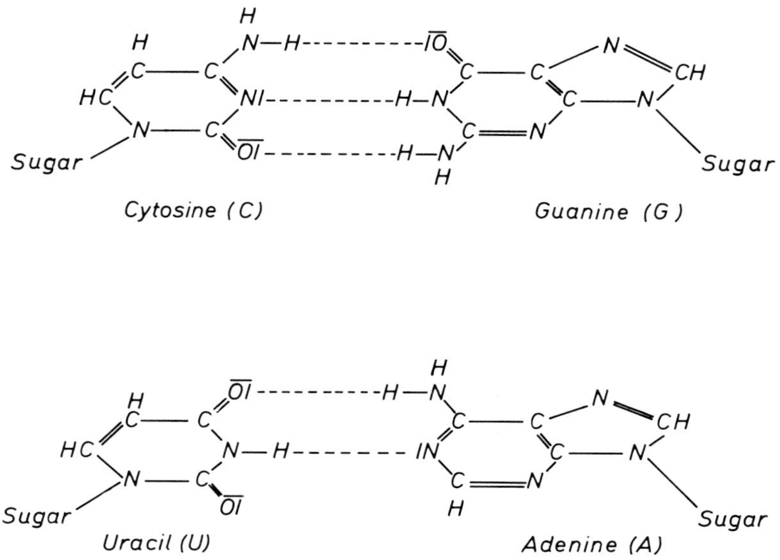

FIGURE 14. Pairing of complementary bases in RNA replication.

their length. These would then have condensed to the complementary strand. Guanine formed hydrogen bonds with cytosine, and uracil with adenine (Figure 14). There are experimental data to support the assumption that *RNA replication* could proceed in the *absence of enzymes.*[40] The alternative to enzyme-free replication would be to assume the presence of abioenzymes.

As the temperature rose above 20°C in summer or during the day, the permeability of the abiomembranes would have changed gradually. Only smaller or fat-soluble molecules could pass it at these temperatures, and the dissociation of H^+ also increased, so that the pH value of the abioid internal milieu must have gone from alkaline or neutral to mildly acidic. This stopped the hydrolysis and caused the double helices to dissociate (melt) into single strands. During the cooling phase, conditions would have been favorable for replication of the individual strands, as described above. When the temperature sank markedly below 20°C, the membrane again became permeable for larger hydrophilic molecules, which provided the needed building materials. The pH value again turned from the acidic range to neutral or slightly alkaline, and the milieu favored hydrolysis. Only the accurately replicated double helices survived to dissociate and replicate again in the following phases.

The incorporation of one non-complementary nucleotide or of one nucleotide with a sugar of the wrong optical configuration would have reduced the cooperativity and thus the formation of intact double helices. Such double helices would not have survived the hydrolysis phase. In this way, the formation of non-uniform RNA was gradually suppressed. The double helices with D-ribose then gradually displaced the L-systems, which probably developed more slowly. Only D-ribose is found in modern biogenic systems.

B. Protobionts

As long as the population of plasma and nucleo abioids existed separately, evolution was in a divergent phase. Further selection could not occur, for plasma abioids lacked a self-reproducing protein-nucleic-acid-lipid system, and the nucleo abioids lacked a productive energy metabolism. Through chance encounters and fusions of plasma and nucleo abioids to *protobionts,* which would have had to happen sooner or later, however, the prerequisites

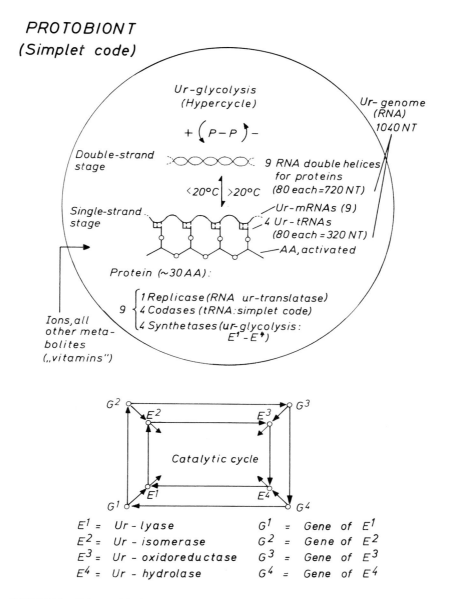

FIGURE 15. Scheme of the hypothesized protobiont functions: nucleic acid and protein systems are coupled as hypercycles. The amino-acid-group-specific simplet code (groups of three bases, of which the middle base is specific; Table 3) contains the information for the synthesis of 4 enzymes for ur-glycolysis, another 4 for the group-specific coupling of amino acids (codases; see Table 2) to 4 different cloverleaf RNA molecules (ur-transfers), and one replicase for error-free replication of RNA (quantitative gene count, see Table 5).

for the coupling of anabolism and catabolism and their genetic control by a coding mechanism were achieved (Figure 15). Then evolution again entered a strongly selective, convergent phase. Before explaining this process, however, we discuss the structure of the internal medium of the protobiont.

The combination of plasma and nucleo abioids applied not only to their two membrane types, but also to the milieu types. However, the milieus were not completely homogeneously mixed. Instead, proteins, nucleic acids and lipids would have collected into a more viscous,

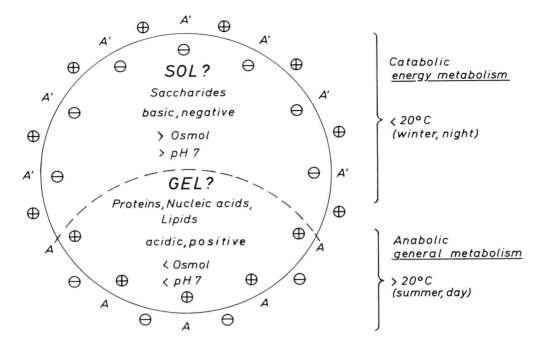

FIGURE 16.   Scheme of the assumed formation of two phases after the fusion of the plasma and nucleo abioids to form protobionts: a basic, negatively charged, osmotically active, watery, sol phase containing dissolved sugars and the corresponding parts of the membrane and an acidic, positively charged, osmotically less active, viscous, colloid or gel phase containing dissolved proteins, nucleic acids, lipids, and the appropriate portion of membrane (see Figure 8). The catabolic energy metabolism is active at night or in winter, since the anabolic processes occur mostly in the day or in the summer.

acidic colloid phase (gel?), leaving the more watery, alkaline phase (sol?) to the sugars (Figure 16). The colloid particles would have been mostly present as anions, which would have attracted an equal number of cations, including H^+, to preserve electrical neutrality. The more diffusible anions, including OH^-, would have been displaced from this phase. The result would be small dynamic, osmotic and pH differences between a more acidic, positive, osmotically active *gel gradient* and a more basic, negative, less osmotically active *sol gradient* (see also the Donnan distribution: Karlson[24]). Thus the protobiont combined the hydrolytic, catabolic milieu of the plasma abioid with the condensation-favoring anabolic milieu of the nucleo abioid, both in a certain sense as an antagonistic gradient system and membrane-specifically separated (see Figures 9 and 12). It was therefore possible for the energy metabolism to develop further in the aqueous sol-like phase, and the genetic system in the gel-like colloid area, while both phases interacted with each other. If we also take into account the periodic changes in the environment, then we must assume that in winter or at night the catabolic energy metabolism was dominant, while in summer or in the daytime, the anabolic building metabolism was in operation.

1. Ur-Messenger and Ur-Transfer RNA

In the colloid region, the double helices taken over from the nucleo abioid duplicated themselves further by way of condensation, using the energy-rich pyrophosphates delivered from the sol region, until they reached a size of about 80 nucleotides. This represents four duplications, with reference to the original 10-nucleotide RNA homomers (10/20/40/80). Longer polymers were apparently not stable, presumably because the chance that they

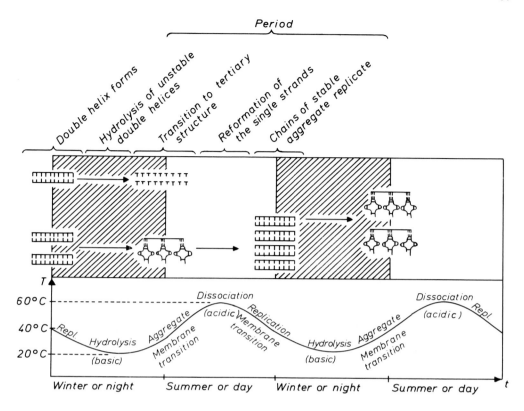

FIGURE 17. Hypothetical representation of the effect of periodic fluctuations on the development of a genetic system in the protobiont (adapted from Kuhn[32]). Toward the end of the hydrolysis phase, as the transition to single-stranded RNA occurs, only those RNAs which form stable aggregates of collector strands and cloverleaf-like associate molecules (tertiary structure) survive. In the dissociation phase, the stable aggregates form single strands which replicate. These thus increase in number and eventually displace the less stable aggregates, which are hydrolyzed during the hydrolysis phase. The contact with the membrane of the protobiont possibly plays an essential role in the formation of the aggregates.

contained a replication error was too great. They were broken down during the hydrolysis phase before they had a chance to replicate and thus increase their numbers.

The 80-nucleotide RNA molecules also were only able to survive when they adopted hydrolysis-resistant tertiary structures, which then unfolded again to open-chained, replicable molecules before the beginning of the replication phase. This sort of transformation has been experimentally observed.[32,51] RNA forms with different tertiary structures were probably also differentially resistant to hydrolysis. Compact forms may be more resistant than others and are more likely to survive the hydrolysis phase. However, selection of such RNA forms purely on the basis of their tertiary structures would not have been possible, due to replication errors. Careful analysis has shown an error probability of 1/100 per base.[6] Thus, on an average, one in 100 bases would be incorrectly replicated, so that for each replication of a 80-nucleotide RNA molecule, approximately one error appeared. For successful selection, however, there could not have been more than one error per 1000 bases, according to computer simulation calculations.[14,15] As a consequence, the hypothetical system went into a phase of divergent evolution of RNA molecules with various tertiary structures. In the process, new forms arose: aggregates between tertiary structures and open-chained forms (Figure 17). There is experimental evidence for formation of aggregates by self-organization (Chapter 3, Section III).

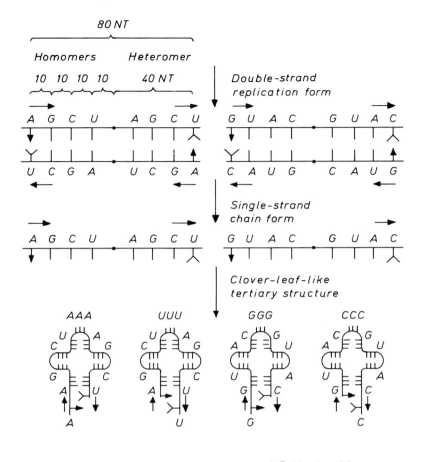

FIGURE 18.   Possibly preferred sequences of the 80-member RNA molecules which form especially stable, base-paired cloverleaf structures of the associate molecules. Such molecules have, for steric reasons, a free base triplet on their closed end. NT = nucleotide.

The RNA aggregates were certainly more resistant to hydrolysis than the individual molecules. Furthermore, they could be reproduced with fewer errors, because imperfectly replicated subunit molecules would no longer fit, due to their changed structure, and so would diffuse away and be hydrolyzed. The aggregate thus chanced to serve a new purpose. The conditions for selection were fulfilled by the aggregates, but not, however, for the individual molecules. The transition began to a strongly selective, convergent phase of evolution.

The result of the convergent phase was hypothetical RNA molecules which were able to form stable *tertiary structures* with specific contact regions through intramolecular pairing of complementary bases. Their chain forms may be at the same time able to serve as *collector strands* for the association of the tertiary structures. Those molecular sequences which were able to form *cloverleaf structures* by intramolecular base-pairing, as they are now found in transfer RNAs, were favored (Figure 18). These could have arisen by end-to-end condensation of identical 40-nucleotide double helices with those sequences which were taken over from the nucleo abioids (Section A.3).

For steric reasons, there were always three unpaired bases on the closed end of the

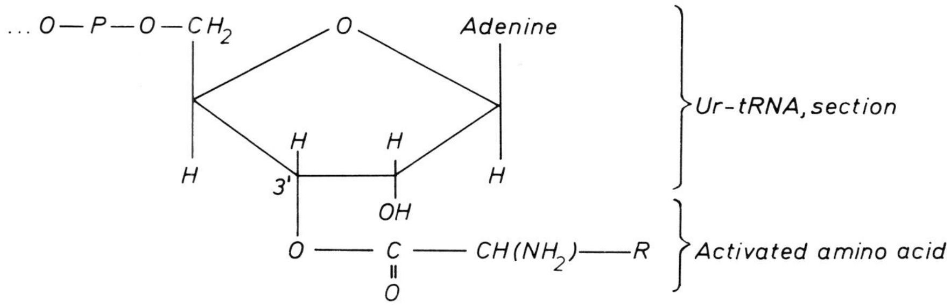

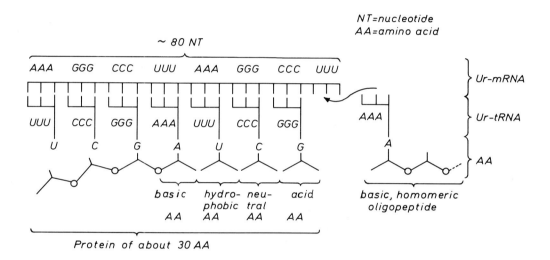

FIGURE 19.   Representation of the probable group-specific coupling of an amino acid to the open end of the cloverleaf molecules (ur-transfers) by a specific codase. The amino acid is preferentially coupled to the 3' position, and thus activated for peptide formation.

FIGURE 20.   Hypothetical model of ur-translation in the protobiont. The ur-transfers, group-specifically charged with amino acids, line up along the collector strand (ur-messenger) by means of base pairing, and thus make possible a genetically controlled polymerization of the coupled amino acids to proteins (hypercycle).

cloverleaf molecule, more or less as ur-triplets (see the wobble hypothesis suggested by Crick and proved by Söll[1]). Since the intramolecular pairing of the molecules was not symmetric, possibly also for steric reasons, there would have been, in the ideal case, and given the sequences discussed above, four different ur-triplets: AAA, UUU, GGG, CCC. These formed the stable associations by pairing with three complementary bases of the collector strand. It may be that at first only the middle bases were complementary to each other, which would be the same as a simplet code (see Table 3). The amino acids or (usually homomeric) oligopeptides taken over from the nucleo abioid could couple to the unpaired, open end of the clover-leaf structure (Figure 19). Such activated amino acids or peptides could easily polymerize to proteins after the carrier molecules had lined up with the collector strand, each with its anticodon triplet matching the codon triplet of the collector strand. For lack of space, carriers with more than one coupled amino acid (oligopeptides) could only have associated at the ends of the strands. In this way the collector strand became the *ur-messenger RNA* (mRNA), and the associating or clover-leaf molecules became the *ur-transfer RNAs* (tRNAs; Figure 20).

2. Hypercycle

Each of the four nucleotide bases has a specific affinity to one of the four groups of amino acids with similar side chains. These four groups are made up of amino acids with acidic, basic, hydrophobic, or intermediate side chains (Table 2). Five of the hydrophobic amino acids (Phe, Leu, Ile, Met, and Val) have a selective affinity for the base uracil. This corresponds to the modern codons xUx which code for precisely those hydrophobic amino acids (Table 3). Intermediate amino acids (e.g. Ser, Pro, Thr, Ala) correspond partly to the base cytosine or the codon series xCx. Hydrophilic amino acids, on the other hand, have a definite preference for adenine and guanine, or the codon series xAx and xGx. However, it is not possible to assign clearly the basic (Lys, Arg) or the acidic (Asp, Glu) amino acids to one or the other base. Probably the modern irregularity arose later through evolutionary incorporation of more amino acids (His, Trp, Gln, Met ?) and the start and stop codons. The amino acid glycine, which was abundant in prebiotic times, also occupies a special position. This may be due to its simple structure, which does not contribute a side chain to the protein. These observations suggest the existence of an early *primitive code* (see also Kaplan[25]). Probably, this primitive code was only specific for the first base of the anticodon triplet on the ur-transfer and the codon of the ur-messenger. According to the nature of their side chains, the amino acids would have been coupled to one of four different types of ur-transfers. This assumption is supported by the observation that the fourth base in modern transfer RNA molecules is specific for the functional group of the amino acid coupled to that species of tRNA (the first three bases are always ACC). The specific coupling of four amino acid groups to four types of transfer RNAs would require four *primitive codases*. Since the code was not unequivocal, eleven to sixteen amino acids would have been incorporated into the primitive proteins. However, since it was specific for a functional group of amino acids, the enzymatic and structural properties of these proteins were probably not greatly influenced by the lack of complete specificity.

The proteins synthesized by this system were no longer of random sequence. Since the folding and function of a protein depend on the order of hydrophobic, neutral, and polar amino acids in its primary structure, the ur-enzymes must have had a modest degree of specificity. However, the proteins with useful properties could not be selected, because the base sequence of the ur-messenger was too rapidly changed by errors in replication. A simplet code for four amino acid groups was thus developed, but it could not become established. This phase of evolution was thus divergent. Sooner or later, however, an enzyme which catalyzed RNA replication and reduced the number of errors from 1/100 to about 1/1000 must have arisen. Now the information on the amino acid sequence required for the synthesis of this *ur-replicase* was conserved. At the same time, the replicase increased the speed of replication, so that the protobiont which possessed it had the ability to supplant all others. The simplet code now gradually became fixed. Any individual with a defect in the anticodon triplets of the ur-transfer RNA produced the wrong proteins and was eliminated.

It follows that the protobiont was a system in which the information molecules, or genes (nucleic acids) were catalytically and self-instructively coupled to the functional molecules or enzymes (proteins; Figure 15). In this system, only one enzyme served as replicase for the reproduction of the various genes on the RNA molecules. Such systems, in which there is only one replicase, and the other elements of the system are only functionally coupled, have been shown by computer simulation to be likely to be unstable (see Section II.D).

Therefore it is logical to postulate a system in which every gene is synthesized by its own replicase. Thus many of these replicases are needed, so that every component of the system has its own specific coupling factor. Eigen[6] proposed such a cyclization in the form of the *hypercycle*. Without going into the arguments in detail, we shall give a short sketch of the evolution of the hypercyclic coupling of the nucleic acids with proteins, as suggested by Eigen and Schuster.[8-10]

Table 2

THE FOUR AMINO ACID GROUPS AND THEIR SELECTIVE AFFINITY TO THE FOUR NUCLEOSIDE BASES

Characteristic \ *AA-Groups*	*Acid AA with ionized groups : COO⊖*	*AA with polar, nonionized groups: -SH, -CO, -NH$_2$, -OH*	*AA with nonpolar residues : -R*	*Basic AA with ionized amide groups : NH$_3$⊕*
H$_2$O solubility	*Hydrophilic*	*Intermediate*	*Hydrophobic*	*Hydrophilic*
Charge	*Negative*	*Neutral*	*Neutral*	*Positive*
Binding	*Ionic*	*H-bridges*	*Hydrophobic*	*Ionic*
Abiogenic AA	*Asparate (Asp)* *Glutamate (Glu)*	*Asparagine (Asn)* *Cysteine (Cys)* *Proline (Pro)* *Serine (Ser)* *Threonine (Thr)* *Tyrosine (Tyr)* — *Alanine (Ala)*	*Glycine (Gly)* *Isoleucine (Ile)* *Leucine (Leu)* *Phenylalanine (Phe)* *Valine (Val)*	*Arginine (Arg)* *Lysine (Lys)*
Presently known only as biogenic AA		*Glutamine (Gln)* *Methionine (Met)*	*Tryptophane (Trp)*	*Histidine (His)*
Rel. binding frequency in copolymerisation of AA-AMP (from Calvin 1969)	*e.g. Asp-Asp : 55% Asp-His : 8%*	*e.g. Ala-Ala : 47% Ala-Gly : 15% Ala-Asp : 2%*	*e.g. Val-Val : 52% Val-Ala : 13%*	*e.g. His-His : 44% His-Asp : 12%*
Selective base affinity (see also Fig. 4.)	*Guanine*	*Cytosine*	*Uracil*	*Adenine*

Table 3
MODERN GENETIC CODE AND HYPOTHETICAL ORIGINAL CODE

A   Modern genetic code

Codon	AA	Codon	AA	Codon	AA	Codon	AA
UUy	Phe	UCx	Ser	UAy	Tyr	UGy	Cys
UUz	Leu			UAz	Stop	UGA	Stop
						UGG	Trp
CUx		CCx	Pro	CAy	His	CGx	Arg
				CAz	Gln		
AUy }		ACx	Thr	AAy	Asn	AGy	Ser
AUA } Ile				AAz	Lys	AGz	Arg
AUG	Met						
GUx	Val	GCx	Ala	GAy	Asp	GGx	Gly
				GAz	Glu		

x : U,C,A   o   G

y : U   o   C

z : A   o   G

B Hypothetical original code (simplet code)

xUx (group I)	xCx (group II)	xAx (group III)	xGx (group IV)
Phe, Leu Ile, Val	Ser, Pro Thr, Ala	Glu, Asp	Arg Lys (?), Gly(?)
hydrophobic AA	intermediate AA	acid AA	basic AA

Note: The most recent studies by Eigen and Winkler-Oswatitsch[11-13] on the evolution of the genetic code also suggest an ur-code in which the four bases G, C, A, and U, when found in the middle position of a triplet G–C,* coded for the then four most prevalent amino acids valine (GUC), alanine (GCC), asparagine (GAC) and glycine (GGC). The results of Miller's simulation experiments (see Chapter 3, Figure 1), indicate that these four amino acids were among the most abundant in the ur-soup.

The code which has recently been discovered and analyzed in mitochondria differs from that of other organisms — some researchers regard it as the more primitive code; it may open up new possibilities of reconstructing the ur-code (cf. Chapter 7, II).

According to Eigen, at the beginning of the biogenetic phase there must have been many protein-like substances of indeterminate sequence and few RNA-like polymers. The RNA polymers became capable of self-replication, either with or without catalysis by abioid ur-enzymes (Table 4A). GC (Guanine-cytosine) rich compounds would have formed the longest reproducible sequences, but AU (adenine-uracil) substituents would have been just as necessary. They allow for structural flexibility, which would promote rapid reproduction. Re-

---

*   G≡C base pairs predominate in the DNA molecules of primitive genetic systems (archaebacteria) and must also have been predominant in primitive systems (protobiont, eobiont, ur-procyte), since they are, as a result of their triple bond, more stable than the A=U pairs with only a double bond.

Table 4
HYPOTHETICAL REPRESENTATION OF THE EVOLUTION OF THE HYPERCYCLE

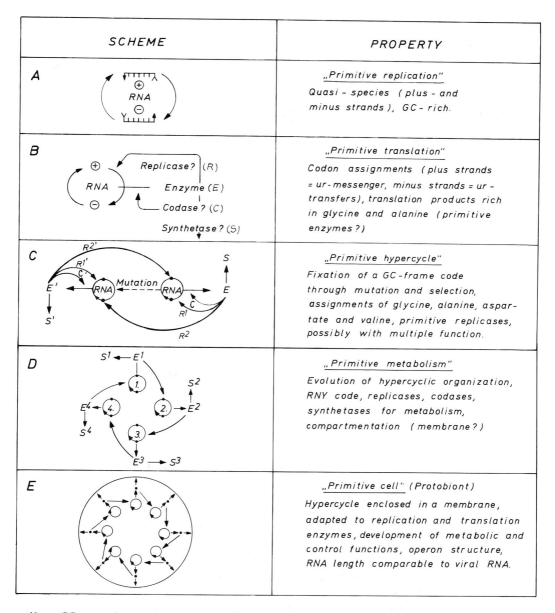

SCHEME	PROPERTY
A	„Primitive replication" Quasi - species (plus - and minus strands), GC - rich.
B	„Primitive translation" Codon assignments (plus strands = ur-messenger, minus strands = ur-transfers), translation products rich in glycine and alanine (primitive enzymes?)
C	„Primitive hypercycle" Fixation of a GC-frame code through mutation and selection, assignments of glycine, alanine, aspartate and valine, primitive replicases, possibly with multiple function.
D	„Primitive metabolism" Evolution of hypercyclic organization, RNY code, replicases, codases, synthetases for metabolism, compartmentation (membrane?)
E	„Primitive cell" (Protobiont) Hypercycle enclosed in a membrane, adapted to replication and translation enzymes, development of metabolic and control functions, operon structure, RNA length comparable to viral RNA.

Note: GC = guanine-cytosine, R = purine, Y = pyrimidine, N = purine or pyrimidine.

Adapted from Eigen and Schuster;[9,10] details, see text.

producible sequences in the form of the plus and minus strands make up *quasi-species*, which would be subject to Darwinian evolution.

In the course of this evolution, the plus strands apparently became chain-like messengers, and the minus strands, perhaps through intramolecular base-pairing, may have become ur-transfers (Table 4B). The amino acids would have been attached to the ur-transfers, in frequencies determined by the availability of the ur-transfers, and the ur-transfers paired with the commaless pattern of the primitive messenger. The products of this translation would have been rather monotonous at first, since they, like the proteinoids, were composed mostly of glycine and alanine. Like the abioproteins, they might have had slight catalytic properties. If one of the possible translation products was able to catalyze its own messenger, this messenger would come in time to dominate in the population of quasi-species (Table 4; C,D). Mutants of the dominant messenger may have been coupled to a reproductive cycle or hypercycle whenever they provided a further advantage, for example in the form of multiple catalytic capacities enabling them both to replicate the messenger and to serve as a codase to couple amino acids to the transfer RNA or as a synthetase for metabolic products.* In this way, an RY (R = purine, Y = pyrimidine) structured code might have been selected which already distinguished four group qualities of amino acids (see Figure 20; Tables 2 and 3) or four amino acids (glycine, alanine, asparagine, valine).

More and more less divergent mutants would have been integrated in the hypercycle (Table 4; D,E). The steadily increasing accuracy of replication made possible a lengthening of sequences. Gene duplication and subsequent divergent development produced from a common ancestor the various functional and structural proteins, i.e. replicases, codases, operons, synthetases, ribosomal proteins, membrane proteins, and the enzymes for a primitive energy metabolism (primitive glycolysis?). Eventually, the RNY (R = purine, Y = pyrimidine, N = purine or pyrimidine) code arose. Eigen proposes that the modern translation and transcription apparatus could have arisen when the individual genes were joined to form a single strand and the information was gradually transferred to DNA (see section D); every enzyme then had its own specific function.

3. Primitive Glycolysis, Primitive Biomembranes

The formation of the hypercycle produced another convergent phase of evolution in which the protobionts which produced proteins with determined amino acid sequences evolved. Those systems which happened to synthesize proteins with useful properties would have prevailed. One can imagine that first ur-codases (aminoacyl synthetases) for the four ur-transfer RNAs were selected which increased the specificity of binding of the amino acids to the RNAs. A coupling of oligopeptides to the tRNAs was no longer possible.

In addition, selection would have favored enzymes which catalyzed the hydrolysis of sugars by the protobionts (see Figure 10). One can speculate that a *primitive fermentative metabolism* (ur-glycolysis) must have included four steps, and thus four different primitive enzymes. Sugar bisphosphates** had to be hydrolyzed, isomerized, and oxidized before they could be hydrolyzed to obtain pyrophosphate (Figure 21). The enzymes catalyzing these steps gave the protobiont a genetically controlled energy metabolism, which was needed to produce the pyrophosphate required for a rapid and accurate production of nucleic acids and proteins. It is possible that the protobiont was also capable of the enzymatic condensation of lipids from abiotic glycerol, fatty acids and phosphoric acid.

Once the incorporation of relatively defined proteins and probably also lipids was possible, the structure and function of the abiomembrane must have been greatly improved. Previously, the passage of molecules across the membrane must have occurred primarily by diffusion. Relatively large molecules could pass in both directions. The internal concentration of a given substrate was equal, at best, to the exterior concentration. An accumulation of substrates against the concentration gradient was not possible. The genetically controlled synthesis of membrane proteins transformed the abiomembrane into a primitive semi-biomembrane *(ur-*

---

*   Such multiple functions of one and the same enzyme could have been influenced or induced by periodic changes in environmental conditions (see Figure 17).

** The abiotic development of sugar phosphate is not easy to understand, but it must be assumed for nucleic acids.

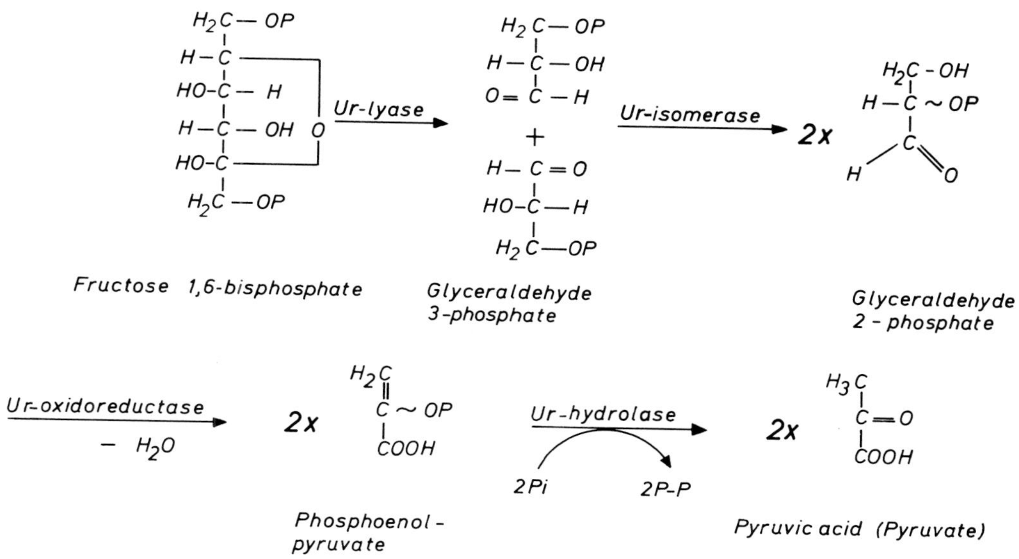

FIGURE 21. Representation of the genetically controlled primitive glycolysis (ur-glycolysis) assumed for the protobionts. At least four enzymes would have been required to lyse, isomerize, oxidize, and finally to hydrolyze the phosphorylated sugar to form pyrophosphate (Glycolysis, see Chapter 6, Figure 6). This diagram of primitive glycolysis is purely hypothetical and is intended to stimulate further discussions. It is open to discussion, whether the primitive energy production might have been realized by oxidation of acetaldehyde (e.g. in form of thioester or acetylphosphate) into activated acetic acid (Karlson, personal communication). It is furthermore questionable, whether a uroxidoreductase was already active at this primitive level. One can speculate that an energy-rich compound was produced by a simple dehydration of glyceraldehyde-2-phosphate, without a simultaneous oxidation of the aldehyde to carbonic acid, particularly since the origin of the oxidation equivalents, as well as the fate of the hydrogen which is detached during dehydration remain unsolved problems. The latter process possibly represents a later stage of energy production. However, cells must have learned at a rather early stage to bind chemical energy in the form of organic phosphate, in particular as ATP.

biomembrane). Tunnel proteins may have been responsible for the beginnings of an *active transport*. The necessary energy would have been provided by pyrophosphate from ur-glycolysis. Larger molecules were no longer able to diffuse away through the membrane. It could be that sugars were only able to pass the membranes in the non-phosphorylated form, and were phosphorylated upon entry by means of pyrophosphate. In this condition they could no longer escape by diffusion. In this way the pyrophosphate could be used to increase the internal substrate concentration.

Thus the first precyte, with about 15 genes, might have arisen (Table 5). This organism had a primitive anabolism (RNA, protein, lipid), the beginnings of an energy metabolism (pyrophosphate), and elementary membrane function (active transport). The protobiotic system was thus the first system which was capable of nearly identical self-replication.

C. Eobionts

The further evolution of precytes was now dependent on the permanent presence of all those proteins which catalyzed essential metabolic steps or had particular structural functions. The formation of ''nonsense'' proteins, which occurred when bases on the ur-messenger were skipped during translation (frame shift), had to be avoided. However, the protobiont lacked a mechanism for starting protein synthesis at a particular point on the ur-messenger. To mark the starting position for protein synthesis a very specific tertiary structure, a sort of *nucleation center*[32,33] was needed, and it could only be formed from long-chain RNA

Table 5
THE HYPOTHETICAL NUMBER OF MACROMOLECULE TYPES IN THE REPRODUCTIVE APPARATUS OF PRE-CELLS AND PRIMITIVE CELLS

System	Proteins (enzymes)	RNA	Genes:RNA	Nucleotides
Protobiont	1 Replicase (RNA)		~1	80
	4 Codases		~4	80 each 320
	5 Synthetases		~5	80 each 400
		4 tRNAs	~4	80 each 320
	10	4	~14	~1120
Eobiont	1 Replicase (RNA)		~1	80
	16 Codases		~16	80 each 1280
	13 Synthetases (Glycolysis, Lipids etc.)		~13	80 each 1040
		16 tRNAs	~16	80 each 1280
	Translatases and Ribosomal proteins	2 tRNAs	~2	160,320 480
	30	18	~48	~4160

System	Proteins (enzymes)	RNA	Genes:DNA	Nucleotides
Ur-procyte	1 Replicase (DNA)		~1	300
	1 Transcriptase (DNA→RNA)		~1	300
	1 Start-stop protein		~1	300
	20 Codases		~20	300 each 6000
	60 Synthetases (Fermentation, PP-path, Lipids) Translatases, Ribosomal proteins (including peptidase)		~60	300 each 18000
		20 tRNAs	~20	80 each 1600
		3 rRNAs	~3	160,1600, 3200 4960
	83	23	~106	~31460

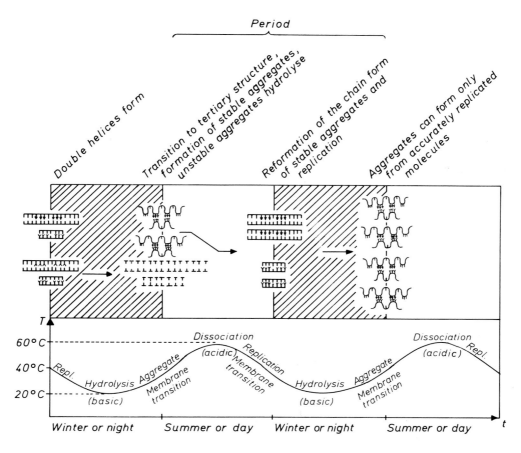

FIGURE 22. Periodic fluctuations in the internal milieu of more highly evolved RNA abioids. Only long-chained RNA molecules can survive the hydrolysis phase by forming stable isometric aggregates. Only these aggregates reproduce themselves during the replication phase after they have dissociated to single strands. Less stable aggregates are either hydrolyzed during this phase or diffuse away through the membrane (adapted from Kuhn[32]).

molecules. However, the formation of long-chain RNA molecules was not possible in the hydrolytically active internal milieu of the protobiont. Evolution had again reached a dead end, i.e. had entered a divergent phase. Development could not continue until a protobiont had combined with a nucleo abioid which in the course of parallel evolution had developed long-chain RNA molecules with the corresponding tertiary structure in its relatively hydrolysis-stable internal milieu. From the union of a protobiont with such a nucleo abioid (see Figure 24) arose the eobiont (a term suggested by Pirie), with a functioning protein synthesis apparatus.

1. Ribosomal Nucleo Abioids

In particularly hydrolysis-stable nucleo abioids, the 80-nucleotide RNA double helices had duplicated by further condensations (80/160/320). These long RNA molecules could survive the hydrolysis phase as single strands only if they formed hydrolysis-resistant, surface-reducing globular tertiary structures (Figure 22). Such tertiary structures, like the RNA types consisting of collector strand and associate molecules (Section B), were subject to selection for specific, in this case compact, aggregates. During the replication phase these aggregates dissociated, and each molecule replicated separately. In the following hydrolysis

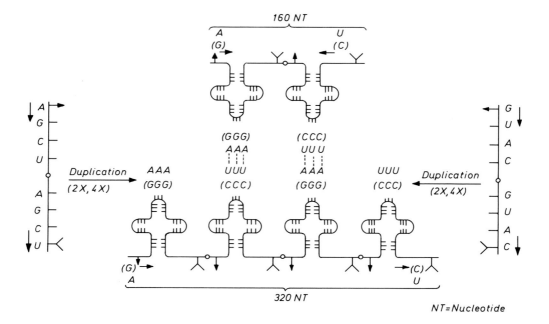

FIGURE 23.   Possibly preferred sequences for the formation of longer RNA molecules in the further evolved RNA abioids. These would have formed several sequential clover-leaf structures by intramolecular base pairing. The individual tertiary structures can fold together by base pairing to form the hydrolysis-resistant isometric aggregates.

phase, each molecule again took on its characteristic tertiary structure, and the complementary molecules associated to form new aggregates. Defectively replicated individual molecules no longer fit together, due to their altered tertiary structure, and were hydrolyzed, or diffused away (through the membrane?). The individual components of the aggregates were thus those which had replicated with relatively few errors. Contact with the abiomembrane might have played a role in the formation of aggregates.

A selection for better fitting forms set in. The result was the formation of *isometric aggregates*,[32] which could have been composed of two RNA molecules, with 160 and 320 nucleotides, respectively (Figure 23). The specific shape of the parts may have arisen through intramolecular base pairing, with the cloverleaf form as a model, and with additional spatial folding. The specific contact region between the two parts may have been formed by complementary triplets.

2. Primitive Translation

One can imagine that a union of a ribosomal nucleo abioid and a protobiont to eobiont must have occurred sooner or later (Figure 24). Purely random complexes between the two parts of the isometric aggregate and the ends of the ur-messenger (RNA single strand stage) could have been formed and could have begun to serve a new purpose (Figure 25). They encouraged the binding of a corresponding ur-transfer at the start position, and that of a second ur-transfer in the neighboring position, thus serving as a kind of *nucleation center*.[32] The two parts of the isometric aggregate could have held the two ur-transfers in position, as in a vise, so that the peptide bond could be formed exactly. Then the second ur-transfer would have taken the place of the first, which diffused away, and a third ur-transfer could take the place of the second, so polymerization could proceed. This system would practically eliminate ''nonsense'' readings of the ur-messenger, thus giving the eobiont which possessed

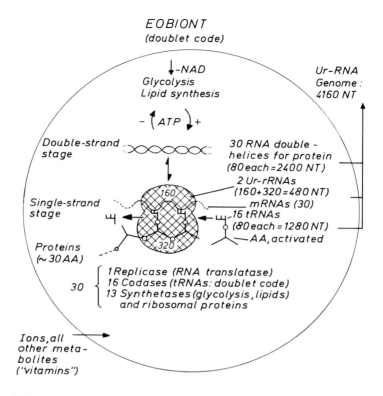

FIGURE 24. Hypothetical functions of the eobionts, which arose from the fusion of the protobiont with the highly evolved RNA abioids. The isometric aggregates developed into ur-ribosomes, and protein synthesis came under their control. The now selectively favored doublet code (1st and 2nd bases pairing specifically) would have been capable of synthesizing an RNA replicase, 16 codases for coupling up to 16 different abiogenic amino acids to the ur-transfers, and 13 synthetases for glycolysis, lipid metabolism, and stabilization of the ribosomal RNA to the ur-ribosome (quantitative listing of the genes, see Table 5). NT = nucleotide.

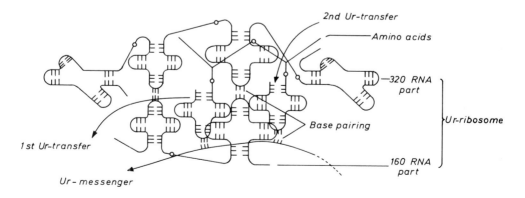

FIGURE 25. Hypothetical arrangement of the ur-messenger and amino-acid-charged ur-transfers in protein biosynthesis on a primitive ribosome composed of a 160-nucleotide and a 320-nucleotide cloverleaf tertiary structure.

it an enormous selective advantage. Gradual changes in the isometric aggregates would lead to growing precision in the meshing of the groups at the site of peptide formation, and the approach of the amino acid charged ur-transfers would be increasingly smooth. The isometric

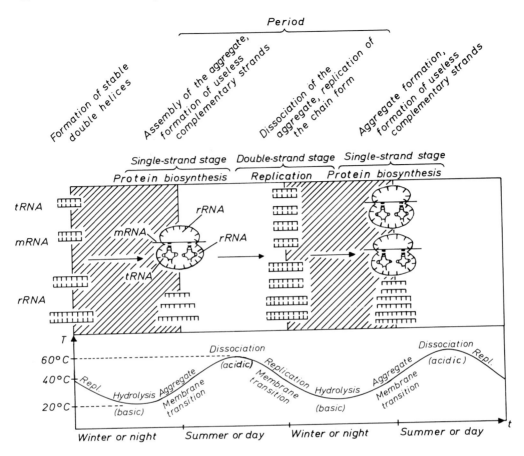

FIGURE 26.   Diagram of the periodic fluctuations in the eobiont. During the hydrolysis phase, only the stable complexes consisting of ur-messenger, ur-transfer, amino acids, and ur-ribosome survive. They also provide the strict control needed for the first exactly regulated protein biosynthesis (adapted by Kuhn[32]).

aggregate had become the primitive ribosome *(ur-ribosome; rRNA)*. Together with the ur-messengers and ur-transfers, which had already been present, the ur-ribosomes formed the primitive translation apparatus *(ur-translation)*.

The system again entered a strongly selective convergent phase. The eobionts with the best enzymes survived (Figure 26). Good enzymes, however, have a precisely determined amino acid sequence, so a selection pressure arose which favored proteins with exactly determined amino acid sequences. This, however, could not be provided by a simplet code. Therefore, eobionts were selected in which not only the first, but also the middle base of the triplet had to be complementary. Thus a doublet code was developed. Now proteins with fixed sequences, containing at least 16 different amino acids, could be synthesized.

As enzymes developed further, the coupling with ions or micromolecules as *pre-coenzymes* (protoprosthetic groups) must have led to an enormous increase in effectiveness.[4] For example, in the reaction

$$2\ H_2O_2 \xrightarrow{\ Fe^{2+}\ } 2\ H_2O + O_2$$

the iron atom has a specific activity of 10^{-5} μ.* Bound to a porphyrin molecule (such as heme), it already has a specific activity of $10^{-3}\mu$. Coupling the heme to a protein increases the activity to a value of 10^5 μ. Other (trace) metals such as Sn, Zn, Mg, Co, Cu, Pb, etc. must have been of similar importance as cofactors for the evolution of enzymes (e.g. Orgel[44-46]).

The specificity and activity of the enzymes of the material and energy metabolism were greatly improved in this way. ATP began to be synthesized. This indirectly favored the rapid synthesis of new enzymes, including 16 codases, and of proteins to stabilize the special structure of the primitive ribosome. With the synthesis of lipid precursors (glycerol, fatty acids), lipid synthesis also came under genetic control, which improved the quality of the membrane. Errors in base pairing and selection were constantly improving the proteins and adapting them to each other, so that their properties increasingly supplemented each other. The function of the whole became more and more dependent on the presence of an entire complement of proteins. The eobiont may have possessed more than 50 different genes (Table 5). It was a relatively well functioning cell, which was capable of slowly reproducing itself.

D. Ur-Procytes

Further development of the eobiont was not possible, since the RNA helices of the individual proteins were not united in a single strand. When the system divided, essential parts could be lost all too easily. Such individuals were then no longer able to reproduce, and dropped out of the evolutionary process. Furthermore, the *complementary strands of the RNA* double helices, which arose during the replication phase along with the RNA strands of the translation apparatus, mRNA, tRNA, and rRNA, were useless for protein synthesis; they were waste products. The eobiont, however, was neither able to combine the individual RNA matrices into a single strand, nor to eliminate the complementary strands of the RNA. The condensation of longer RNA chains was limited by deviations from the 3′ to 5′ nucleotide linkages, which produced side chains. Even a single wrong nucleotide coupling interfered with the winding of the helix. For lack of cooperativity, such strands could no longer be completely replicated, so that essential information was lost. The unification of the information in a single strand only became possible with the introduction of DNA, which cannot form branched chains, since its sugar moiety has no hydroxyl group on the 2′ carbon. The divergent phase gave way to a new convergent phase of evolution. This development was introduced by the combination of the eobiont with a DNA nucleo abioid to form an *ur-procyte* or *archaebacterium* (ur-prokaryotic cell = archuekaryote; Figure 27).

1. DNA Nucleo Abioids

DNA was either less often synthesized abiotically, or its evolution presumably started later than that of RNA. RNA and DNA have essentially the same structure, except that DNA contains the sugar 2-deoxyribose instead of ribose, and the methylated form of uracil, thymidine, instead of uracil. Presumably DNA nucleo abioids arose in much the same way as RNA nucleo abioids, but later. They became enriched by nucleotides containing both D- and L-deoxyribose. As in the RNA eobionts (Section A.3) DNA double helices containing only D-deoxyribose were selected. This development occurred parallel to that of the eobiont, so that sooner or later the eobionts and the DNA nucleo abioids must have come into contact with each other. Their union then made possible the transition into a new phase of evolution.

2. Primitive Transcription

One can imagine that after the union of the eobiont and the DNA nucleo abioid, two

---

* μ = μmol substrate per min and mg protein.

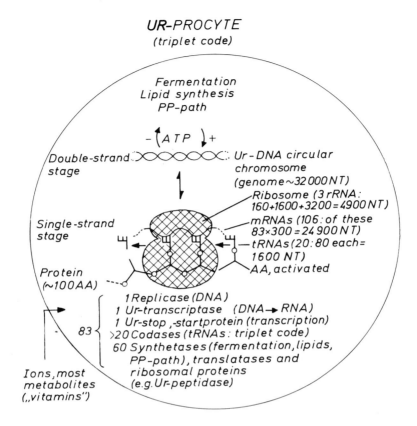

FIGURE 27.   Hypothetical function scheme for the ur-procyte, which evolved from the union of the eobiont with a DNA nucleo abioid. In the ur-procyte, the interfering complementary RNA strands are replaced by DNA, which incorporates the entire genetic information in a single, membrane-bound ring molecule. A kind of primitive transcription has developed. The now complete triplet code is able to synthesize at least 1 DNA replicase, 1 RNA transcriptase, 1 primitive stop or start protein, 20 codases and about 60 synthetases for glycolysis, the pentose phosphate pathway, lipid synthesis, translatases, and ribosomal proteins (including a primitive peptidase; quantitative summary of the genes, see Table 5).

enzymes E I and E II evolved (possibly by mutation of the RNA replicase; Kuhn[32]). E I produced DNA complementary strands from RNA templates, while E II catalyzed the production of RNA from the complementary DNA. This avoided the formation of useless RNA complementary strands.

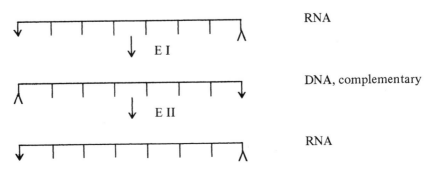

Later a protein L, which coupled all open-chained RNAs together at the beginning of each replication phase, must have arisen as a random change, which given enough time was sure to have occurred, and then to have been perfected in time by selection.

The coupled RNA strand would then have served as a template for the polymerization of DNA with the enzyme E I. The DNA formed in this way would have contained all the genetic information of an individual in a single strand. Sooner or later a DNA-dependent polymerase would also have arisen. DNA replication then became a second form of reproduction. After each replication, the L proteins would have bound to the points on the DNA strand at which the synthesis of a RNA molecule began or ended, as a kind of start and stop indicator. They would have located these points with their recognition regions, with which they could distinguish the complementary DNA strands. The increased reproduction effect, as well as the incorporation of all the genetic information into a single strand, would have brought the new system a considerable selective advantage. The loss of individual RNA matrices from the whole organized system was avoided by having all the information on one DNA strand, which became the *ur-chromosome* as selection pressure produced an ever more accurate DNA replication apparatus. E I degenerated and E II became the *ur-transcriptase*. Thus the essential elements of the transcription apparatus evolved along with translation.[32]

Evolution again entered a convergent phase. The third base in the codon triplets gradually came to be meaningful, although to a lesser extent than the other two. The modern triplet code with its start and stop signals arose. From this point on, further evolution seems a necessary consequence. Longer proteins, including enzymes, with higher molecular weights and thus higher specificity, were formed. For the replicase, this meant a reduction in the frequency of errors to 10^{-6}. On the other hand, new proteins with definite functions, which together resulted in a highly organized structural and functional unit, now arose relatively quickly. The ribosome was improved to an accurately working complex, the replication and translation apparatus were perfected, and the complete enzyme complement for fermentation, the pentose-phosphate pathway (PP-path*), lipid synthesis, and the construction of the plasma membrane with various proteins for transport was realized. A *primitive procyte,* which in the essential elements of its metabolism had become independent of the periodic changes in its environment (Figure 27), had emerged. It may have contained more than 100 genes (Table 5, see also Kaplan[26]).

The ur-procyte must have had such a tremendous selective advantage that its numbers and kinds increased explosively. It was aided by the closing of its DNA chromosome to a more stable ring molecule. When several copies of the circular chromosome were attached to the plasma membrane, the ur-procyte could divide without producing nonviable division products, as was the case in the protobionts and the eobionts.

E. Viruses

The ur-procyte at first still took up all the metabolites necessary for reproduction as a kind of "vitamin", such as sugars, bases, amino, fatty, and other organic acids, from the surrounding "primal soup", by active or passive transport. The much smaller abioids were threatened with elimination, as the ur-procytes used them as cheap sources of raw materials. Possibly, only a few nucleo-abioids escaped this fate by appropriating some of the genes from the procytes which had incorporated them. These genes gave them the ability to synthesize, on the procyte's own ribosomes, a stable coat of protein, sometimes also including lipids and saccharides, and to replicate, also with the procyte's replication apparatus. If this is true, then the modern RNA and DNA viruses are the descendants of the RNA and DNA nucleo abioids (Figure 28). Their relatedness to parts of the procyte or eucyte (Chapter 7)

---

* Pentose phosphate path, see Chapter 6, Figure 7.

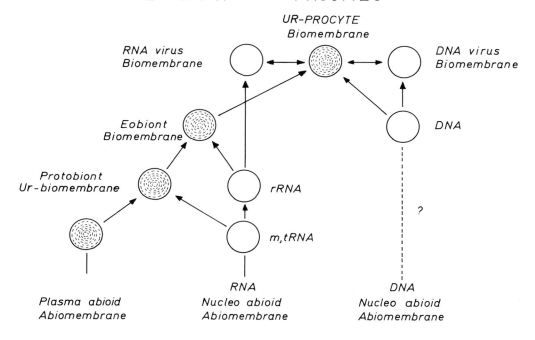

FIGURE 28. Presumed evolutionary development of the protobionts (plasma abioids and m, tRNA nucleo abioid), the eobionts (protobionts and RNA nucleo abioids), and the ur-procytes (eobiont and DNA nucleo abioid). According to this scheme, the RNA and DNA viruses are the descendants of the RNA and DNA nucleo abioids, which through genetic exchange with the procytes, or later, with eucytes, secondarily acquired the capacity to form biomembranes and to benefit from cell metabolism.

reproduction apparatus would consequently be a result of the descent of procytes from nucleo abioids, and also of a constant exchange of genes between the two. The genome of the nucleo-abioid may have grown to the modern molecular weight of viruses, which may be up to 10^7, by integration of parts of foreign genomes (compare with episomes, plasmids*) and by internal duplication. The ribosomes, which also represent the remains of RNA nucleo abioids, have by comparison a molecular weight of at most 10^6.

There is, to be sure, another explanation for viruses, namely that they represent cellular genes which have become independent, like the *episomes* or *plasmids*. Other researchers hold them to be highly degenerate bacterial cells which have adapted to their parasitic way of life.[52]

IV. MECHANISMS OF PRECYTE EVOLUTION

The prerequisites for chemical evolution on the primitive earth were a spatially varied environment and periodic changes in it. These conditions were fulfilled by the variability of the primitive atmosphere, hydrosphere, and lithosphere and by the earth's rotation. This rotation, in turn, is the result of conservation of angular momentum generated by instabilities in the condensing cosmic matter. Chemical evolution was thus a direct continuation, on the molecular level, of cosmic evolution.

---

* Plasmids or episomes are circular extrachromosomal DNA strands which can be integrated into the bacterial DNA, and replicate either in the integrated state or independently. They determine properties such as resistance to antibiotics or conjugation behavior.

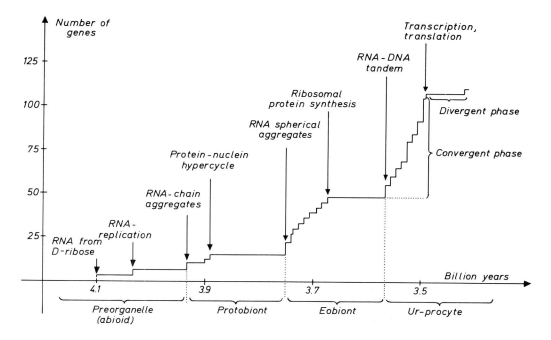

FIGURE 29. Scheme of the evolution of the reproductive system of abioids, protobionts, eobionts, and ur-procytes (archaebacteria), showing the most important convergent and divergent phases in relationship to time (billions of years) and the level of organization (number of genes see Table 5; adapted from Kuhn[32,33]).

A property which the two systems have in common is that free energy is added to them against the energy gradient. This leads to an increase in their degree of order or complexity. The driving force for the chemical system is the self-organization of the molecules according to their various charge distributions, possibilities for non-covalent binding, and structure. The forces of the environment acting on them as selection in the broadest sense also determine their organization. A prerequisite for selection is, according to Kuhn[32,33] a periodically changing environment, as provided by the day-night, summer-winter, and tidal rhythms. This causes a constant repetition of an entire complex of interacting processes, so that the chemical system is pushed back and forth between two states (energy dissipation). In this way it attains a degree of variability on which selection can act. Selection itself then occurs in often repeated variations from divergent to convergent phases and back again. In the divergent phases, many molecular variants with similar chances of survival arise. No decisive changes occur until a mechanism suddenly comes to serve a new purpose. This turning point is followed by a strongly selective convergent phase.

With the development of the precytes, bioevolution followed directly from chemical evolution. The mechanisms were the same. The individual evolving precytic systems were again the result of the coevolutive confrontation between the internal and external milieu, i.e. the products of self-organization and selection by a periodically changing environment. Here too, the process of evolution as such went through divergent stationary phases and convergent, progressive phases (Figure 29). "Chance accidents of Nature", as the creative element of evolution, thus determine the time at which decisive events occur and the details in the formation of self-organizing systems, but not the general direction.[32] If it were possible to recreate the conditions of the primitive earth, then in principle the same forms of life would again have to arise. The unavoidable steps to higher stages of evolution, i.e. formation of an outer covering, replication, formation of aggregates, development of a code, concen-

Hypothetical scheme of structure, function, and evolution of ur-procytes

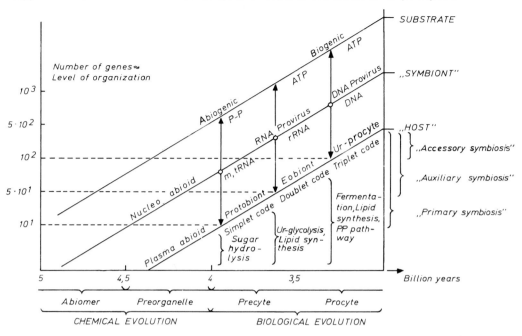

FIGURE 30.   Hypothetical mechanism of precyte evolution to the ur-procyte stage (archaebacterium). The procyte is thought to have developed in three steps. The first step ("primary symbiosis") occurred when a plasma abioid (1st "host") assimilated a provirus-like mRNA, tRNA abioid (1st "symbiont"), forming the protobiont with a simplet code and genetically controlled, elementary ur-glycolysis and pyrophosphate synthesis. The second step ("auxiliary symbiosis") was the combination of the protobiont (2nd "host") with the rRNA abioid (2nd "symbiont") to form the eobiont with a doublet code, normal glycolysis and ATP and lipid synthesis. The third step ("accessory symbiosis") finally produced, by integration of a provirus-like DNA abioid (3rd "symbiont") into the eobiont (3rd "host"), the ur-procyte, with a triplet code, glycolysis, the pentose-phosphate pathway, lipid synthesis, etc. The driving force behind the development of the primary, auxiliary, and accessory symbiosis may have been the necessity to synthesize substrates which had been depleted in the primal soup (see Table 5).

tration of information, and the principle of fermentative metabolism, would occur again. On the other hand, it would be reasonable to expect that all the steps for subordinate molecular "organizational problems", such as the formation of the individual elements of the genetic apparatus or the reproductive apparatus, that is transcription and translation, would take a different form if the "experiment" could be repeated.

Another mechanism of chemical evolution, which continued in the bio-evolution of precytes, is the principle of preformed elements (modular construction principle). Each more highly evolved level is composed of representatives (variants, mutants) of the next lower level. Atoms are composed of elementary particles, molecules of atoms, macromolecules of molecules, pre-organelles (abioids) of macromolecules, and finally, the ur-procytes were composed of various precytes.

The evolution of ur-procytes from precytes appears to have occurred in three characteristic stages, whereby the genetic information increased through gene duplication (Figure 30; Haapala and Sorsa[21]). The protobiont arose from the union and further common evolution of plasma and nucleo abioids. Therefore one could, figuratively speaking, call the protobiont the "primary symbiosis"* of the procyte, since it was the basis for the development of the

---

* Symbioses are, in general, permanent associations of organisms which yield mutual benefits.

procyte. The eobiont, which arose from the combination of protobiont and ribosomal nucleo abioid, would in the same sense represent the ''auxiliary symbiosis'' of the procyte. The last stage, the acquisition and assimilation of the DNA nucleo abioid by the eobiont, would represent a sort of ''accessory symbiosis'' of the procytes (see Chapter 7.I: the endosymbiont theory of eucyte evolution[36,54-56]). This three-layered system had such an enormous selective advantage that it increased explosively.

The model for precyte evolution developed here from the multilist and multistep hypothesis could be labeled the *''modular construction''* hypothesis. It corresponds in principle to the endosymbiont theory of eucyte evolution.[54-56] The modular construction hypothesis has been worked out in sufficient detail to serve as the impulse for a detailed experimental testing.

REFERENCES

1. **Bresch, C. and Hausmann, R.,** *Klassische und Molekulare Genetik,* Springer Verlag, New York, 1970.
2. **Calvin, M.,** *Chemical Evolution,* Clarendon Press, Oxford, 1969.
3. **Calvin, M.,** Biopolymere; Entstehung, Chemie und Biologie, *Angew. Chemie,* 86(3), 111, 1974.
4. **Calvin, M.,** Chemische Evolution, *Naturw. Rdschau.,* 29(4), 109, 1976.
5. **Dose, K. and Rauchfuss, H.,** Chemische Evolution und der Ursprung lebender Systeme, *Wissenschaftliche Verlagsgesellschaft,* Stuttgart, 1975.
6. **Eigen, M.,** Selforganization of matter and the evolution of biological macromolecules, *Naturwissenschaften,* 58, 465, 1971.
7. **Eigen, M. and Winkler, R.,** *Das Spiel. Naturgesetze steuern den Zufall,* Piper Verlag, Munich, 1976.
8. **Eigen, M. and Schuster, P.,** The Hypercycle. A principle of natural self-organization. Part A: Emergence of the Hypercycle, *Naturwissenschaften,* 64(11), 541, 1977.
9. **Eigen, M. and Schuster, P.,** The Hypercycle. A principle of natural self-organization. Part B: The abstract Hypercycle, *Naturwissenschaften,* 65(5), 7, 1978.
10. **Eigen, M. and Schuster, P.,** The Hypercycle. A principle of natural self-organization. Part C: The realistic Hypercycle, *Naturwissenschaften,* 65(7), 341, 1978.
11. **Eigen, M. and Winkler-Oswatitsch, R.,** Transfer-RNA: The early adaptor, *Naturwissenschaften,* 68, 217, 1981.
12. **Eigen, M. and Winkler-Oswatitsch, R.,** Transfer-RNA, an early gene? *Naturwissenschaften,* 68, 282, 1981.
13. **Eigen, M. and Winkler-Oswatitsch, R.,** Ursprung der genetischen Information, *Spektrum der Wissenschaft,* 6, 37, 1981.
14. **Försterling, H. D. and Kuhn, H.,** *Physikalische Chemie in Experimenten,* Verlag Chemie, Weinheim, 1971.
15. **Försterling, H. D., Kuhn, H., and Tews, K. H.,** Computermodell zur Bildung selbstorganisierender Systeme, *Angew. Chemie,* 84, 862, 1972.
16. **Folsome, C. E.,** Synthetic organic microstructures and the origins of cellular life, *Naturwissenschaften,* 63, 303, 1976.
17. **Fox, S. W.,** *The Origin of Prebiological Systems and Their Molecular Matrices,* Academic Press, New York 1965.
18. **Fox, S. W.,** Selfordered polymers and propagative cell-like systems, *Naturwissenschaften,* 56, 1, 1969.
19. **Fox, S. W. and Dose, K.,** *Molecular Evolution and the Origin of Life,* Freeman, San Francisco, 1972.
20. **Garret, R. A. and Wittmann, H. G.,** in *Aspects of Protein Biosynthesis,* Anfinsen, T. B., Ed., Academic Press, New York, 1971.
21. **Haapala, O. and Sorsa, V.,** Evolution of eukaryotic chromosome organization, *Biol. Zbl.,* 95, 317, 1976.
22. **Harpold, M. A. and Calvin, M.,** AMP on an insoluble solid support, *Nature,* 219, 486, 1968.
23. **deJong, H. G. B., Decker, W. A., and Swan, O. S.,** Zur Kenntnis der Komplex-Koazervation. III. Komplex-Koazervate unter physiologischen Milieubedingungen, *Biochem. Z.,* 221, 392, 1930.
24. **Karlson, P.,** *Biochemie,* Georg Thieme Verlag, Stuttgart, 1970.
25. **Kaplan, R. W.,** Lebensursprung, einmaliger Glücksfall oder regelmässiges Ereignis? Probleme der Beschaffenheit von Urorganismen und der Wahrscheinlichkeit ihres Entstehens, *Naturw. Rdschau.,* 30(6), 197, 1977.

26. **Kaplan, R. W.,** *Der Ursprung des Lebens. Biogenetik, ein Forschungsgebiet heutiger Naturwissenschaft.* *2. Auflage,* Georg Thieme Verlag, Stuttgart, 1978.

27. **Klima, J.,** Die Entstehung des Lebens, *Biologie in unserer Zeit,* 3(1), 9, 1973.

28. **Krampitz, G., Baars, S., Haas, W., and Kempfle, M.,** Zur Kondensation von Aminoacyl-adenylaten. Ein Modell für die abiogene Proteinsynthese, *Naturwissenschaften,* 56, 416, 1969.

29. **Küppers, B.,** The general principles of selection and evolution at the molecular level, *Progr. Biophys. Molec. Biol.,* 30(1), 1, 1975.

30. **Küppers, B.,** Towards an experimental analysis of molecular self-organization and precellular Darwinian evolution, *Naturwissenschaften,* 66, 228, 1979.

31. **Küppers, B.,** Evolution im Reagenzglas, *Mannheimer Forum 80/81. Hrsg. Boehringer Mannheim GmbH,* 47, 1981.

32. **Kuhn, H.,** Selbstorganisation molekularer Systeme und die Evolution des genetischen Apparates, *Angew. Chemie,* 84(18), 838, 1972.

33. **Kuhn, H.,** Model consideration for the origin of life, *Naturwissenschaften,* 63, 68, 1976.

34. **Kuhn, H. and Kuhn, C.,** Evolution of a genetic code simulated with the computer, *Origins of Life,* 9, 137, 1978.

35. **Lacey, J. C., Jr. and Pruitt, K. M.,** Origin of the Genetic Code, *Nature,* 223, 799, 1969.

36. **Margulis, L.,** *Origin of Eukaryotic Cells,* Yale University Press, New Haven, 1970.

37. **Miller, S. L.,** A production of amino acids under possible primitive earth conditions, *Science,* 117, 528, 1953.

38. **Mills, D. R., Kramer, F. R., and Spiegelmann, S.,** Complete nucleotide sequence of a replicating RNA molecule, *Science,* 180, 916, 1973.

39. **Nagy, B.,** Organic chemistry on the young earth, *Naturwissenschaften,* 63, 499, 1976.

40. **Naylor, R. and Gilham, P. T.,** Studies on some interactions and reactions of oligonucleotides in aqueous solution, *Biochemistry,* 8, 2722, 1966.

41. **Nöll, G.,** Modellmembranen — Membranmodelle, *Biologie in unserer Zeit,* 6(3), 65, 1976.

42. **Oparin, A. I.,** *Der Ursprung des Lebens. 1. Ausgabe* (russ. Proiskhozdenic Zhizny), Moskowskiy, Rabochii, Moskau, 1924.

43. **Oparin, A. I.,** *Genesis and Evolutionary Development of Life,* Academic Press, New York, 1968.

44. **Orgel, L.,** *An Introduction to Transition Metal Chemistry,* Methuen, London, 1961.

45. **Orgel, L.,** Evolution of the genetic apparatus, *J. Mol. Biol.,* 38, 381, 1968.

46. **Orgel, L.,** *In Origins of Life. 1st Interdisciplinary Communications Program,* L. Margulis, Ed., New York, Gordon and Breach, 1970.

47. **Paecht-Horowitz, M., Berger, J., and Katchalsky, A.,** Prebiotic synthesis of polypeptides by heterogeneous polycondensation of aminoacid adenylates, *Nature,* 228, 636, 1970.

48. **Paecht-Horowitz, M.,** Die Entstehung des Lebens, *Angew. Chemie,* 85, 422, 1973.

49. **Rahmann, H.,** *Die Entstehung des Lebendigen,* Gustav Fischer Verlag, Stuttgart, 1972.

50. **Ring, K.,** The effect of low temperatures on permeability in *Streptomyces hydrogenans, Biochem. Biophys. Res. Commun.,* 19, 576, 1965.

51. **Scheffler, I. E., Elson, E. L., and Baldwin, R. L.,** Helix formation by dATP oligomers. I. Hairpin and straight-chain helices, *J. Mol. Biol.,* 36, 291, 1968.

52. **Schlegel, H. G.,** *Allgemeine Mikrobiologie,* Georg Thieme Verlag, Stuttgart, 1974.

53. **Schuster, P.,** Vom Makromolekül zur primitiven Zelle — die Entstehung biologischer Funktion, *Chemie in unserer Zeit,* 6, 1, 1972.

54. **Schwemmler, W.,** Zikadenendosymbiose: ein Modell für die Evolution höherer Zellen. Zur Verifikation der Endosymbiontentheorie der Eukaryonten-Zelle, *Acta Biotheoretica,* 23, 132, 1974.

55. **Schwemmler, W.,** Allgemeiner Mechanismus der Zellevolution, *Naturw. Rdschau.,* 28(10), 351, 1975.

56. **Schwemmler, W.,** Die Zelle: Elementarorganismus oder Endosymbiose? *Biologie in unserer Zeit,* 7(1), 7, 1977.

57. **Schwemmler, W.,** Die Rekonstruktion der Präzytenevolution als experimenteller Forschungsanreiz, *Courier Forschungsinstitut Senckenberg,* 28, 1, 1978.

58. **Schwemmler, W.,** Evolution der Urzeller: Rekonstruktion des Lebensursprunges, *Natur und Museum,* 108(2), 49, 1978.

59. **Sitte, P.,** Biomembran: Struktur und Funktion, *Ber. Dtsch. Bot. Ges.,* 82(5,6), 329, 1969.

60. **Sparrow, A. H. and Naumann, A. F.,** Evolution of genome size by DNA doublings, *Science,* 192, 524, 1976.

61. **Spiegelmann, S., Mills, D. R., and Peterson, R. L.,** An extra-cellular Darwinian experiment with a selfduplicating nucleic acid molecule, *Proc. Natl. Acad. Sci.,* 59, 217, 1967.

62. **Spiegelmann, S.,** *The Neurosciences,* 2nd study program (Hrsg. F. O. Schmidt), Rockefeller University Press, New York, 1970.

63. **Wagenführ, W.,** *Biogenese — theoretische und experimentelle Aspekte,* Staatsexamensarbeit, Freie Universität Berlin, 1973.

64. **Wisnieski, B. J. and Fox, C. F.,** Correlations between physical state and physiological activities in eukaryotic membranes, especially in response to temperature, Hastings, I. W. and Schweiger, H.-J., Eds., *Dahlem Workshop on the Molecular Basis of Circadian Rhythms,* S. 247, Abahon Verlagsgesellschaft, Berlin, 1976.

Chapter 6

EVOLUTION OF THE PROCYTES

The procytes (protocytes) are generally unicellular and potentially immortal, in so far as the parent cell lives on after cell division in its daughter cells. The formation of tissues with specialized cells occurs only in exceptional cases, and only then do the cells die, leaving corpses. One example of such a development is the differentiation of nitrogen-assimilating heterocysts in filamentous colonies of certain blue-green algae, and another is the formation of the multicellular fruiting bodies of myxobacteria (slime bacteria). Since procytes are generally unicellular and short-lived, their morphological development (ontogenesis) is slight, and therefore difficult to detect. Haeckel's Principle (1903)* is thus not applicable in the usual way to beings "without a history". Furthermore, there are no mechanisms for genuine sexual reproduction, and thus no defined boundaries (inability to cross-breed) between the species. This explains the difficulty of developing a natural, i.e. phylogenetic, system of classification for the confusing variety of procytic forms. Hypotheses must first be invented to make the development of the procytes understandable. "One could compare it with the task which an inhabitant of one of the planets of Sirius would face, if it should try to reconstruct the development of the automobiles now present on the earth, purely by observation, and with no direct knowledge of the older models and their development".[1] The criteria for a reconstruction of procyte evolution are (1) thermodynamic possibilities, (2) biological usefulness (selective advantage), (3) mechanical plausibility of their metabolic processes, and (4) sequence analyses. One of the main results of such deductive studies is the conclusion that all presently existing forms have descended from the same ancestral strain. All procytes, like other organisms, consist of proteins and nucleic acids which are constructed from the same subunits with comparable sequences. The various metabolic processes can also be derived from basic types, such as fermentation, chemoautotrophy, respiration, photergy, and photosynthesis, which appear to derive from a basic ancestral procytic metabolism, with ATP as the general energy currency, and NADH as the key substance in hydrogen transport. The metabolic types differ with respect to their source of energy, the substrates for their metabolism and their source of carbon. In essence, two alternative hypotheses are currently being discussed for the evolution of procytic metabolic types. The two hypotheses can be designated the *"conversion"* hypothesis and the *"splitting"* hypothesis.

I. WORKING HYPOTHESES

The basic idea of the *conversion hypothesis* is already very old, but it was only logically and precisely formulated in the last decade (e.g. Broda[1,2,7,10,21]). According to this hypothesis, the development of the procytes began from the anaerobic fermenters (Figure 1, A). These organisms obtain their energy from chemical processes (chemotrophy, see Figure 3). In addition, they require organic compounds, e.g. organic substrate (organotrophy). In particular, they require a source of organic carbon, and are thus dependent on other living organisms (C-heterotrophy). The photosynthesizers** are held to have developed from the fermenters.

---

* *The Basic Principle of Biogenetics*, formulated by Ernst Haeckel (1834—1919), holds that ontogeny (development of the individual) is, to a certain extent, a short recapitulation (repetition) of phylogeny (evolution of the species).

** Photosynthesizers are cells or organisms which are capable of photosynthesis. Photosynthesis is used here in Broda's sense[1] as a synonym for photoassimilation.

A CONVERSION HYPOTHESIS

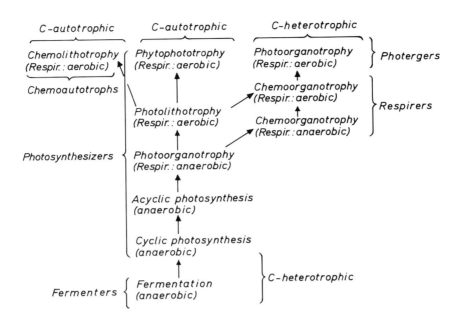

B SPLITTING HYPOTHESIS

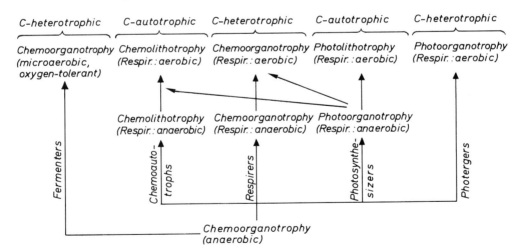

FIGURE 1.   Schematic diagram of the two alternative hypotheses on the evolution of procyte metabolism. According to the ''conversion'' hypothesis (A), photosynthesis must have arisen before (anaerobic) respiration, while the ''splitting'' hypothesis (B) suggests that it was the other way around (see text). According to the latter hypothesis, certain aerobic respirers have also developed from aerobic photosynthesizers which lost their photosynthetic apparatus (see Section II).

They use the much more efficient sunlight as a source of energy (phototrophy). The anaerobic photosynthesizers obtain their reducing equivalents for synthesis by oxidizing organic substrates (organotrophy), while the aerobic forms obtain theirs from inorganic compounds

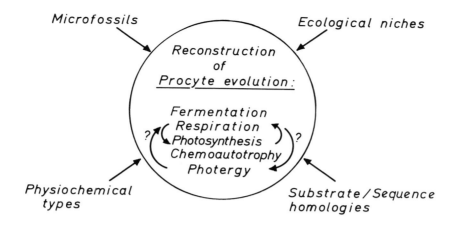

(lithotrophy: lithos, Greek = stone). Both forms use CO_2 from the air as a source of carbon, so that they are in this respect independent of other organisms (C-autotrophy). The respirers are then thought to have developed by conversion, or loss of the photosynthetic apparatus, from the photoorganotrophs (for example, from cyclic and acyclic photosynthesizers, Section III.D), and the chemoautotrophs* from the photolithotrophs. The respirers have a chemoorganotrophic, C-heterotrophic metabolism, while the chemoautotrophs, as their name suggests, have a chemolithotrophic, C-autotrophic metabolism. The photoorganotrophic, C-heterotrophic photergers,** finally, are supposed to have derived from the aerobic respirers. The photergers generate ATP from sunlight, although in a different way from the photosynthesizers.

In contrast to the classical conversion hypothesis, the more recent *splitting hypothesis* suggests (e.g. Margulis,[16,18] Schwemmler,[28,29] Müller[19]) that the anaerobic respirers evolved directly from the fermenters (Figure 1, B). These diverged into four branches: purely respiring, C-heterotrophs; respiring, photosynthetic C-autotrophs; respiring, C-autotrophic chemoautotrophs; and finally, the C-heterotrophic, respiring photergers. All four branches are thought to have developed aerobic forms from anaerobic intermediates. Certain aerobic respirers have developed from aerobic photosynthesizers which lost their photosynthetic apparatus (see Section II.E).

The main difference between the two hypotheses lies in the derivation of the respirers. According to the conversion hypothesis, the aerobic and anaerobic respirers, or their respiratory chain, have descended from the photosynthesizers. The splitting hypothesis suggests, in contrast, that the photosynthesizers, or their photosystems I and II, derive from anaerobic respirers. Since only one of these hypotheses can be correct, it is reasonable to examine the available data to determine whether they support one or the other hypothesis. It more or less follows that data which support one hypothesis in the same measure cast doubt on the other.

II. EXPERIMENTAL RESULTS

Essentially five different methods have been used to elucidate the order of development of the different types of protocyte physiology:

---

* Chemoautotrophs are all those cells which have a chemoautotrophic metabolism.

** Photergers are cells which are capable of photergy, which is, according to Kaplan,[13] any process of light-driven ATP production (without NAD(P)H production), in this case, the synthesis of ATP with the help of protein-bound retinal (rhodopsin), a derivate of carotene.

- analysis of microfossils
- study of ecological niches
- comparison of the physiochemical* composition of procyte internal milieus
- comparison of catabolism and anabolism
- analysis of sequence homologies

A. Microfossil Discoveries

One field of micropaleontology is concerned with examining ancient sedimentary layers (see Chapter 3, I) for *chemical microfossils,* using extremely sensitive methods such as gas chromatography. In this context, it is understandable that we are primarily interested in those organic molecules (biomers) which are indicative of biotic systems. However, one must also take into consideration the formation of degradation products. For example, the discovery of isoprenoid may or may not indicate the presence of chlorophyll, which has an isoprenoid derivative (phytol) as a side chain. The presence of chlorophyll, in turn, could be correlated with the rise of photosynthetic mechanisms (Table 1). If the ratio of the carbon isotopes ^{12}C to ^{13}C is greater in the fossil than in the surrounding rock, which would be the result of selective photoassimilation of $^{12}CO_2$ (see also $^{34}S/^{32}S$), this is also interpreted as an indication of photosynthetic activity. Finally, n-alkanes with an odd number of carbon atoms could well be the result of extraorganic decarboxylation of fatty acids with even numbers of fatty acids. Unfortunately, the interpretation of the data is complicated by the possibility of contamination of the rock layers by organic materials from later periods (epigenetic sources) which might simulate a syngenetic formation.

A further source of micropaleontological data is the microscopic examination of thin sections of sediments which reveals the structures of *organismic microfossils.* These studies have, however, yielded the rather disappointing result that the time in which the basic procyte types must have evolved, from 4 to 2 billion years ago, produced very few fossils. This era, after all, represents 50% of the biological history of the earth! However, the few discoveries which have been made (summarized in Table 1) do allow us to draw a few conclusions. The oldest chemical and biological finds come from the Onverwacht Formation of South Africa (3.4 to 3.2 billion years ago). They suggest the existence of eobionts or even primitive fermenting bacteria at this time. The next layer, the Fig Tree (3.2 to 3.1 billion years ago), contains microstructures *("Spherotype B")* which were, with a probability near to certainty, fermenting organisms, and also anaerobically respiring, lithotrophic or photergic organisms which are similar to sulfur or halobacteria, the so-called *"Eobacterium isolatum".* In even younger Fig Tree layers (2.7 billion years ago: Rahmann[22]), microorganisms comparable to the green sulfur bacteria have been found *("Archaeosphaeroides barbertonensis"),* suggesting that by this time the mechanism of anaerobic photosynthesis was present. This is confirmed by the contemporary finds from the Bulawayo layer of South Africa and the Soudan layer of North America (2.5 to 2.3 billion years old). Structures corresponding to the modern blue-green algae were present, at the latest, by 2.5 to 2.3 billion years ago, as shown by the finds in the Witwatersrand layer. These organisms were probably capable of aerobic photosynthesis. The analysis of the Gunflint formation of North America (2 to 1.9 billion years ago) has yielded the widest spectrum of procyte types to date. It is highly probable that at this time several kinds of oxidizing and thus aerobically respiring bacteria *("Eostrion")* and *Oscillatoria*-like blue-green algae were present.

The above finds, with their suggested interpretation, indicate that metabolic types arose in the following order:

---

* The word physiochemical is used here in the sense of physiological (pH, osmotic pressure) and chemical properties (type and concentration of ions, molecules), as in Landureau (Chapter 8, Reference 15).

Table 1

CHEMICAL AND CELLULAR FOSSILS RELEVANT TO DEVELOPMENT OF PHYSIOLOGICAL TYPES AMONG THE PROCYTES

Geological Formation	Onverwacht (S. Africa)	Fig Tree (S. Africa)	Bulawayo (S. Africa) Soudan (N. America)	Witwatersrand (S. Africa)	Gunflint (N. America)
Estimated age (x 10⁹ years)	3.4 — 3.2	3.2 — 3.1	2.8 — 2.7	2.5 — 2.3	2 — 1.9
Chemical fossils	- Porphyrins (Cytochrome, Chlorophyll, Membrane?) - Sporopellenin (Carotenoids?) - n-Alkanes	- Isoprenoids: Phytane, Pristane (Quinone, Chlorophyll?) - Porphyrins (Cytochrome, Chlorophyll, Membrane?) - Amino acids: 0.1 $\mu g/g$ Val, Ala, Gly (Proteins?) - Sporopellenin (Carotenoids?) - $^{13}C/^{12}C$: + (Photoassimilation of CO_2?)	- n-Alkanes C (2n+1) (Fatty acids?) - Isoprenoids: Phytane, Pristane (Quinone, Chlorophyll?) - Stromatolith: Chalk (Blue-green algae?) - $^{12}C/^{13}C$: + (Photoassimilation of CO_2?)	- Amino acids: 1$\mu g/g$ (Proteins?) - $^{12}C/^{13}C$: + (Photoassimilation of CO_2?) - $^{34}S/^{32}S$: + - Stromatolith: Chalk (Blue-green algae?) - n-Alkanes - Porphyrin (Cytochrome, Chlorophyll, Membrane?)	- Stromatolith: Chalk (Blue-green algae?) - $^{12}C/^{13}C$: + (Photoassimilation of CO_2?) - Isoprenoids: Phytane, Pristane (Quinone, Chlorophyll?) - n-Alkanes
Cellular fossils	- Spheroid microfossils: $\emptyset$ 10-30 μm (Eobiont?, fermenting bacterium?)	- Spherotype B $\emptyset$ 5-30 μm (Fermenting bacterium?) - Eobacterium isolatum: 0.6/0.25 μm (Sulphur bacterium?, Halophilic bacterium?)	- Single microfossils or chains: $\emptyset$ 0.8-1.5 μm (Blue-green algae?) - Archaeosphaeroides barbertonensis: $\emptyset$ 1.9 μm (Green sulphur bacterium?)	- Microfossils: individuals, clusters and chains, some with skeletons: $\emptyset$ 20-60 μm (Blue-green algae)	- Eosphera: $\emptyset$ 30 μm - Kakabekia umbellata $\emptyset$ 10-30 μm } Blue-green algae: Oscillatoria? - Eostrion 0.5/11 μm - Various bacteria } Iron bacteria: Sphaerotilus, Siderocapsa, Siderococcus?
Possible metabolic types	Fermentation →	Fermentation Respiration, anaerobic → Photergy → Photosynthesis, anaerobic (Chemoautotrophy, anaerobic?)	Photosynthesis, anaerobic anaerobic (Photosynthesis, aerobic?) → Photosynthesis, aerobic →		Photosynthesis, aerobic Respiration, aerobic Chemoautotrophy, aerobic

Note: The possible or suggested interpretations of each result are given in parentheses. Data from Rahmann,[22] Dose and Rauchfuss,[6] and Kaplan.[13]

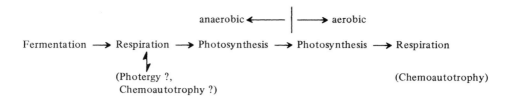

Thus, photosynthesis emerged after anaerobic respiration, which is in agreement with the splitting hypothesis but contradicts the conversion hypothesis.

B. Ecological Niches

Geology provides another means of direct examination of materials from the time of procyte evolution. It has been found that the atoms in the oldest rocks are primarily present in reduced form, i.e. they exhibit their lowest chemical valence. However, in younger rocks (for example, the layered marine iron ores), starting about 2.8 billion years ago, the oxidized forms of the elements predominate. It has therefore been assumed that iron which was originally dissolved in the primitive oceans as Fe^{2+} was gradually oxidized to Fe^{3+} by oxygen released from aerobic photosynthesizers* and was sedimented in layers, due to periodic precipitation, as a mixture of $Fe_3^{2+/3+}O_4$. In this way the oxygen released by photosynthesis was at first prevented from entering the atmosphere. However, after the iron and other oxidizable elements in the oceans had been exhausted, the O_2 entered the atmosphere and oxidized the iron exposed to it to form the red terrestrial iron ores of $Fe_2^{3+}O_3$. The last layered marine ores are about 1.8 billion years old, which is also the approximate age of the first oxidized iron ores on land.

Several interesting conclusions may be drawn from these facts with regard to the development and occupation of *ecological niches* for procytes with various types of metabolism. It must first be assumed that there was little free oxygen in the hydrosphere about 3 billion years ago and that there was essentially none in the atmosphere. Therefore, the primitive anaerobic, fermenting procytes could only have developed in the dark depths of the reducing hydrosphere at least 20 m below the surface (Table 2). Here they were protected from the UV radiation of the sun which, for lack of an ozone layer, fell directly onto the surface of the earth. The more than 3 billion year old fossils of procytes like *"Eobacterium isolatum"*, which were possibly capable of anaerobic respiration, must also have evolved in the deep, anaerobic, dark niches of the hydrosphere. The colonization of higher water layers, which were flooded with UV light, was only possible after the development of protective pigments in the cell membrane. Only the precursors of photergers, retinal (carotenoid with protein: bacteriorhodopsin) in their cell membranes, and those of the photosynthesizers, with chlorophyll, possessed such pigments, so that they must have been the first to conquer the higher, lighted niches. A light-powered metabolism later developed from these protective pigments and other metabolic components. The photosynthesizers released O_2, starting about 3 billion years ago, first into the water and, from about 2 billion years ago onward, into the atmosphere.** This must have produced a strong selective pressure for development from anaerobiosis to aerobiosis. The aerobic respirers and chemoautotrophs were able, under the protection of the developing ozone layer, to move into the unoccupied, dark aerobic niches of the surface water and moist land surfaces. Later, the aerobic chemoautotrophs and photosynthesizers, which were completely independent of organic substrates, were able to settle

---

* Fe^{2+} can also be oxidized to Fe^{3+} by chemoautotrophs (see Figure 19).
** The conversion of the reducing atmosphere into an oxidizing one lead to the disappearance of energy-rich abiotic compounds. These are unstable in the presence of oxygen and quickly decompose through oxidation. As a result, chemical evolution was brought to a complete standstill.

the light-flooded, moist niches of the lithosphere. However, all procyte types must have been dependent on an aqueous or moist medium at least for part of their life cycles.

On the basis of these geological facts, we can argue that the ecological niches for anaerobic respiration could only have developed before photosynthesis. The geological facts are too uncertain, however, to support one hypothesis or the other.

C. Physiochemical Types

The comparison of certain physiochemical data on the intracellular milieu of various procytes yields further evidence which can be used to clarify the phylogeny of the procytes. The data are determined for soluble procyte fractions. The concentrations of the most frequent inorganic ions (Na^+, K^+, Mg^{2+}, Ca^{2+}) and of organic molecules (sugars, amino acids, and other organic acids) are measured in these fractions. The ratios of these concentrations within various types of procytes appear to follow certain rules,[28] and they can be arranged in a system of characteristic fluctuations (Figure 2). In this system, the total concentration of the inorganic components decreases as the organic components increase. In particular, the concentrations of sugar, Cl^- and Na^+ decrease as the concentrations of amino acids and other organic acids, PO_4^{3-}, K^+, Mg^{2+} and Ca^{2+} increase.

The chemical parameters appear to be correlated with the physiological parameters. As the concentration of organic acids increases, so does the H^+ concentration, from about pH 8 to about pH 6. At the same time, the osmotic pressure sinks from about 450 mosmol ($\sim \Delta t°C - 0.8$) to about 250 mosmol ($\sim \Delta t°C - 0.4$), due to the replacement of smaller inorganic ions with larger organic molecules. On the basis of these positive correlations, even those procytes for which not all the physiochemical data are known can be roughly fit into the system.

The scale on which the ratios of the total inorganic components to the total organic components are plotted can be divided into three regions. In the first, the inorganic fraction dominates over the organic, in the second, the two are in balance, and in the third, the organic fraction predominates over the inorganic. These correspond to the *physiochemical types I, II, and III*, which have also been found in the cell fluids of plants and animals. According to Sutcliffe,[33] they are the result of a phylogenetic development which runs from physiochemical type I to type III (review of literature, also Schwemmler;[26,27] see also Chapter 8, III).

Physiochemical type I includes, for example, all fermenters. Their anabolism is still highly dependent on the substrate, which may be related to the dominance of inorganic substances over the organic ones in type I. Glycolytic activity and the associated storage of starch account for the fact that here the sugar concentrations are higher than organic acid concentrations. The pH and osmotic pressure of the fermenters correspond to the values postulated for the primal milieu (Chapter 3, I).

Certain anaerobic respirers belong to the transition type I/II or type II. Their increased synthesis of organic building materials should explain the increase in the organic fraction of this type. Amino acid synthesis and the citrate cycle account for the increase in the concentration of organic acids. The sugar concentration decreases, due to the reduced rate of glycolysis and the correspondingly lower requirement for substrate. The formation of new metallo-organic intermediates and end products (e.g. Ca^{2+} and HPO_4^{2-} for ATP and ATPase; Ca^{2+} and Mg^{2+} for stabilization of membranes of the respiratory chain) may have been responsible for the decrease in Na^+ and Cl^- concentrations in favor of other ions. The pH rises with the increase in organic acids, whereby the osmotic pressure sinks accordingly.

Photosynthesizers and aerobic respirers, finally, can be placed in the transition type II/III or in type III. The more or less complete independence of their metabolism from the supply of organic material, which is made possible by the synthesis of new organic structures and compounds, explains the dominance of the organic fraction over the inorganic and also

Table 2

PRESUMED DEVELOPMENT OF MEDIUM, SUBSTRATE, AND METABOLISM OF BASIC PHYSIOLOGICAL PROCYTE TYPES

Note: For the sake of clarity, the group of chemoautotrophic organisms has not been integrated into the table. Data in part from Rahmann,[22] Broda,[1,2] Kaplan,[13] Schön.[25]

	Fermenter (anaerobic)	Respirer (anaerobic)	Photerger (anaerobic)	Photosynthesizer (anaerobic)	Photosynthesizer	Respirer (aerobic)
Time (years ago,×10⁹)	3.6 – 3.2		3.2 – 2.8		2.8 – 2.4	2.4 – 1.8
Atmosphere	Reducing				Oxidizing	
Biotope	Dark, anaerobic		Light, anaerobic		Light, aerobic	Dark, aerobic
	Deep water		Deep water		Water surface	Water or moist land areas
Developmental trend	Dark → Light → Anaerobiosis → Aerobiosis → Water → Land					
Procyte type	Fermenter	Respirer (anaerobic)	Photerger	Photosynthesizer (anaerobic)	Photosynthesizer	Respirer (aerobic)
Carbon source	C-Heterotrophy (e.g. sugars, acids)		C-Autotrophy (e.g. sugars, acids, CO_2)			C-Heterotrophy (e.g. sugars, acids)
Energy source	Chemotrophy		Phototrophy			Chemotrophy
Energy substrate	Organotrophy		Organotrophy or Lithotrophy (inorganotrophy)		Lithotrophy	Organotrophy
H-Donor	e.g. Sugars	e.g. Sugars, Organic acids (NADH)	e.g. Sugars, Organic acids, H_2O (NADH)	e.g. Sugars, Organic acids, H_2, H_2S, S^0, $S_2O_3^{2-}$, (NADH)	H_2O (NADPH)	e.g. Sugars, Organic acids (NADH)
H-Acceptor	e.g. Glycolysis intermediates NAD(P)H	Aldehyde(?) CO_2(?) $SO_4^{2\ominus}$ $NO_3^{\ominus}$	CO_2 (?) O_2 (later)	CO_2 (?) (later)	NAD(P)H	NAD(P)H	O_2
ATP Synthesis	Substrate phosphorylation (glycolysis)	Oxidative phosphorylation (incomplete anaerobic respiratory chain)	Photophosphorylation (light driven proton pump)	Cyclic/acyclic photophosphorylation (photosystems I,II)	Cyclic and acyclic photophosphorylation (photosystems I,II)	Respiratory chain phosphorylation (complete aerobic respiratory chain)
New synthetic abilities (compared to pre-decessors)	Decarboxylation, Catalases, Isoprenoids, CO_2 fixation, N_2 fixation	NAD, FAD,(FMN) Quinones, Ferredoxine, Cytochromes (b,c) ?Partial citrate-cycle	Carotenoids, Complete citrate cycle?, Partial Calvin cycle	Chlorophyll, H*production, Complete Calvin cycle	H_2O Lysis, O_2 Production	Cytochrome oxidases (Cytochrome a)

Metabolism

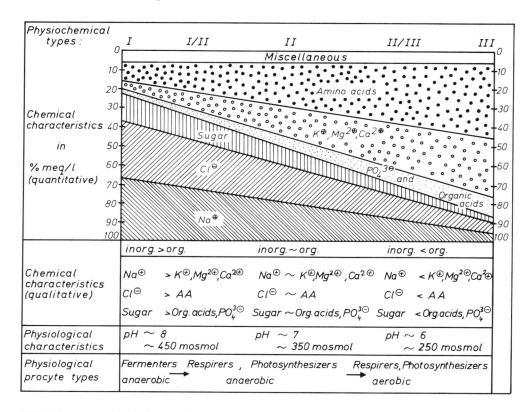

FIGURE 2.   Presumed chemical and physiological trends in the development of the intracellular milieu of procytes. There are three main physiochemical types, with characteristic physiological properties (see Schwemmler;[26-28] details, see text).

the further reduction of Na^+ and Cl^- in favor of the other ions. The osmotic pressure is even lower, due to the increase in the organic fraction, whereas H^+ concentration is higher, as a result of the increase in the concentration of organic acids. Respiring or photosynthesizing bacteria which shift to another form of metabolism undergo a corresponding change in physiochemical character. For example, when the well-known enterobacterium *Escherichia coli* shifts from fermentation to respiration, its physiochemistry changes from type I to type II/III.[8] In the purple bacterium *Rhodospirillum rubrum,* the K^+ concentration rises during the shift from respiration to photosynthesis, at the expense of the Na^+ concentration, as also occurs in the transition from physiochemical type II to type III.[31]

In summary, it appears that on the basis of their physiochemical type and the apparent phylogenetic trend (see Sutcliff[33]), the anaerobic respirers should be placed before the photosynthesizers phylogenetically. This evidence, like that previously discussed, supports the splitting hypothesis and contradicts the conversion hypothesis, but cannot unequivocally prove its validity.

D. Substrate Homologies

Since the procytes lack morphological structures which could be compared in the sense of the homology criteria,[23] it is only possible to compare them with one another on a chemical basis (Table 2). One possibility is to compare the various procyte physiological types on the basis of their source of energy, the substrates for their catabolism or their sources of reducing equivalents and carbon (Table 2). For this purpose we must discuss the metabolism

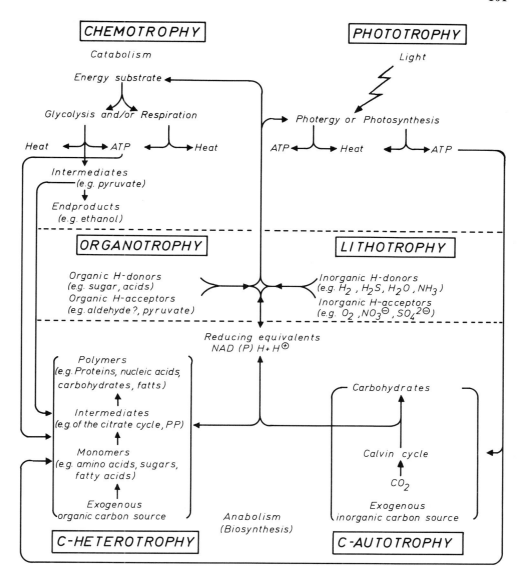

FIGURE 3. Summary of procyte metabolism (details, see text).

of the procytes in somewhat more detail (Figure 3; see Schön[25]). The *metabolism* of cells has two main functions (compare Chapter 4.I.): to build new cell components for growth and reproduction (*anabolism*) and to provide energy for these syntheses and other energy-dependent processes *(energy metabolism, catabolism).*

Anabolism (biosynthesis) includes the synthesis of about 150 different micromolecular components (e.g. amino acids, nucleotides, sugars, and fatty acids) and some macromolecular components (e.g. proteins, nucleic acids, carbohydrates, and fats) of the cell. Both groups are derived from a few basic precursors, of which the most important are sugar phosphates, pyruvate, acetate, oxaloacetate, succinate, and α-ketoglutarate. The biosynthetic mechanisms are also similar in all cells. There are differences, however, in the nature of their carbon sources. If the carbon used for biosynthesis is obtained from organic substrates, the procytes are *C-heterotrophic* — the fermenters, photergers and respirers belong in this category. If, on the other hand, CO_2 fixed in the Calvin cycle is used for the synthesis of carbohydrates, the organisms are *C-autotrophic* — chemoautotrophs and photosynthesizers make up this group.

Energy metabolism supplies the energy for the anabolic processes and for the maintenance of the complex cell structure. Energy can be obtained in two completely different ways. In *chemotrophic* energy metabolism (catabolism), chemical energy released by the degradation (oxidation) of organic or inorganic substrates is stored in the form of ATP. In *phototrophic* energy metabolism (photergy, photosynthesis), energy is obtained by transforming light energy into chemical energy, again in the form of ATP. Both types of metabolism can be further subdivided.

Most *chemotrophic* procytes obtain their energy by oxidizing or dehydrating organic energy substrates, such as glucose (glycolysis, respiration). They transfer the hydrogen released from the substrate (H-donor) to H-acceptors, such as NAD(P) (see below). In the simplest case, as in the fermenters, H-donors and H-acceptors are organic compounds. In the respirers, however, only the H-donors are organic, while the H-acceptors are almost without exception inorganic (anaerobes: SO_4^{2-}, NO_3^-; aerobes: O_2). Fermenters and respirers have a *chemoorganotrophic* metabolism, as do all animals. The group of chemoautotrophs, on the other hand, have a *chemolithotrophic* (chemoinorganotrophic) metabolism. They utilize only inorganic energy substrates, i.e. both their H-donors (H_2, H_2S, NH_3, Fe^{2+}) and their H-acceptors (O_2, NO_3^-, SO_4^{2-}, Fe^{3+}) are inorganic.

The *phototrophic* procytes, which, like the green plants, use sunlight as a source of energy, also oxidize chemical compounds. The hydrogen atoms or electrons released are also transferred to H (electron)-acceptors. The two most important cellular H (electron)-acceptors are two pyridine nucleotides, nicotinamide adenine dinucleotide *(NAD)* and nicotinamide adenine dinucleotide phosphate *(NADP):*

$$NAD^+ \underset{- 2H^+ - e^-}{\overset{+ 2H^+ + 2e^-}{\rightleftharpoons}} NADH + H^+$$

$$NADP^+ \underset{- 2H^+ - 2e^-}{\overset{+ 2H^+ + 2e^-}{\rightleftharpoons}} NADPH + H^+$$

These hydrogen carriers serve as reducing agents in biosyntheses, such as the reduction of CO_2 in the Calvin cycle to carbohydrates. Depending on whether the hydrogen they use for the formation of reducing equivalents comes from inorganic (H_2, H_2S, H_2O etc.) or organic H-donors (succinate, malate, etc.) the procytes are classified as *photolithotrophs* (= photoinorganotrophs) or *photoorganotrophs*. The photergers are photoorganotrophs, whereas there are both photoorganotrophic and photolithotrophic members of the group of photosynthesizers.

If we now systematically evaluate these results and those found in Table 2 and attempt to find the greatest common denominator, we arrive at the hypothetical relationships among the most important procyte physiological types shown in Figure 4. This assumed evolutionary relationship among the physiological types is another indication that the photosynthesizers arose from anaerobic respirers. This argumentation on the basis of substrate homologies cannot, however, produce evidence pertaining to the central question, i.e. whether the cytochrome chain was used first for respiration or photosynthesis. This problem is treated by sequence and conformation analysis of nucleic acids and proteins.

E. Sequence Homologies
Many nucleic acids and proteins are a kind of fossils in the sense that their structures

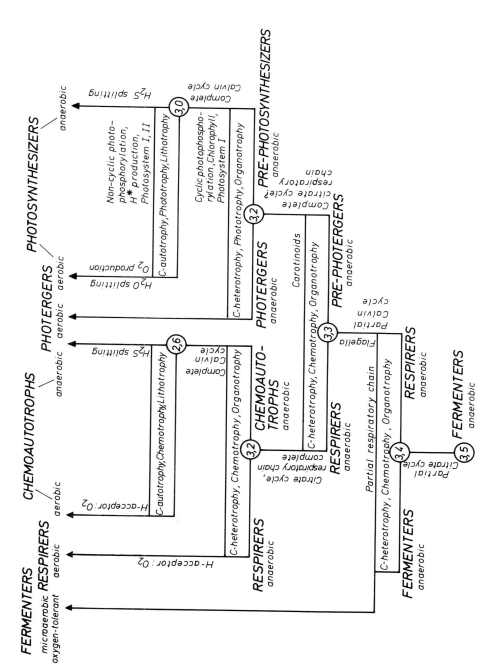

FIGURE 4A. Evolution of the physiological procyte types as suggested by comparison of substrates and basic metabolic pathways. H* = activated hydrogen; "partial Calvin cycle" means here the (heterotrophic) CO_2 fixation which was presumably not yet cyclic; the numbers in the circles indicate time in billions of years.

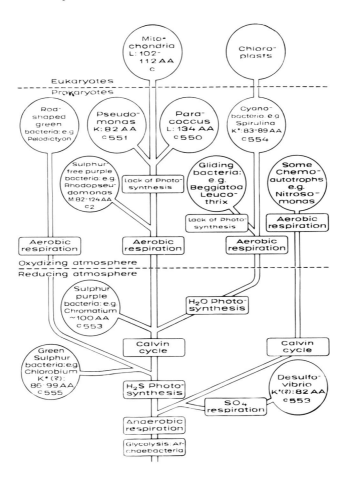

FIGURE 4B.   Essential aspects of the evolution of basic procytic metabolic types can be deduced from the sequences and structures of cytochrome c molecules and various nucleic acid molecules, as well as from other available data[18] (see Section III and Figure 20). According to this diagram, anaerobic respiration developed before anaerobic photosynthesis, whereas aerobic respiration developed independently at the same time in different bacteria phylla. Some of today's aerobically respiring bacteria must be derived from phototrophic bacteria which have lost their photosynthetic capability. Some of these bacteria resemble, according to the criteria studied here, the mitochondria, whereas the chloroplasts resemble the blue-green algae (details see Chapter 7).

   L, M, K, K* = classes of cytochrome c molecules.

   AA = number of amino acids in each cytochrome c molecule.

have been dynamically conserved by evolution over billions of years. Thus, the degree of relatedness of two organisms can be expressed in terms of their mutual similarity. We can assume that there has been a steady accumulation of mutations in the genome of each group of organisms. Since these mutations are random, each species begins to accumulate different mutations as soon as it diverges from a sister species. There is therefore less and less agreement between their DNAs or RNAs as time goes on. Thus forms whose common ancestor lies far back in time have many more differences in DNA or RNA sequence than forms which diverged from one another recently. Since the amino acid sequence of the proteins is also indirectly correlated, via transcription and translation, with the base sequence of the genome, the same is true for proteins. Dayhoff and co-workers have compiled a computer archive with which they can compare sequences.[9,30] They were thus able to investigate genetic similarities of biological polymers on a large scale and to propose phylogenetic trees showing the relatedness of homologous nucleic acids and proteins in different species.

With regard to nucleic acids, the homologous sequences of the tRNAs and the single small and two large rRNAs of various species have been compared in particular (e.g. reviewed by Dayhoff et al. in Schwemmler and Schenk[30] and in Fredrick[9]). These data are of particular significance with regard to the homologies between mitochondria and aerobic, phototrophic bacteria and between plastides and blue-green algae; furthermore, they offer evidence of differences between DNA-containing cell organelles and prokaryotes as compared with the nucleocytoplasmic system. Therefore, these data will be presented and discussed in greater detail in connection with the question of the origin of the eukaryotic cell (see Chapter 7, Figure 11).

Of the proteins, homology analyses have been carried out on ATP-synthetases, ferredoxins, and cytochrome molecules of the various species. ATP-synthetases occur in all prokaryotes. Ferredoxins are a class of iron-sulfur proteins that act as electron carriers in a number of biochemical processes, including nitrogen fixation, nitrite reduction, and photosynthetic electron transport (Figure 16). Cytochrome c acts as electron carrier in respiration and photosynthesis processes (Figure 8, B). Although much of this data and its interpretation is subject to discussion, it can shed light on the evolution of some essential elements of the basic metabolic types of the prokaryotes. In the words of Dickerson,[5] " cytochrome c can lead us through the labyrinth of bacterial evolution like Ariadne's thread".

Sequence analyses substantiate the subdivision of cytochrome c molecules in the four classes L, M, K and K*, based on their X-ray structure. This can be shown with the help of an identity matrix. Such a matrix is constructed by comparing the amino acid sequences of two homologous cytochromes and counting the number of chain positions at which the same amino acid is found in both. However, the abbreviated identity matrix reproduced here shows only the averages for the cytochrome c molecules of the four size classes (according to Dickerson[5]). One can recognize that the number of homologies between cytochromes

	L(6)	M(9)	K(8)	K*(7)	$C_{555}(2)$	$C_{553}(1)$
$L(6:C_2,C_{550})$	49,7	38,5	17,8	17,5	15,6	12,0
$M(9:C_2,C)$	38,5	51,6	18,9	19,1	15,1	11,9
$K(8:C_2,c_{551})$	17,8	18,9	43,8	17,6	13,8	14,4
$K*(7:c_{554},f)$	17,5	19,1	17,6	43,6	18,4	6,4
$c_{555}(2)$	15,6	15,1	13,8	18,4	47,0	12,0
$c_{553}(1)$	12,0	11,9	14,4	6,4	12,0	—

of the same class (numbers in the rectangles) is much higher than the homologies between different classes. The M and L cytochromes (dotted rectangles) are an exception and would build a single class if their structure were not quite distinct. The numbers in parentheses are the number of investigated cytochromes of each class. Dickerson does not place cytochrome c_{555} (green sulfur bacteria) and c_{553} (*Desulfovibrio*) in the K group, since they have a unique amino acid sequence. On the basis of structure and sequence analyses of cytochrome c molecules and investigations of nucleic acids, the phylogenetic tree seen in Figure 4, B can be constructed. This diagram summarizes several essential aspects of procytic metabolic types and again indicates that anaerobic respiration must have developed before anaerobic photosynthesis. Furthermore, it is evident that aerobic respiration (cytochrome-oxidase) developed independently in different bacteria much later, after the bacteria phylla had developed in different directions. This is also indicated by the fact that the bacterial cytochrome c reacts easily with the corresponding cytochrome reductases in the mitochondria but only with difficulty with the mitochondrial cytochrome oxidases. This is to be expected if one assumes that cytochrome reductase is a common evolutionary legacy of all prokaryotes but that cytochrome oxidase is of polyphyletic origin. Nevertheless, the alternative possibility of an early transfer of the genes which code for oxidase in the form of episomes or plasmides from one basic bacteria form to another cannot definitely be discarded.

III. HYPOTHETICAL RECONSTRUCTION OF PROCYTE EVOLUTION

The facts and conclusions discussed above all make the descent of the photosynthesizers from anaerobic respirers seem more probable than an evolutionary development in the opposite order. They also indicate that the descent of the photergers from primitive anaerobic respirers is highly probable (see also Thauer and Fuchs[34]). Thus it can practically be concluded that the *splitting hypothesis* holds in principle for the evolution of procytes, although the conversion hypothesis may apply to certain cases (e.g. *Paracoccus, Pseudomonas,* and *Beggiatoa* or *Leucothrix,* formerly photosynthetic blue-green algae). The following outline of the metabolic development of procytes is thus based on the splitting hypothesis.[1,2,13,16,17,24,29] It is intended only as a short, hypothetical summary. Details of the metabolic pathways mentioned are to be found in the literature (see above). The following discussion of the individual metabolic paths is based on this literature, in particular on Schön,[25] although here the phylogenetic aspects are stressed.

A. Fermenters

The modern *fermenters* must have evolved about 3.5 billion years ago from the fermenting *ur-procytes* postulated in Chapter 5, III.D. (and Chapter 5, Figure 27). They developed a complete *fermenting metabolic apparatus* (Figure 5), for which carbohydrates are the main energy sources. Glucose is one of the most important of these. The most widespread form of glucose degradation is glycolysis (also fructose 1,6-biphosphate pathway, or Embden-Meyerhof-Parnas pathway = EMP pathway; Figure 6). Since it is found in all groups of organisms, glycolysis is probably the original metabolic process. The energy balance of glycolysis yields two ATP molecules per C_3 molecule, or four ATPs per glucose. Since two ATPs are consumed in the preparatory reactions, the net yield is only two ATP molecules per glucose molecule.

NADH produced in fermentation must be reoxidized in order to be re-used (Figure 8,A). One possibility for the regeneration of NAD^+ is to transfer the hydrogen atoms (or electrons) to intermediates from fermentation. One important intermediate of this process is *pyruvate,* from which the various end-products of fermentation are derived by enzymatic transformation to ethanol, lactic acid, propionic acid, formic acid, acetic acid, etc.

Another degradative pathway frequently used by fermenters is the *pentose phosphate pathway* (= PP pathway, Figure 7). In the oxidative branch of the PP-pathway, glucose is degraded to glyceraldehyde and three CO_2 in a cyclic, partially reversible series of reactions. The significance of this pathway lies in its role as a supplier of ribose 5-phosphate for the synthesis of nucleic acids and NADPH for many synthetic reactions in the procyte. In addition to other pathways of glucose degradation, some of the fermenters also have a partial *citrate cycle** (glyoxylate cycle; Figure 18). Other important metabolic pathways flow into the simple citrate cycle or derive from it, for example, the beginnings of independent fat and protein metabolism.

Examples of recent fermenters are the *mycoplasms, rickettsiaes, lactobacilli, spirochaetes*

---

* The partial and complete citrate cycles are described in Section III.D.

FERMENTERS

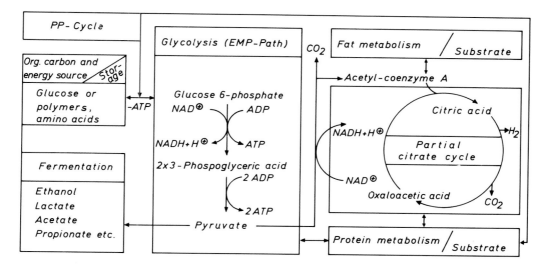

FIGURE 5. Fermentative metabolism of fermenters, such as *Clostridia,* and the cytoplasma and possibly amyloplasts of eucytes. Organic substances serve as sources of both energy and carbon. Phosphorylated sugars are hydrolyzed and fermented by glycolysis (EMP path = Embden-Meyerhof-Parnas pathway) or other pathways (PP = pentose-phosphate pathway), and the energy released is bound in the form of energy-rich ATP. The incomplete citrate cycle is supplied with acetyl-coenzyme A obtained from pyruvate. The beginnings of an independent fat and protein metabolism have developed, and these are coupled to the citrate cycle, which is here only anaple(u)rotic (supplying intermediate products). (From Schwemmler;[68] details see text.)

and most *Clostridia* (see also Figure 20). Some fermenters have even become tolerant of oxygen or microaerophilic (microaerobic). The *propionic acid bacteria* are an example of oxytolerance, while certain *lactobacilli* are microaerophilic. Fermenters are found in moist ecological niches as decomposers.

B. Anaerobic Respirers

The activity of the rapidly multiplying fermenters must have led gradually to a scarcity of abiogenic organic substrates. These compounds were only formed at a constant, low rate. The fermenting procytes appear therefore to have experienced a severe "substrate and energy crisis" about 3.4 billion years ago. This led to selection of forms which, as *anaerobic respirers,* had acquired through mutation of their genetic information the ability to utilize new energy sources. This ability was based on the utilization of *redox systems,* which are characterized by electrochemical potential differences. The redox energy is transformed into the chemical energy of energy-rich phosphate bonds in ATP.

These redox systems could only have developed in the course of a long evolutionary process. At first, presumably, individual redox pairs were embedded in the membrane and bound to proteins. The energy released in the transfer of electrons from the redox partner with higher potential to that with lower potential was now used to synthesize ATP. This was the beginning of the *respiratory chain** (electron transport chain, Figure 17, A and B). The respiratory chain gradually grew through the successive incorporation of more electron transport pairs. The order of the redox pairs in the modern chain may also indicate the order of their evolution. *NAD* (Figure 8, A) or similar coenzymes then stood at the

---

* For the complete and shortened respiratory chain, see Section III.E.

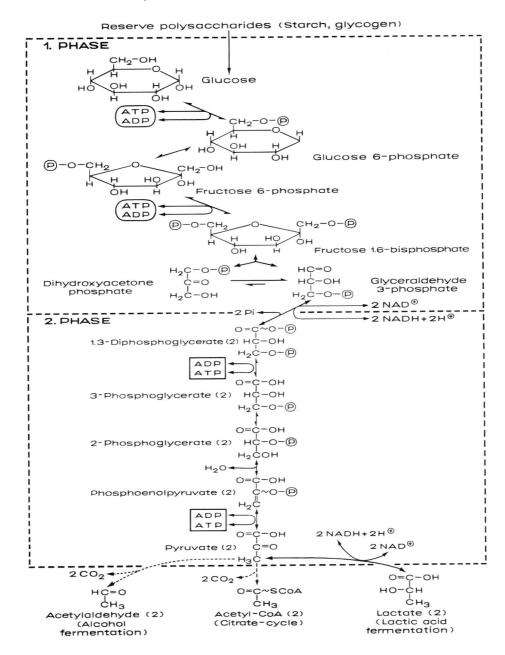

FIGURE 6.   Glycolysis occurs in two phases. In the preparatory phase, a molecule of glucose is transformed into fructose bisphosphate (a process which consumes ATP) and split into two triose phosphate molecules, dihydroxyacetone phosphate and glyceraldehyde phosphate, by the enzyme aldolase. Dihydroxyacetone phosphate is almost completely isomerized to glyceraldehyde phosphate. In the second phase of glycolysis, the latter is oxidized and phosphorylated by uptake of inorganic phosphate (P$_i$) to the energy-rich diphosphoglycerate. The hydrogen released by the oxidation is accepted by NAD$^+$, which is reduced to NADH + H$^+$. The high-energy phosphate group on the diphosphoglycerate is now transferred to ADP, forming ATP. The 3-phosphoglycerate formed as the other product of this reaction is isomerized to 2-phosphoglycerate, which on losing H$_2$O becomes the very energetic phosphoenolpyruvate. This compound can also donate its phosphate group to ADP, forming ATP and pyruvate. Since the ATP synthesis described here occurs at the substrate level, it is called substrate phosphorylation. Pyruvate can be further transformed to ethanol or lactic acid (or other products) or can be converted to acetyl-coenzyme A which enters the citrate cycle.

Balance of glycolysis: 1 Glucose → 2 Pyruvate + 2 ATP + 2 NADH.

P P PATH

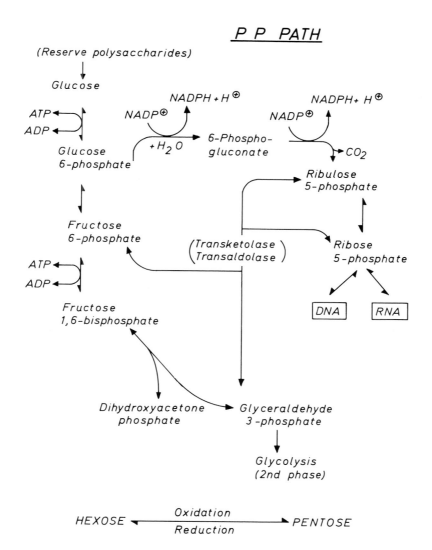

FIGURE 7. In the oxidative pentose phosphate path (P P path), glucose is degraded in a cyclic reaction series to three CO_2 and glyceraldehyde phosphate, which can be further degraded by glycolysis to yield ATP. The entire process can be divided into two phases: the actual oxidation process, and the transformation reactions. In the oxidation process, glucose is first phosphorylated to glucose 6-phosphate. This is combined with water and is simultaneously dehydrogenated to 6-phosphogluconate. The hydrogen is transferred to $NADP^+$ (not NAD^+). The 6-phosphogluconate is itself dehydrogenated, with the hydrogen again being transferred to $NADP^+$ and immediately decarboxylated to form ribulose 5-phosphate. This is in equilibrium, via an isomerase, with ribose 5-phosphate, the sugar required for RNA and DNA synthesis. The following transformation reactions serve to regenerate hexoses (fructose 6-phosphate, glucose 6-phosphate) from the pentoses in a cyclic process involving transaldolases and transketolases. The P P path thus resembles to a certain degree the Calvin cycle (reductive P P path: Figure 13), except that it proceeds in the opposite direction. In contrast to glycolysis (see also the citrate cycle, Figure 18A), the P P path produces primarily NADPH. The hydrogen atoms released are thus used for reductive syntheses, rather than for energy yield.

Balance of the P P path: 1 Glucose 6-phosphate → 1 pentose 5-phosphate + 1 CO_2 + 2 NADPH.

beginning, and the *cytochromes* (Figure 8, B) at the end of this development. If this is so, then photosynthesis could not have been the beginning, but rather the necessary result of

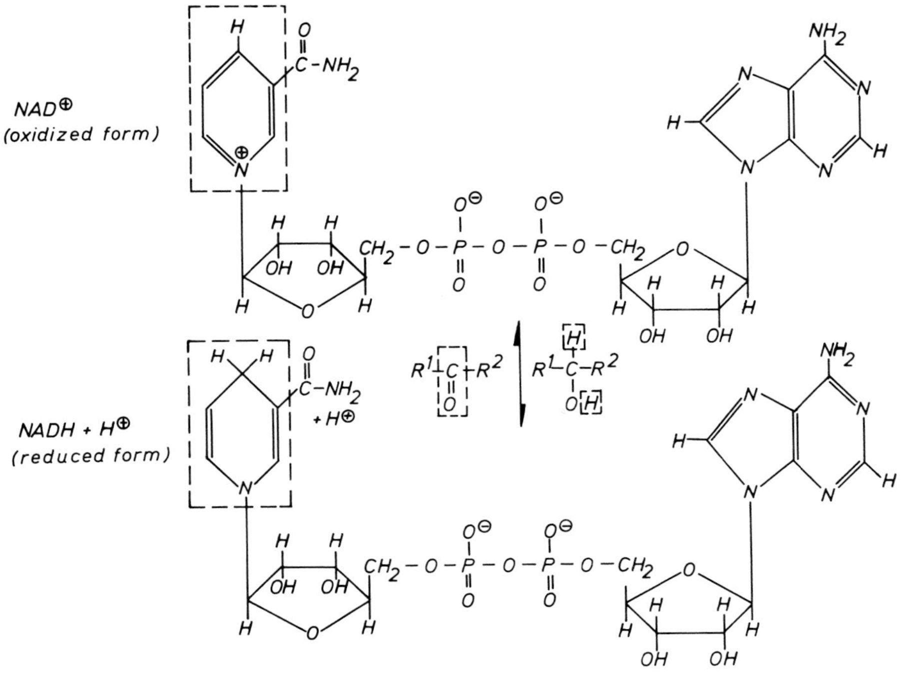

FIGURE 8A. Basic structure of the oxidized and reduced forms of nicotinamide adenine dinucleotide. NAD(H) is the general hydrogen carrier, serving both as a reducing equivalent and as the substrate for respiration. The related compound nicotinamide adenine dinucleotide phosphate NADP(H), on the other hand, is primarily involved in reductive processes and biosyntheses.

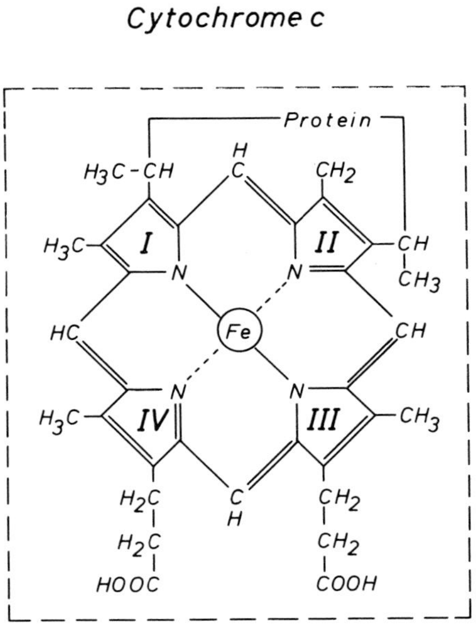

Cytochrome c

FIGURE 8B. The basic structure of the cytochromes. Cytochromes are electron carriers which contain an iron-heme ring. The iron atom can undergo a reversible change of its oxidation state:

$$Fe^{3+} \xrightleftharpoons[- e^-]{+ e^-} Fe^{2+}$$

RESPIRER , anaerobic

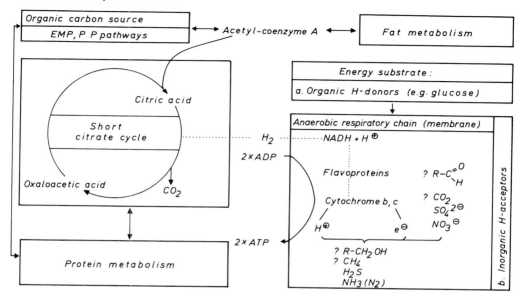

FIGURE 9. Fermentative, oxidative metabolism of the anaerobic respirers such as the nitrate and sulfate bacteria. Organic compounds serve as the carbon sources, organic H-donors and inorganic H-acceptors as the energy substrates. Hydrogen electrons from the H-donors and from the citrate cycle are transferred along an enzyme chain of redox systems (anaerobic respiratory chain) to a lower energy state, and the energy released in the process is bound in the form of ATP. The oxidized hydrogen is thought to have been transferred to aldehyde (?) or CO_2 (?) in the hypothesized ancient forms, and is transferred to SO_4^{2-} or NO_3^- in modern forms. Less ATP is recovered from the incomplete anaerobic respiratory chain, since some of the cytochrome oxidases (cytochrome a_3) are lacking.

the evolutionary process; the cytochromes are in the middle and the chlorophylls are at the end of the photosynthetic potential gradient (see Figure 12, 14, and 17, C).

Cytochrome a_3 *(cytochrome oxidase)* could not yet have been present as the last member of the respiratory chain, for lack of O_2 as the terminal electron acceptor (Figure 9). Probably the electrons were at first transferred to aldehydes (?) or, later, to inorganic CO_2 (?) or SO_4^{2-} and NO_3^-. CO_2 must have already been present, partly as a result of fermentation, and SO_4^{2-} could have been produced by volcanism before free oxygen appeared. NO_3^- could possibly only have formed after free oxygen appeared, and was then available as an H-acceptor. Sulfate can be reduced to hydrogen sulfide, nitrate to molecular nitrogen or ammonium. These forms of reduction are also called sulfate or nitrate respiration (Figure 17, B).

With their short respiratory chain, the anaerobic respirers were able to extract much more energy from organic substrates than could be yielded by fermentation. The fermentation of a molecule of glucose produces two molecules of ATP, while its anaerobic respiration by procytes yields at least 24 molecules of ATP (aerobic respiration by procytes: 38 ATP). This enabled the anaerobic respirers to increase their synthetic activity enormously. In addition to their fermentative metabolism (EMP and PP pathways), they might have had an incomplete heterotrophic, acyclic CO_2 fixation as a possible precursor of the Calvin cycle* (Figure 13, B), a simple partial citrate cycle* (Figure 18, B), and protein and fat metabolism independent of organic substrates.

---

* The complete description of the respiratory chain and the citrate cycle is given in Section III.F, that of the Calvin cycle in Section III.E.

PHOTERGERS

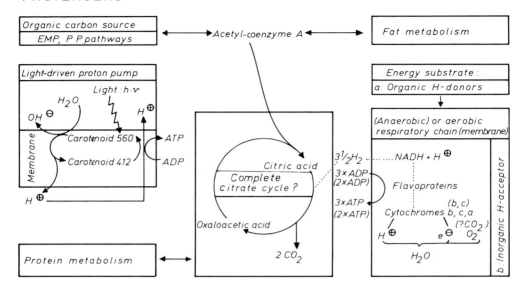

FIGURE 10.   Fermentative, oxidative photergic metabolism of photergers, such as modern halobacteria, and possibly also the cones of the retina and chromoplasts in plants (?). Organic compounds are used for synthesis and energy production. However, electrons and hydrogen ions released by the oxidation of substrates are separated by a proton pump and then used to drive ATP synthesis. The proton pump is powered by sunlight trapped by carotenoids. The citrate cycle is now perhaps complete, and the syntheses of protein and fats are for the most part independent of external precursors. Data for the anaerobically respiring photergers, which are probably phylogenetically old, are given in parentheses.[20,28,32]

The nitrate respirers, such as the *denitrificants,* and the sulfate respirers, like *desulfovibrionts* or *Desulfotomaculum* species, can be regarded as possible descendants of the hypothetical anaerobic respirers (see also Figure 20). Nitrate respirers can be easily isolated from the soil, and sulfate respirers from bottom muds.

The anaerobic respirers presumably then split into four main groups about 3.3 billion years ago. The modern representatives of these are the aerobic respirers, the aerobic chemoautotrophs, the aerobic photergers, and the aerobic photosynthesizers. All four branches first went through an anaerobic stage (see Figures 2, 4, and 20).

C. Photergers

The development of photoorganotrophic, C-heterotrophic photergers from anaerobic respirers must have come about through the selection of forms which had happened to acquire through mutation the ability to synthesize *carotenoids* from isoprene molecules. The yellow to red pigments were embedded in the membrane and bound to proteins. Such a chromoprotein (chroma, Greek = color), when it occurs in the halobacteria is called *bacteriorhodopsin.* It is similar to the purple vision pigment, rhodopsin, in animal visual cells. Bacteriorhodopsin presumably served the first photergers as a protection against UV radiation. They could now conquer the higher, light-flooded areas of the primitive seas.

In time, the purple complex attained another function; it developed into a light-driven *proton pump*[20] (Figure 10). Upon irradiation, the purple complex reversibly changes its absorption maximum from 560 to 412 nm; at the same time, protons H^+ are taken from the cytoplasm and released outside the membrane. In this way, a proton gradient is formed across the plasma membrane. The electrochemical energy stored in this gradient can be used for ATP synthesis when the protons flow back into the cell.

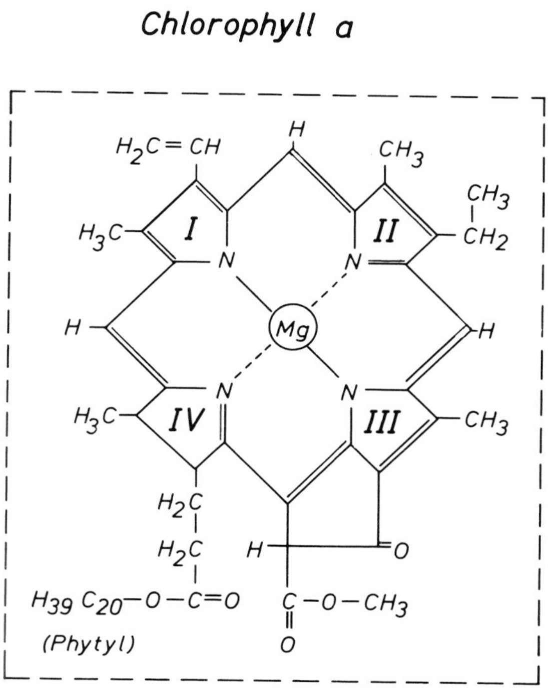

Chlorophyll a

FIGURE 11.   The basic structure of chlorophyll a.

The photergers were thus able, in the absence of oxygen, to carry out *photophosphorylation* in light, although with structures which completely differ from those of the phototrophic bacteria. The originally anaerobic photergers could also have possessed a nearly complete citrate cycle (Figure 18) and both fermentation and anaerobic respiration. Their H-acceptors may have been similar to those used by the anaerobic respirers.

The only known modern descendants of such anaerobic photergers are the extremely salt-loving *halobacteria* from salt lakes and evaporating sea water. They are normally strictly aerobically respiring bacteria. Only when they grow in conditions with limited oxygen do they revert to anaerobic photergy. The above derivation indicates that they developed the ability for aerobic respiration in the further course of evolution (see also Figure 20).

D. Photosynthesizers

The *photosynthesizers* must have evolved parallel to the photergers, but beginning somewhat later, i.e. about 3.2 billion years ago. They too developed membrane pigments, including the much more effective chlorophyll (Figure 11), which they synthesized from porphyrins. Like the carotenoids, the chlorophylls were only effective in combination with protein. They also have the ability to absorb light by being excited to a higher electronic state. In this state they were strong reducing agents. Unlike the carotenoid system, the chlorophyll system transferred the excited electrons to a system of redox partners which was probably developed from the respiratory chain. The redox partners here belong to the same class of chemical compounds as those in the respiratory chain. Here too, the energy released by electron transfer was fixed in the form of ATP. The electrons were finally cycled back to the chlorophyll, thus regenerating it. The process is therefore called *cyclic photophosphorylation* (Figure 12). Aside from the original excitation of the chlorophyll by light, the process can occur in the dark. At present, no recent photosynthesizers with *purely* cyclic photophosphorylation are known.

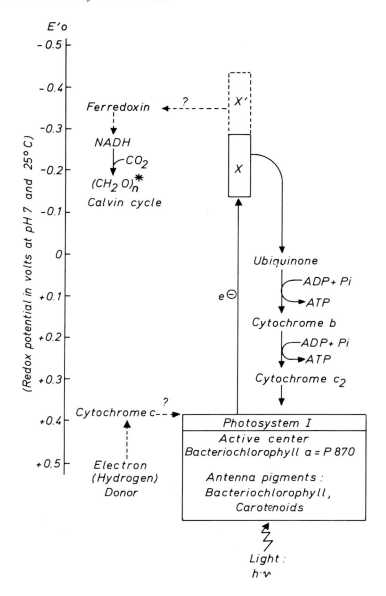

FIGURE 12.   ATP synthesis is achieved by cyclic photophoshorylation (Photosystem I) in the anaerobic photosynthesis of phototrophic bacteria. Antenna pigments (carotenoids, bacterial chlorophyll) absorb light and transfer the energy of excitation to the active center of the bacterial chlorophyll a (P 870). This releases an electron with a high reducing potential in the active center which reduces a suitable acceptor X. The electron hole in the active center is filled by an electron from cytochrome c_2, so that the cytochrome molecule is oxidized. The electron from the reduced acceptor X flows via ubiquinone and cytochrome b back to the oxidized cytochrome c_2. ATP is produced as the electron falls back to the level of cytochrome c_2 (compare also Figure 17C). There is presumably a non-cyclic electron transport for the formation of NADH (broken line), which draws its electrons both from organic (e.g. succinate) and inorganic (e.g. H_2S) substrates. Here, no ATP is synthesized. The NADH formed is used primarily for CO_2 reduction in the Calvin cycle (Figure 13.; adapted from Schön[25]).

($CH_2O)_n$* does not denote formaldehyde here, but rather a unit amount of carbohydrate.

The cyclic process produced energy, but no electrons which could be used for the reduction of CO_2 or bicarbonate (HCO_3^-). This inorganic source of carbon was first made available by the process of *non-cyclic photophosphorylation* (Figure 12). This reaction, which occurs in the light, yields both ATP and NADPH (nicotinamide adenine dinucleotide phosphate in reduced form). NADPH is usually the reducing agent for CO_2, which is reduced by simultaneous expenditure of ATP and NADPH to carbohydrate. The process by which this occurs, the *Calvin cycle* (Figure 13A), takes place in the dark. The source of electrons needed to reduce the CO_2 can be either organic compounds like sugar or organic acids, or inorganic, e.g. reduced sulfur compounds (H_2S, S^-, $S_2O_3^{2-}$). However, only in the latter case is there a net increase in biomass. The phototrophic bacteria are modern examples of this kind of obligate anaerobic photosynthesizers. They are found principally in anaerobic waters. There are several kinds of phototrophic bacteria, i.e. *sulfur-free, red purple bacteria,* which use organic H-donors (photoorganotrophy), and *green sulfur bacteria* and *sulfur-containing red purple bacteria,* which have mostly inorganic H-donors (photolithotrophy). Many phototrophic bacteria also have various fermentative pathways, and the sulfur-free purple bacteria are even capable of aerobic respiration (Figure 15; see also Figure 17, C). The photosynthetic pigments are not synthesized under aerobic conditions.

Sooner or later, however, the decreasing availability of H_2S limited the evolution of the purple bacteria. The limitation was first overcome by the *blue-green algae* (cyanobacteria), which replaced H_2S with H_2O. However, the oxidation or cleavage of H_2O is energetically much more difficult than that of H_2S (Figure 14). In comparison we can consider the greatly simplified net reactions for purple bacteria and blue-green algae:

$$2\,H_2S + CO_2 = (CH_2O)^* + H_2O + 2S; \quad \Delta G = 50.4^{**}\ kJ/mol$$

$$2\,H_2O + CO_2 = (CH_2O)^* + H_2O + O_2; \quad \Delta G = 483.0^{***}\ kJ/mol$$

The C-autotrophic, photolithotrophic blue-green algae then had become completely independent of organic substrates for energy or carbon atoms (Figure 15). They had become independent of the primal soup, at least for some of the time. On the other hand, their activity released oxygen. This was first absorbed by the hydrosphere, and later by the atmosphere. Starting about 2 billion years ago, photochemical reactions of the atmospheric oxygen formed O_3 (ozone) from O_2, which then collected in the outer regions of the atmosphere as a thin layer, where it absorbed a large part of the deadly short-wavelength UV radiation from the sun. The conditions for the colonization of the land and air had been created. The original blue-green algae were probably not much different from their modern descendents. In principle, they must have been equally capable of anaerobic fermentation and aerobic respiration, as are the modern forms.

E. Aerobic Respirers

Aggressive O_2 forced the anaerobic bacteria which had existed up to that time either to withdraw into the oxygen-free parts of the biosphere (e.g. bottom muds) or to adapt gradually. In all probability, the *anaerobic respirers* achieved the latter. In time they developed the ability to oxidize the reduced hydrogen carriers, for example the NADH formed by dehydration of the organic substrate in the citrate cycle, by transferring the hydrogens to O_2 thus forming water (Figure 16). In this biochemical reaction, however, the components H and O_2 do not react directly. Instead, the hydrogen (proton + electron) is transferred in steps along the redox systems of the anaerobic respiration chain and finally combined with the O_2. This mechanism conserves some of the energy released by the oxidation reaction and

---

* (CH$_2$O) as used here does not denote formaldehyde, but a unit of carbohydrate.
** = 12 kcal/mol
*** = 115 kcal/mol

A   <u>*Calvin cycle*</u>
Cyclic, autotrophic CO_2 fixation

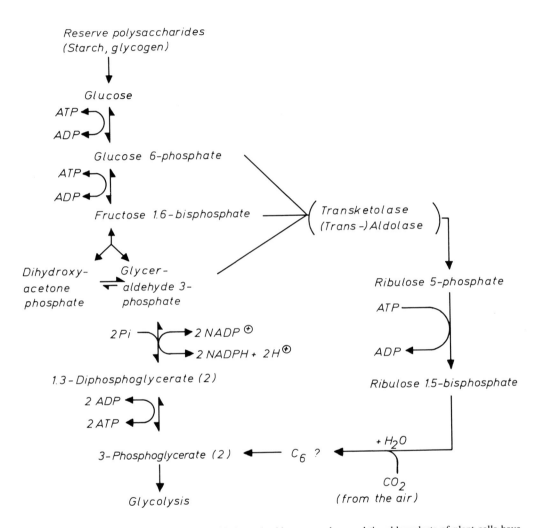

FIGURE 13A.   Photo- and chemoautotrophic bacteria, blue-green algae and the chloroplasts of plant cells have the enzymes of the Calvin cycle (Calvin-Bassham cycle, or reductive pentose phosphate path). The Calvin cycle can be subdivided into three phases: CO_2 uptake by an acceptor, reduction of the fixed CO_2 to carbohydrate, and regeneration of the CO_2 acceptor. The CO_2 acceptor is ribulose 1,5-bisphosphate. Incorporation of CO_2 produces an unstable, 6-carbon intermediate, which has not been identified. This immediately dissociates into two identical C_3 bodies (3-phosphoglycerate), which are reduced by the reducing equivalents of NADPH and phosphorylated by ATP to glyceraldehyde phosphate and dihydroxyacetone phosphate. These can combine to form fructose 1,6-bisphosphate. These reactions are familiar from glycolysis (Figure 6) and the pentose phosphate path (Figure 7), where, to be sure, they run in the opposite direction. Ribulose 5-phosphate, the precursor for the CO_2 acceptor, is finally regenerated by means of the enzymes transaldolase and transketolase (see the pentose phosphate path). Both the hexoses and the intermediates with fewer C atoms are withdrawn from the cycle for biosyntheses.

Balance of the Calvin cycle:
$6\ CO_2\ +\ 18\ ATP\ +\ 12\ NADPH\ +\ 6$ ribulose 1,5-bisphosphate $\rightarrow$ 1 hexose + 6 pentoses

B Non-cyclic, heterotrophic CO₂ fixation
(possible phylogenetic precursor of the Calvin cycle)

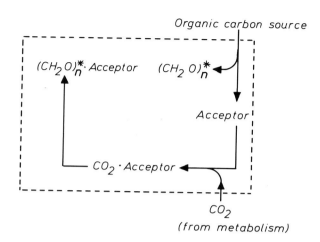

FIGURE 13B. In addition to the cyclic, autotrophic CO_2 fixation of the Calvin cycle, there is a non-cyclic, heterotrophic CO_2 fixation. The latter is found in connection with gluconeogenesis. The C-heterotrophic organisms do not take the CO_2 from the air, however, but bind CO_2 produced in the cell. The CO_2 acceptors (e.g. pyruvate) in this process are different from those in the autotrophic fixation. There is no net fixation of CO_2, i.e. de novo synthesis of biomass. Furthermore, the CO_2 acceptor is not regenerated, but must be newly synthesized, In some respects, the heterotrophic CO_2 fixation can be regarded as a phylogenetic precursor of the autotrophic CO_2 uptake, the former being not yet incorporated into a cycle. (A, B adapted from Schön.[25])

$(CH_2O)_n{}^*$ = unit quantity of carbohydrate.

stores it in the form of energy-rich compounds, such as ATP, instead of releasing all of it as heat. The drop in redox potential, and thus the change in free energy, is large enough at three stages in the process (steps I, II, and III; see Figure 17, A) to drive the synthesis of one molecule of ATP each. The production of ATP in the aerobic *respiratory chain,* as in the anaerobic, is called *oxidative phosphorylation* (respiratory chain or electron transport chain phosphorylation), as opposed to the photophosphorylation of the photosynthesizers.

The aerobic respirers differ from their hypothetical anaerobic ancestors in that they have an additional redox step probably the *cytochrome oxidase,* which has a sufficient redox potential to transfer the hydrogen directly to the oxygen. With this step, the aerobic forms obtain one ATP more per electron than the anaerobic forms.

Furthermore, the aerobic respirers have a complete *citrate cycle,* in contrast to the anaerobes (Figure 18, A). This is the main pathway for the complete degradation of organic substrates in anaerobic organisms, and it delivers hydrogen atoms or electrons (NADH + H⁺) for respiration. The cycle also produces many intermediary products for anabolic processes, so that even strictly anaerobic procytes possess most of the enzymes of this cycle.

The aerobic respirers were now able, thanks to their oxygen-adapted metabolism, to colonize the surface waters and later the moist terrestrial areas, as soon as the photosynthesizers had produced enough organic substrate for their C-heterotrophic nutritional requirements. The modern descendants of aerobic respirers are the *Eubacteria,* some *Micrococci, Mycobacteria* and *Pseudomonads* (see Figure 20). They live in various ecological niches. Some of the aerobic respirers must be derived from aerobic photosynthesizers which have lost their photosynthetic apparatus secondarily (Figure 17, C). Among the representatives

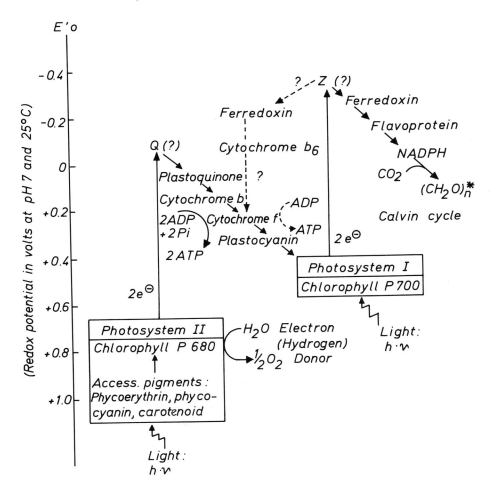

FIGURE 14.   Blue-green algae and the chloroplasts of plant cells use H_2O for the hydrogen donor to form NADPH reducing equivalents in aerobic photosynthesis, in contrast to the anaerobic phototrophic bacteria. Two successive pigment systems are required: photosystem I, which corresponds to the photosystem of the bacteria (Figure 13) and photosystem II, which accomplishes the photolysis of water.

Accessory pigments of photosystem II (phycoerythrin, phycocyanin, carotenoids, etc.) absorb light and transfer the excitation energy to the active pigment chlorophyll a_{II} (P 680), which is thus excited and passes electrons to the still unknown component Q. The electrons are replaced from water, which is oxidized to oxygen in the process. The electrons then fall from Q via plastoquinone, cytochrome b, cytochrome f, and plastocyanin (electron transport chain; compare with the respiratory chain, Figure 17) to the chlorophyll a (P 700) of the photosystem I. ATP is generated in the course of this electron passage, which is referred to as non-cyclic photophosphorylation.

A cyclic electron transport (cyclic photophosphorylation) other than the above pathway is postulated (dotted line). This leads electrons excited by light from photosystem I over ferredoxin, cytochrome b_6, cytochrome f, and plastocyanin back to chlorophyll a of photosystem I. Here too, possibly in contrast to the phototrophic bacteria, ATP is formed. It is not clear to what extent the capacity for ATP formation outside non-cyclic photophosphorylation is utilized under natural conditions.

The electrons in photosystem I (chlorophyll P 700) can be again activated by light. This raises them to the energy level of the as yet unidentified component Z, from which they descend via ferredoxin to $NADP^+$. NADPH is then available for the reduction of CO_2 in the Calvin cycle. (Adapted from Schön.[25])

$(CH_2O)_n$* = unit quantity of carbohydrate.

of these respirers are *Pseudomonas, Paracoccus, Beggiatoa,* and *Leucothrix* (compare also Section II.E, Figure 4, B; Dickerson[5]).

A   PHOTOSYNTHESIZERS

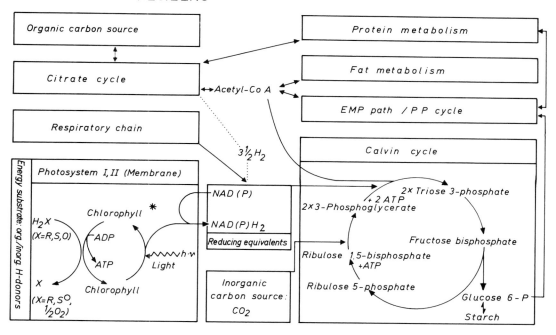

B

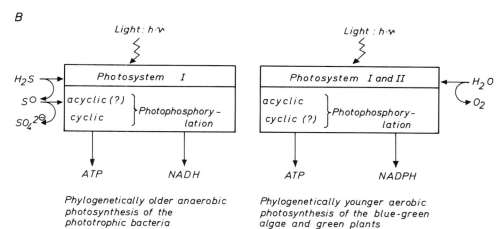

FIGURE 15.   Fermentative, oxidizing, photergic metabolism of photosynthesizers e.g. certain purple bacteria, blue-green algae, and, under some circumstances, the chloroplasts of green plant cells. The substrates for synthesis can be either organic compounds or CO_2 from the water or air (see also CO_2 in the citrate cycle). The energy donors are sunlight (h·ν) and organic or inorganic hydrogen sources. Chlorophyll which has been activated by sunlight reverts to a less energetic state, and the energy released is bound in the form of ATP. Unlike the anaerobic photosynthesizers, the aerobic photosynthesizers are able to use water as a hydrogen donor. Both use the hydrogen obtained from their hydrogen donors to form reducing equivalents (e.g. NADH) which reduce or fix CO_2 in the Calvin cycle. The oxygen released by the lysis of H_2O escapes into the atmosphere. (Figure 15A from Schwemmler;[18] 15B adapted from Schön.[25])

H_2R = organic compound like sugar or organic acid.

Chlorophyll* = excited chlorophyll.

F. Chemoautotrophs

The adaption from anaerobic to aerobic metabolism must also have occurred in the fourth great branch of the physiological procyte types. This meant a transition from anaerobic to

RESPIRERS, aerobic

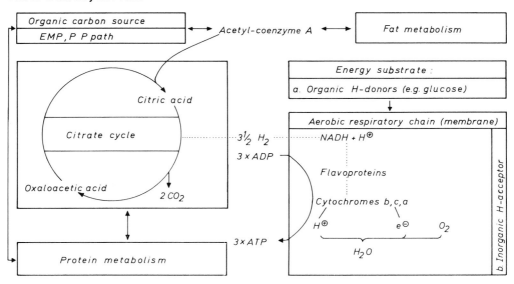

FIGURE 16.   Fermentative, oxidizing metabolism of the aerobic respirers, such as enterobacteria, to some extent mitochondria or perhaps etioplasts of eucytes. Organic material serves as a carbon source, while inorganic material can serve as an energy substrate. Hydrogen electrons from the citrate cycle and from the energy substrate are passed along an enzyme chain (respiratory chain) to a lower energy level. The energy released is bound in the form of ATP; the oxidized hydrogen is transferred to O_2. The fat and protein metabolism have become completely independent of substrate (from Schwemmler[28]).

aerobic respiration. The *anaerobic chemoautotrophs* may, like the anaerobic respirers, have expanded their short respiratory chain to include cytochrome oxidase (cytochrome a_3). They would then have been able to extract more ATP per electron, and to transfer the electrons to O_2 as terminal acceptor, instead of to SO_4^{2-} (?) or NO_3^- (Figure 19). Both anaerobic and aerobic chemoautotrophs, however, like most photosynthesizers, obtained their carbon by reduction of CO_2 in the Calvin cycle (Figure 13). Through their complete independence of organic substrate and their adaption to an aerobic metabolism, the aerobic chemoautotrophs are able to leave the water, at least in some phases, and to colonize the moist areas of the land.

Modern descendants (see Figure 20) of the anaerobic chemoautotrophs might be certain sulfur-oxidizing bacteria such as *Thiobacillus denitrificans*. This bacterium can oxidize sulfur compounds, even in the absence of air, with NO_3^- as electron acceptor. However, it requires ammonium as a source of nitrogen, since it is unable to carry out assimilatory nitrate reduction. The recent aerobic chemoautotrophs are represented by the *nitrificants, iron bacteria, ammonium oxidizers,* colorless *sulfur bacteria* and *hydrogen bacteria.* All of these are of great importance for the material circulation of the biosphere.

Like animals and plants, bacteria are named by a *binary nomenclature.* The *taxonomy* of bacteria groups individual species according to particularly important characteristics. The characteristics on which such a *classification* is based are the morphological, physiological, biochemical, serological, and genetic properties of the various procytes. The standard reference work for bacterial classification is *Bergey's Manual of Determinative Bacteriology.*[3] This artificial classification, however, generally has nothing to do with the phylogenetic relationships of the procytes. On the other hand, since there are only hypothetical proposals as to the phylogenetic descent of the various bacterial groups, it is not presently possible to

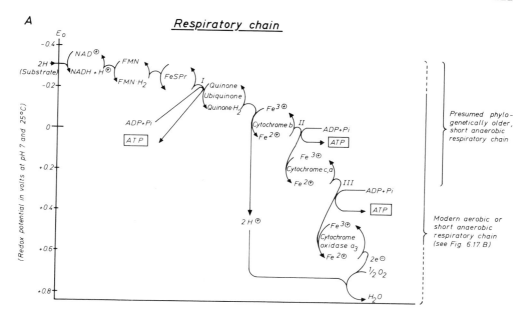

A

Respiratory chain

FIGURE 17A.   Many procytes and the mitochondria of eucytes are able to generate ATP in the course of moving electrons from NADH along the respiratory chain, a system of redox components. The first step in the electron transport is the transfer of hydrogen ($H^+ + e^-$) to another nucleotide FMN (flavin mononucleotide). From here the hydrogen is transferred to FeSPr, an iron-sulfur protein, which passes it on to ubiquinone. The following redox components, the cytochromes, transfer only the electrons. At the end of the cytochrome chain stands cytochrome oxidase (cytochrome a_3), which was presumably added to the system by the aerobic respirers. It transfers the electrons to O_2, the terminal electron acceptor, which joins the two hydrogen ions from ubiquinone to form H_2O. Three moles ATP are formed per one mole of water (steps I, II, and III). The formation of ATP by the respiratory chain is called oxidative phosphorylation (respiratory chain phosphorylation).

B

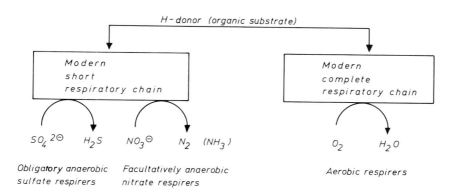

FIGURE 17B.   Modern aerobic respirers have a complete respiratory chain at their disposal, while some modern anaerobes have a chain which is incomplete at the beginning or end. The respiratory chain of the hypothetical phylogenetically ancient anaerobic respirers was presumably always incomplete at the end. The respirers probably lacked the cytochrome oxidase (cytochrome a_3), and thus step III. They thus generated one fewer ATP per electron passage. (Figure 17B adapted from Schön.[25])

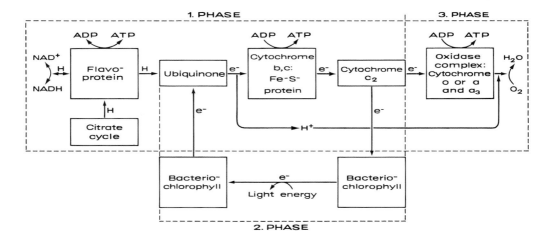

FIGURE 17C.   Sulfur-free purple bacteria such as *Rhodopseudomonas* are capable of both aerobic respiration and aerobic photosynthesis. The part of the electron transport chain involved in respiration is only slightly different from the respiratory chain in *Paracoccus* or in the mitochondria. However, there is also a secondary pathway. Cytochrome c_2 can transfer its electrons not only to cytochrome oxidase, but also to bacteriochlorophyll. The electrons which are excited by sunlight flow back to bacteriochlorophyll by way of ubiquinone and cytochrome b and c_2. On the basis of these facts and conclusions, we must assume that, in the course of evolution, first the shortened, anaerobic respiratory chain (phase 1) developed, then the anaerobic, followed by the aerobic photosystem (phase 2), and, finally, the aerobic respiratory chain of normal length (phase 3). Later, some descendents of the photosynthesizers (e.g. *Pseudomonas, Paracoccus, Beggiatoa, Leucothrix*) lost their photosynthetic apparatus and thus formed some of today's aerobically respiring prokaryotes. (Adapted from Dickerson.[5])

arrange the bacteria in a *natural phylogeny*. The phylogeny of the prokaryotes (Monera) given in Figure 20 is thus only a preliminary, hypothetical attempt to illustrate schematically the relationships worked out in this chapter. The most problematic part of this system is the polyphyletic derivation of the cytochrome-oxidase in the aerobic respiratory chain.

IV. MECHANISMS OF PROCYTE EVOLUTION

The evolution of the precytes resulted from the response of self-organizing molecules to a periodically changing environment. Analogously, the evolution of the *procytes* resulted from the confrontation between their metabolism and the available substrates or milieu. The result of this confrontation is the systematic exploitation of mutations to colonize previously uninhabited parts of the biosphere. There were several characteristic stages in this process. First was the adaptation to light, which led to an oxygen-producing metabolism. Then an adaptation to oxygen became necessary. The O_2 escaping into the atmosphere eventually produced an UV shield around the earth, making the adaptation to land and air possible. The procytes and the biosphere were thus constantly influencing each other: each changed the other, and in return was itself changed.

Just as the evolution of protobionts and eobionts reached a dead end, so it would appear that the procytes also reached the limits of their development. According to Kuhn,[14] their system became more and more complicated by the addition of new enzymes to the functional repertoire. The amount of information transferred from one generation to the next increased. The probability of an error in base pairing was optimized, leading to a limit in the capacity of the organism for genetic information, at 10^6 mononucleotides per DNA strand. This corresponds roughly to the 10^3 proteins encoded in the DNA of the modern enterobacterium *Escherichia coli*. With larger amounts of DNA, the frequency of defective, nonviable off-

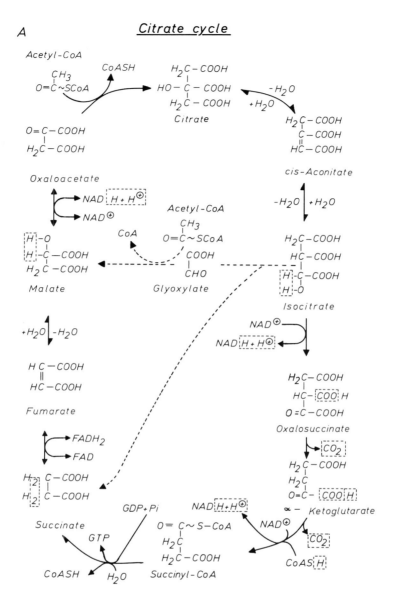

A   *Citrate cycle*

FIGURE 18A.   The citrate cycle* (Krebs cycle) fulfills three essential functions: final oxidation of organic substrates, formation of hydrogen carriers for the respiratory chain, and formation of intermediates for anabolism. In the citrate cycle, oxaloacetate and acetyl-CoA (from carbohydrates, fats, and proteins) join to form citrate and release CoASH. The acetate is then completely oxidized to CO_2 in the course of the cycle, so that two CO_2 molecules are produced per turn of the cycle. The hydrogen released is bound to NAD^+ or FAD, so that three NADH + H^+ and one $FADH_2$ are formed. The former transfer their hydrogens to the respiratory chain (Figure 17). In addition, the citrate cycle produces by substrate-level phosphorylation one high-energy GTP (guanosine triphosphate) which corresponds to ATP. In addition, intermediates of the citrate cycle, such as α-ketoglutarate (→ amino acid), succinic acid (→ heme), etc. are precursors for further biosyntheses.

* Oxalsuccinate does not occur in the citrate acid cycle of mitochondria.

spring would become too great. Evolution therefore entered another divergent phase (Chapter 5, Figure 29).

In the procytes, this limitation can be overcome by exchange of genetic material in

B <u>*Glyoxylate cycle*</u>
*(possible phylogenetic precursor
of the citrate-cycle)*

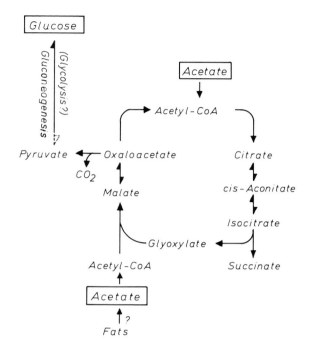

FIGURE 18B. C-heterotrophic procytes use the citrate cycle mainly for the synthesis of intermediates for anabolism (anapleurotic reactions = replenishing reactions), and less or not at all for the formation of energy substrate (NADH). Microorganisms which utilize acetate as a carbon source, or other substances which are degraded directly to acetate without involving pyruvate, such as fatty acids and carbohydrates, display a much shorter version of the citrate cycle, the glyoxylate cycle (see also the dotted line in A). In the short citrate cycle, as in the normal one, pyruvate can be generated from oxaloacetate by decarboxylation. In addition to other products, glucose can be generated by reversal of glycolysis. It is thus possible that the glyoxylate cycle is a phylogenetic relict of the hypothesized anaerobic respirers. They might have possessed, in addition to their fermentation and anaerobic respiration, a glyoxylate cycle (running in reverse, however); even now the glyoxylate cycle is largely reversible. (Adapted from Schön.[25])

parasexual processes, such as *conjugation* in the coliform bacteria. Two bacteria make change contact and exchange genetic material over a plasma bridge (pilus). In this way one individual can receive advantageous genetic material from another. The offspring of these conjugation partners with preferred genetic material eventually displace other bacteria. The pseudosexual process of conjugation makes gene recombination* possible, followed by a rapid collection of advantageous genetic materials, by way of selection. This development introduced another convergent phase of evolution in which the sexual apparatus was constantly improved, so that the capacity for genetic information was in turn increased.

The constant improvement of the genetic apparatus must eventually reach a limit, however, due to the loss of information through thermal collisions. This limit is first reached in the evolution of the eucytes, which will be the subject of the next chapter.

---

* Gene recombination = formation of new genetic combinations, for example by exchange of parts of chromosomes.

CHEMOAUTOTROPHS

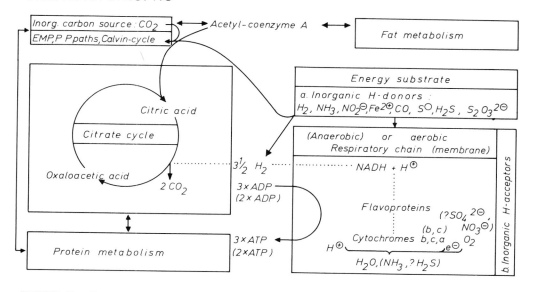

FIGURE 19. Fermentative, oxidizing metabolism of the group of chemoautotrophs, including certain sulfur bacteria, iron bacteria, and nitrifying bacteria. Like most photosynthesizers, they are completely independent of organic substrates for energy and synthesis. They fix CO_2 in the Calvin cycle and are able to utilize the products as their only carbon source (see also the citrate cycle). They obtain energy by extracting hydrogen from H_2S, nitrate, etc. and passing it (or its electrons) along the respiratory chain to produce ATP via oxidative phosphorylation. The hydrogen may also be used to reduce CO_2 in the Calvin cycle. The properties of the hypothetical ancestral anaerobically respiring forms are given in parentheses. (From Schwemmler.[28])

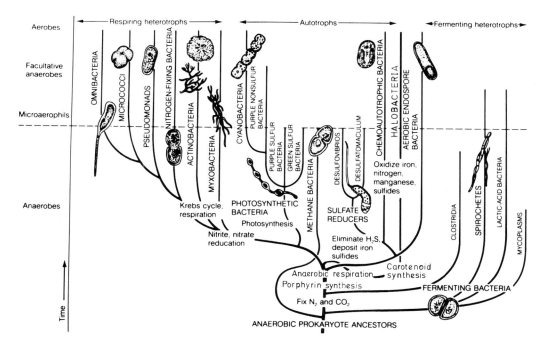

FIGURE 20. Hypothetical phylogeny of procytes. (Modified from Margulis.[18])

REFERENCES

1. **Broda, E.**, *The Evolution of the Bioenergetic Processes,* Pergamon Press, New York, 1975.
2. **Broda, E.**, Die Evolution der bioenergetischen Prozesse, *Biologie in unserer Zeit,* 7(2), 33, 1977.
3. **Buchanan, R. E. and Gibbons, N. A.**, *Bergey's Manual of Determinative Bacteriology,* 8th ed., Williams & Wilkins Company, Baltimore, 1974.
4. **Cramer, W. A.**, Photosynthesis I. In *Encyclopedia of Plant Physiology.* New Series, Vol. V: Photosynthetic Electron Transport and Photophosphorylation, Springer Verlag, New York, 1977.
5. **Dickerson, R. E.**, Cytochrom c und die Entwicklung des Stoffwechsels, *Spektrum der Wissenschaft,* 5, 47, 1980.
6. **Dose, K. and Rauchfuss, H.**, Chemische Evolution und der Ursprung lebender Systeme, *Wissenschaftliche Verlagsgesellschaft,* Stuttgart, 1975.
7. **Duysens, L. N. M.**, Photosynthesis, *Prog. Biophys. Molec. Biol.,* 14, 1, 1964.
8. **Epstein, W. and Schultz, S.**, Cation transport in *E. coli.* V. Regulation of cation content, *J. Gen. Physiol.,* 49, 221, 1965.
9. **Fredrick, J. F.**, Origins and evolution of eukaryotic intracellular organelles, *Ann. New York Acad. Sci.,* 361, 1981.
10. **Gaffron, H.**, In *The Origin of Prebiological Systems and of Their Molecular Matrices,* Fox, S. W., Ed., Academic Press, New York, 1965.
11. **Hartmann, H.**, Speculations on the origin and evolution of metabolism, *J. Mol. Evol.,* 4, 359, 1975.
12. **Kandler, O.**, Archaebakterien und Phylogenie der Organismen, *Naturwissenschaft,* 68, 183, 1981.
13. **Kaplan, W.**, *Der Ursprung des Lebens. Biogenetik, ein Forschungsgebiet heutiger Naturwissenschaft.* 2. Aufl., Georg Thieme Verlag, Stuttgart, 1978.
14. **Kuhn, H.**, Model consideration for the origin of life, *Naturwissenschaft,* 63, 68, 1976.
15. **Kull, U.**, *Evolution,* J. B. Metzler, Stuttgart, 1977.
16. **Margulis, L.**, *Origin of Eukaryotic Cells,* Yale University Press, New Haven, 1970.
17. **Margulis, L.**, Early cellular evolution, In *Exobiology,* Ponnamperuma, C., Ed., North Holland Publishing, Amsterdam, 1972.
18. **Margulis, L.**, *Symbiosis in Cell Evolution,* W. H. Freeman, San Francisco, 1981.
19. **Müller, H. E.**, Evolution und Alter von Bakterien, *Naturwissenschaft,* 63, 224, 1976.
20. **Oesterhelt, D.**, Bacteriorhodopsin als Beispiel einer licht-getriebenen Protonenpumpe, *Angew. Chem.,* 88(1), 16, 1976.
21. **Olson, J. M.**, The evolution of photosynthesis, *Science,* 168, 438, 1970.
22. **Rahmann, H.**, *Die Entstehung des Lebendigen,* Gustav Fischer Verlag, Stuttgart, 1972.
23. **Remane, A.**, *Die Grundlagen des natürlichen Systems, der vergleichen-den Anatomie und der Phylogenetik,* Akademische Verlagsgesellschaft Geest u. Portig, Leipzig, 1952.
24. **Schlegel, H. G.**, *Allgemeine Mikrobiologie,* 4. Aufl., Georg Thieme Verlag, Stuttgart, 1976.
25. **Schön, G.**, *Mikrobiologie,* Herder Verlag, Vienna, 1978.
26. **Schwemmler, W.**, Physiko-chemische Korrelation zwischen Lebewesen, Lebensraum und Symbiose am Beispiel der Hemipteren, *Zool. Anz. Suppl.,* 35 (Verh. d. Dtsch. Zool. Ges. 1971), 144, 1972.
27. **Schwemmler, W.**, Zikadenendosymbiose: ein Modell für die Evolution höherer Zellen. Zur Verifikation der Endosymbiontentheorie der Eukaryontenzelle, *Acta Biotheoretica,* 23, 132, 1974.
28. **Schwemmler, W.**, Allgemeiner Mechanismus der Zellevolution, *Naturwiss. Rdschau.,* 28(10), 357, 1975.
29. **Schwemmler, W.**, *Mechanismen der Zellevolution. Grundriss einer Modernen Zelltheorie.* W. de Gruyter, New York, 1978.
30. **Schwemmler, W. and Schenk, H. E. A.**, *Endocytobiology. Endosymbiosis and Cell Biology. A synthesis of recent research.* W. de Gruyter, New York, 1980.
31. **Stenn, K.**, Cation transport in a photosynthetic bacterium, *Bacteriology,* 96(3), 862, 1968.
32. **Sumper, M., Reitmeier, H., and Oesterhelt, D.**, Zur Biosynthese der Purpurmembran von Halobakterien, *Angew. Chem.,* 88(7), 203, 1976.
33. **Sutcliffe, D. W.**, The chemical composition of hemolymph in insects and some other arthropods in relation to their phylogeny, *J. Comp. Biochem. Physiol.,* 9, 121, 1963.
34. **Thauer, R. K. and Fuchs, G.**, Methanogene Bakterien, *Naturwissenschaft,* 66, 89, 1979.
35. **Trebst, A.**, Energy conservation in photosynthetic electron transport of chloroplasts, *Ann. Rev. Plant Physiol.,* 25, 423, 1974.

Chapter 7

EVOLUTION OF THE EUCYTES

In the course of evolution, a new type of cell, the *eucyte,* developed in addition to the procyte. Unlike the procytes, whose nuclear equivalent is not bounded by a membrane (prokaryote: karyon, Greek = nucleus), the eucyte has a true nucleus (Figure 1, Table 1), which contains a species-specific number of chromosomes. The latter consist of DNA, which is here bound to specific proteins (histones, non-histones). The eucyte is also more highly differentiated than the procyte in other respects, and is compartmented into cell organelles.

The fermentative cytoplasm of the eucyte is permeated by a finely branching membranous canal system, the *endoplasmic reticulum,* which is believed to have transport functions. The *ribosomes,* the site of protein synthesis, are located on the outer side of the reticulum, and also lie unattached in the cytoplasm. The so-called *Golgi apparatus* is also frequently observed. It consists of membrane-bound spaces, which are stacked like plates. They are related to synthetic processes. Another important cell organelle is the *mitochondrion,* which is surrounded by a double membrane. The inner membrane is often folded into the interior to form tubules or cristae, on which the respiratory enzymes, which fix energy in chemical form, are located. The typical plant cell also contains DNA-containing *plastids* the most important of which are the photoassimilating chloroplasts. They are also bounded by a double membrane. Invaginations of the inner membrane form the thylakoids, on which the photosynthetic pigments are localized. The eucyte is thus characterized by morphologically bounded reaction areas with different functions, such as heredity, fermentation, respiration and photosynthesis.

Reproduction of procytes and eucytes also differ. Procytes reproduce by simple lateral division, and, with the exception of the myxobacteria, they do not form multicellular organizations. The eucytes, on the other hand, generally divide by the strictly regulated processes of mitosis or meiosis. Most eucytes are organized in multicellular systems, namely as plants, fungi, animals, and humans. The adult human, for example, is composed of 10^{13} to 10^{14} eucytes, which are all genetically identical because they arose from a single cell by division. Several hundred different types of cells are found in the human body. They vary widely in size and form, and perform many different tasks. Some synthesize secretions and transmit electric impulses, others form muscle strands, still others form hair or bones. However, they are identical in their basic structure, and correspond to the eucytes of other animals, plants and fungi. The various eucyte forms can be traced back to common ancestral forms.

The question now arises, whether these eucytic ancestral forms and the procytes developed independently in the course of evolution, that is polyphyletically, or monophyletically. The correlation of the genetic code and the metabolic processes of fermentation, respiration, and photosynthesis in the procytes and eucytes clearly indicates that they arose monophyletically. This means, however, that the more highly developed eucytes must have emerged from the more primitive procytes. At present there is general consensus on this point; however, the way in which this common evolution of the procytes and eucytes took place is the subject of controversy. The *compartment hypothesis* states that the eucytes developed successively from *one* procyte by intracellular differentiation or compartmentation. The *endosymbiont hypothesis,* on the other hand, holds that the eucytes evolved from endosymbiosis between *several* procytes.

In the following, both possibilities will be briefly presented. It can then be determined which data support one or the other hypothesis. Again, that which supports the one hypothesis should, in principle, contradict the other.

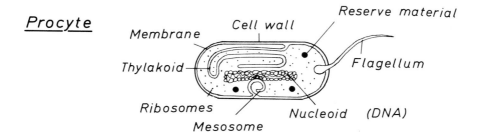

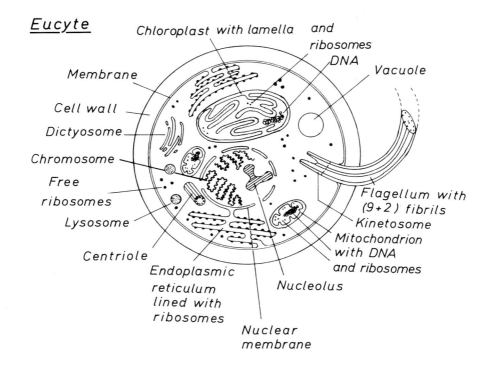

FIGURE 1.   Idealized structure of procytes and eucytes. The most widespread cell organelles are illustrated schematically. (Adapted from Kaplan;[28] details see text.)

I. WORKING HYPOTHESES

The basic idea of the *endosymbiont hypothesis* was suggested by Schimper[61] in the last century and was then precisely formulated at the turn of the century by Altmann,[2] Mereschkowksy and Famintzin.[48] However, the hypothesis was only definitively developed in the last decade (e.g., Margulis,[38,46] Schnepf and Brown,[62] and Schwemmler[67-70]).

According to this hypothesis, the animal cell developed before the plant cell. This was achieved by a fermenting bacterium which first took up and then integrated photergic flagellated bacteria, (Figure 2;* see also (Chapter 6, III.C) then respiring bacteria, and finally photosynthetic blue-green algae (according to Margulis:[44] first respiring, then flagellated,

---

*   Figure 2 is located after page 134.

Table 1
THE MOST IMPORTANT DIFFERENCES BETWEEN
PROCARYOTES AND EUCARYOTES

Criteria	Procaryotes	Eucaryotes
Structure	– Usually small cells: ∅ 1 - 10 μm	– Usually large cells: ∅ 10 - 10^2 μm
	– Little cell differentiation: thylakoids, mesosomes, flagella, inclusions, nucleoids, ribosomes, cell wall, etc.	– Pronounced cell compartmentation: chloroplasts, mitochondria, flagella, centriol, microtubuli, nucleus, ribosomes, vacuoles, lysosomes, glyoxysomes, peroxisomes, cell wall, dictyosomes (Golgi apparatus), endoplasmic reticulum, etc.
	– All are single-celled, often as filamentous or mycelial complexes with fruiting bodies (Myxobacteria)	– Single-celled, but usually multicellular in the form of metazoa or metaphyta
Division	– Simple cell division, usually as equatorial division	– Structured cell division: mitosis (centriole, mitotic spindle)
Chromosome	– One DNA-containing strand of chromatin with polyamines, chromosome characteristically a closed ring and attached to the membrane	– Several chromosomes containing DNA, RNA and proteins (Feulgen staining); characteristically chainlike chromosome not bound to the cell membrane
Sexuality	– Occasional parasexuality, one-sided transfer of genetic information	– Sexuality is the rule: exchange of information in two directions; meiotic production of gametes
Development	– Never a development of multicellular complexes from diploid zygotes; no cell differentiation	– Multicellular complexes develop from diploid zygotes; high degree of cell differentiation
Metabolism	– Anaerobic, facultatively aerobic, microaerophilic and aerobic forms	– Only aerobic metabolism, aside from secondary modifications or certain differentiated cells (see Table 9.1.)
Flagella	– Simple bacterial flagella, if present (one microtubule)	– Complex flagella or cilia, if present (axial structure: (9+2) microtubuli)

Adapted from Margulis.[44,45]

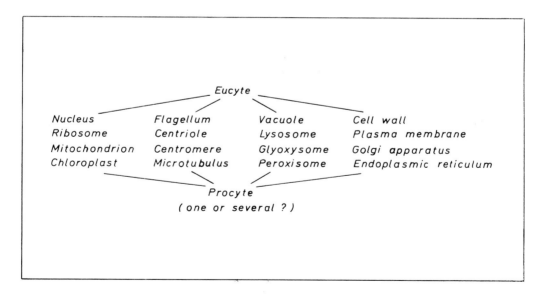

but fermenting bacteria). The incorporation of the microorganisms is thought to have occurred each time in connection with a decisive change in milieu or energy substrate (see Table 2). The integration of the photergers ensued in connection with the change from dark to light or chemotrophy to phototrophy, the integration of the respirers with the change from anaerobiosis to aerobiosis, or from organotrophy to inorganotrophy (= lithotrophy) and the integration of the photosynthesizers was a prerequisite for the change from water to land,

Table 2
COMPARISON OF DNA FROM MITOCHONDRIA OF VARIOUS ORIGIN
FROM PLASMIDS AND BACTERIA

Source of DNA	Circular molecule	DNA: length in μm	$10^6 \times mol.wt.$
Mitochondria from			
Protozoa	no (?)	15 – 25	15 – 50
Plants	no (?)	12 – 30	25 – 60
Animals	yes	5	10
Plasmids from			
Escherichia coli (fertility factor)	yes	25 – 40	50 – 80
Bacterium			
Escherichia coli	yes	~1000	1600 – 2800

(Adapted from Parthier.[50])

or from C-heterotrophy to C-autotrophy. In this way the cytoplasm with the nucleus, flagella, and mitochondria are thought to have developed from bacteria, and the chloroplasts from blue-green algae. Accordingly, the cytoplasm and nucleus of the eucyte represent the host, while flagella, mitochondria, and chloroplasts are its now extremely adapted endosymbionts (endocytobionts).

The *compartment hypothesis* was developed in the last few years as an alternative to the endosymbiont hypothesis (among others, Raff and Mahler,[57] Uzzell and Spolsky,[85] Bogorad,[9] Cavalier-Smith,[14] Parthier[50]). A series of different concepts are included in the compartment hypothesis. All of them, however, are based on the common assumption that the eucytes developed successively, in a direct line, from one procyte. The hypotheses differ only with respect to the type of procyte ancestor and the individual steps of evolution.

If the ancestral bacterial form is assumed to have had a fermenting or respiring metabolism, then it is hypothesized that first the nucleus and flagella, then the mitochondria evolved, thus producing an animal cell (Figure 2), and then finally the chloroplasts and thus plant cells. If it is assumed that a photosynthetic form was ancestral to the eucytes, then the chloroplasts, followed by nucleus and flagella, or by mitochondria, are thought to have evolved in that order. The animal cells would then have arisen by loss of the chloroplasts.

In essence, two different proposals for the intracellular differentiation of mitochondria and chloroplasts are being discussed. The *plasmid hypothesis,* also called the cluster clone hypothesis, was suggested by Raff and Mahler[57] for mitochondria. They postulate that special enzymes were first incorporated, in the course of mutation and selection, into the plasma membrane of the ancestral procyte form. Vesicles were formed when these sections of the membrane invaginated to create a greater surface area and were then pinched off. Plasmids or extrachromosomal DNA carrying the genetic information for certain RNAs and proteins then moved into these vesicles (Table 2). The mitochondria thus arose from membrane vesicles carrying respiratory enzymes, and the chloroplasts from vesicles bearing photosynthetic pigments. The alternative *gene duplication-segregation hypothesis* (transformation hypothesis: Allsopp,[1] Parthier[50]) assumes that the differentiation of the DNA-containing organelles was initiated by a partial doubling of the genome of a procyte ancestor. The duplicate genes developed in different directions through the action of mutation and selection.

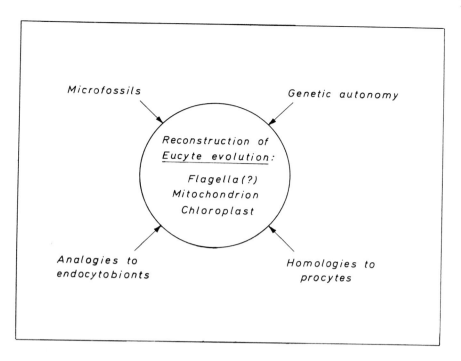

The membrane-bound genes remained rather conservative and did not change significantly, while the non-bound duplicates grew rapidly through further duplications. The cell volume was increased in the process and resulted in a segregation of the chromosomes. Finally, the greatly enlarged chromosomes enclosed themselves in a membrane and became the cell nucleus. The small chromosomes, together with their enzymatically specially equipped membrane structure, were also wrapped in a membrane. The mitochondria thus arose from those membrane segregates with cytochromes, and the chloroplasts from those with chlorophyll.

Uzzell and Spolsky[85] combined the essential elements of the plasmid hypothesis with those of the gene duplication-segregation hypothesis to produce a hypothesis which one might call the *genome duplication-invagination hypothesis*. According to this, eucyte evolution began through genome duplication in respiring and photosynthetic ancestral procyte forms without subsequent cell division. The membrane invaginated, taking the associated duplicates with it, and formed a double membrane around them. The inner membranes of all three segregates originally contained both respiratory enzymes and pigments, but they differentiated in the course of further development. The nuclear membrane lost its enzymes and pigments. The genome, however, freed itself from the membrane and developed rapidly, through further duplications, to protein-packed chromosomes. The inner organelle membranes invaginated again, thus increasing their surface area. Some genes were lost from their genomes, but their DNA remained associated with the membrane and therefore conservative. The mitochondria developed from one segregate by loss of photo-pigments, while the chloroplasts developed from the other by loss of the respiratory enzymes.

For the sake of completeness, we must also mention a sort of *combination hypothesis,* according to which the chloroplasts are of endosymbiotic origin and the other cell organelles developed compartmentally (e.g., Taylor[82]). Taylor[83] refers to an endosymbiotic formation of DNA-containing cell organelles as *exogenous,* whereas their formation by cell differentiation is designated *endogenous.*

Table 3
MICROFOSSIL DISCOVERIES RELEVANT TO EUCYTE EVOLUTION AND SUGGESTED INTERPRETATIONS

Geologic formation	Gunflint (N. America)	Crystal Spring (N. America)	Nonsuch (N. America) Bitterspring (Australia)
Estimated age(x10⁹y.)	2 - 1.9	1.5 - 1.2	1 - 0.8
Cell fossils	Single cells, 20 μm diameter, some with peripheral bodies. Preflagellates? (first flagellate eucytes?)	Filamentous multicellular forms, some with central bodies. Fungal hyphae? (first mitochondria - containing eucytes?)	Multicellular forms, 10 μm diameter, 15 -100 μm long, with central bodies. Green algae? (first chlorophyll - containing eucytes?)

II. EXPERIMENTAL RESULTS

In the final analysis, the above working hypotheses differ in the derivation of the extra-nuclear DNA of eucytes, either from foreign DNA (endosymbiont hypothesis) or internal DNA (compartment hypothesis). Investigations intended to verify one of the two alternative hypotheses naturally concentrate on the DNA-containing organelles, the chloroplasts, mitochondria, and to some extent, also the flagella. A number of methods have been successfully applied, including the study of microfossils, the genetic autonomy of cell organelles and their analogies to endocytobionts or their homologies to procytes.

A. Microfossil Evidence

It is difficult to interpret older cellular fossil discoveries. Due to a lack of sufficiently preserved substructures, procytic and eucytic cell remains can hardly be distinguished, let alone further classified. The presence of organelles is important for the identification of eucytic fossils, and some progress has been made in this direction. The bodies within microfossils which were previously thought to be the result of cell clumping (plasma plus cell wall) have now been recognized as organelles.[63]

The oldest microfossils presumed to be eucytes come from the 2 to 1.9 billion year old Gunflint formation of North America. The fossils represent spherical single cells, about 20 μm in diameter. They have been rather well preserved by silicification, and thin sections can be microscopically examined in three dimensions. The sections occasionally demonstrate central or peripheral bodies. These are probably plasma coagulations, possibly even cell organelles. It is less likely that they represent flagella remnants. Therefore, the finds are only in part considered to be flagellated preeucytes (Table 3).

Microfossils dated between 1.5 and 1.2 billion years ago have been discovered in the Crystal-Spring layer of North America. Some of them represent filamentous multicellular organisms, and in some there are single or multiple central bodies. In part, they are thought to represent primitive *fungal hyphae*. Thus, mitochondria-containing eucytes must have existed since this time at the latest.

The oldest known fossils of chloroplast-containing eucytes come from the Nonsuch formation of North America and from the Bitterspring layer of Australia, which are both about

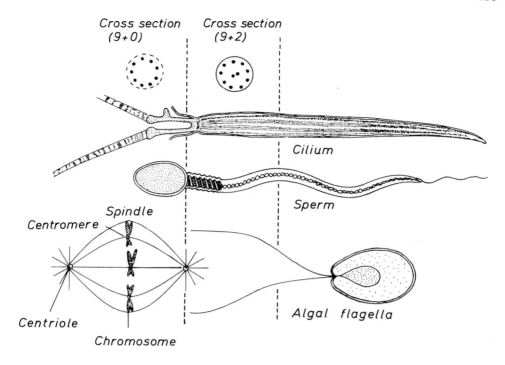

FIGURE 3. The (9 + 2) or (9 + 0) microtubule pattern can be observed in various structures: the flagella of flagellates and sperm, the cilia of protozoa and ciliated tissues, and the centriole of the spindle apparatus which organizes nuclear division during mitosis (From Margulis[39]).

1 billion years old. The fossils are branched or unbranched multicellular filaments, about 10 μm thick and 15 to 100 μm long, or spherical cells. The cells fairly regularly contain at least one central body with a diameter of 0.7 to 2.3 μm. These could be starch grains (Ø 0.8 to 2.1 μm) and/or nuclei (Ø 1.5 to 2.2 μm). The find was therefore interpreted to be a green algae similar to *Vaucheriaceae* or *Siphonaceae*.

The above interpretations of the presently known fossils indicate that the DNA-containing organelles arose in the following order: flagellum, mitochondrion, chloroplast. It will now be interesting to see how the genetic autonomy of the organelles compares to this presumed phylogenetic order.

B. Genetic Autonomy
1. Flagellum

Many protozoa are characterized by the possession of a flagellum *(Flagellata)* or cilia *(Ciliata)*, which are smaller (Figure 3). These motive organelles are also found in animals and many plants, at least in some tissues or in some phases. Male gametes, for example, usually have flagella. All of these organelles consist of two central, single filaments and nine peripheral double filaments or microtubuli (9 + 2) pattern. They are distinctly different from bacterial flagella, which contain only one filament. Flagella and cilia are surrounded by the plasma membrane. Furthermore, they are anchored in the cytoplasm by a basal body (blepharoplast, kinetosome) from which they are also formed. The two central filaments are lacking in the basal body, so that there the axial skeleton has a (9 + 0) pattern. The basal bodies are morphologically and, in some cases, functionally homologous to the centrioles, from which they may arise by division, or vice versa. The centrioles, which therefore also have an axial skeleton with a (9 + 0) pattern, are the cell division organelles of animal and many plant thallophyte cells (Chapter 6, V).

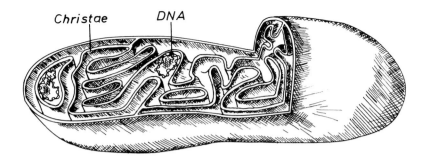

FIGURE 4.   Section (schematic) of the cristae-type mitochondrion. Two DNA molecules are shown (according to Hirsch et al. in Barckhausen et al.[7a]).

Basal bodies and centrioles belong to the class of eucyte organelles for which specific DNA has been postulated but has not been proven. To date, only flagellum-specific RNA (probably coded by nuclear genes) has been demonstrated (see Hartmann[24] and Dodson[19]). Various workers have, to be sure, reported the isolation of a circular DNA from flagella with a molecular weight of about 10^6, which corresponds to a medium-sized virus. However, others have suggested that this DNA was a contamination from mitochondria, since mitochondria are always closely associated with the flagellar apparatus of the zooflagellate examined, *Trypanosoma*. Also, the sequence of bases in the isolated DNA was very similar to that in mitochondrial DNA. Thus the question whether or not basal bodies or centrioles contain endogenous DNA has not yet been definitely answered. Exact analysis of the DNA in pure fractions of basal bodies from the ciliate *Tetrahymena pyriformis* is underway at present. [84] Even if the previous analyses had proven the existence of endogenous flagellar DNA, the flagellum with its 10,000 base pairs would only code for about 20 mRNAs and few proteins with an average molecular weight of 20,000.

One can compare this amount of the possible endogenous flagellar proteins/RNAs with approximately 100 proteins/RNAs calculated by Kaplan[28] to be the minimal complement required for a procyte. On the basis of this comparison, the flagellum controls a little more than 20% of its own properties. If, as a further simplification, we set this percentage equal to the degree of genetic autonomy, then the flagellum has a genetic autonomy of about 20%, and is dependent on the cell nucleus for at least 80% of its properties. Furthermore, the flagella probably have no endogenous protein biosynthesis, since to date no flagellum-specific 70S* ribosomes have been demonstrated.

2. Mitochondrion

Biochemical and fine-structure analyses have proven that mitochondria contain endogenous DNA as circular molecules (Wolf[90]; Figure 4). Each mitochondrion contains several copies, which are almost protein-free and are directly associated with the inner mitochondrial membrane. They have a lower molecular weight than most procytes, about 10^7 daltons (10^{-13} to 10^{-14} mg per molecule). The mitochondria account for 0.1 to 1.0% of the total DNA of a cell. The molecular weights of mitochondrial DNAs from animals and plants differ in some cases greatly (Table 4). The mitochondrial genome thus varies between 15,000 and 30,000 base pairs, which could code for the synthesis of about 10 to 30 proteins of an average molecular weight of 20,000. These include hydrophobic proteins: about 5 to 10% of the proteins of the inner membrane with cristae and tubuli (6 to 10 subunits of a total of 75), including several enzymes (ATPase) or their subunits (2 of the 3 subunits of cytochrome,

---

* S = Svedberg unit, in which sedimentation constants are expressed.

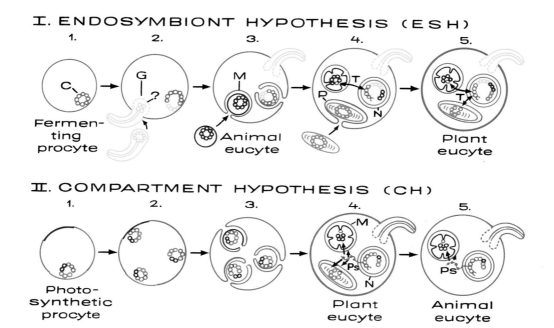

FIGURE 2.   Simplified representation of the alternative hypotheses on eucyte evolution.

I. According to the endosymbiont hypothesis, the cell developed by integration of several procytes: 1. fermenting procyte as host with a membrane-associated chromosome (C); 2. uptake and incorporation of flagella symbiont (G), whose DNA was entirely or partly integrated into the host chromosome; 3. integration of aerobic bacteria as mitochondria (M); formation of an animal eucyte by subsequent membrane invagination to form a nucleus (N); 4. integration of blue-green algae as chloroplasts (P); gene transfer (T) between nucleus and mitochondria; formation of a plant eucyte; 5. gene transfer also between chloroplast and nucleus, formation of a cell wall.

II. According to the compartment hypothesis, the eucyte arose from a procyte by intracellular differentiation or compartmentation: 1. photosynthetic procyte as parent form; 2. duplication of the procyte genome; 3. invagination of the cell membrane with its respiring and photosynthesizing enzymes to form a double membrane around the nucleus (N), the mitochondria (M), and the chloroplasts (P); 4. protein biosynthesis (Ps) dependent on mitochondria, plastids, and nucleus; differentiation of the cell wall and the flagellum, forming the plant eucyte; 5. development of the animal eucyte by loss of plastids and cell wall.

blue = mitochondrial descent
green = chloroplast descent
yellow = flagellum descent
red = parts of the cell derived from the original host cell (endosymbiont hypothesis) or the fermenting part of the cell (compartment hypothesis)

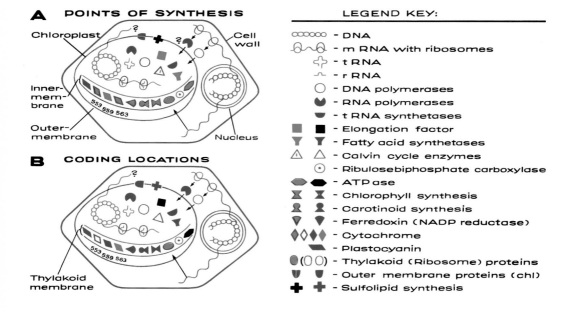

A   POINTS OF SYNTHESIS

Chloroplast

Cell wall

Inner-membrane

553 559 563

Outer-membrane

Nucleus

B   CODING LOCATIONS

553 559 563

Thylakoid membrane

LEGEND KEY:

- DNA
- m RNA with ribosomes
- t RNA
- r RNA
- DNA polymerases
- RNA polymerases
- t RNA synthetases
- Elongation factor
- Fatty acid synthetases
- Calvin cycle enzymes
- Ribulosebiphosphate carboxylase
- ATPase
- Chlorophyll synthesis
- Carotinoid synthesis
- Ferredoxin (NADP reductase)
- Cytochrome
- Plastocyanin
- Thylakoid (Ribosome) proteins
- Outer membrane proteins (chl)
- Sulfolipid synthesis

FIGURE 6.I.   Genetic control or intracellular location of the synthesis of chloroplast-specific nucleic acids and proteins. The proportions of nucleus and plastids are not true to scale.[8,74]

(A) Sites of synthesisof chloroplast components
   Green = chloroplast components synthesized in the chloroplasts or on chloroplast ribosomes
   Red = chloroplast components synthesized on cytosomal ribosomes
   Green and red = chloroplast components synthesized on both chloroplast and cytosomal ribosomes
   Black = chloroplast components whose site of synthesis is unknown.

(B) Sites of coding of chloroplast components
   Green = chloroplast components coded by the chloroplast DNA
   Red = chloroplast components coded by the nuclear DNA
   Green and red = chloroplast components coded by chloroplast and nuclear DNA
   Black = chloroplast components whose site of coding is unknown.

Table 4

DIFFERENCES BETWEEN MITOCHONDRIAL AND BACTERIAL DNA AND RNA

Components	Mitochondria from			Bacteria (Escherichia coli)
	Plants (gymnosperms, angiosperms)	Fungi (ascomycetes)	Animals (vertebrates, invertebrates)	
DNA: length in μm	7 – 30	10 – 25	5	~ 1000
mol. wt. (×10⁷)	2.5 – 6	2 – 5	1	160 – 280
circular molecule	no(?)	yes	yes	yes
Ribosomes: sedimentation const.	78	70 – 73	50 – 55	70
rRNA: mol. wt. (×10⁶)				
23 S	1.15	1.26	0.65 – 0.95	1.10
16 S	0.70	0.70	0.35 – 0.50	0.56
5 S present	no (in some lower plants, yes)	no	no	yes
% Guanine-cytosine: RNA	35	32	60	53
Poly (A) - rich mRNA	?	yes	yes	no
Protein synthesis:				
inhibition by fusidinic acid	no	no	?	yes
inhibition by chloramphenicol	good	weak	no	strong

Note: The mitochondrial nucleic acids are from several different eukaryotes.

Adapted from Parthier,[50] and Mahler in Schwemmler and Schenk.[74]

6 or 4 of the 7 subunits of cytochrome oxidase, both insoluble respiratory chain enzymes) and probably a few ribosomal proteins are coded by mitochondrial DNA. In addition, the mitochondrial DNA codes for the three ribosomal RNAs of the mitocondrial 70S ribosomes (5S, 16S, 23S) and for more than 20 transfer RNAs.

All other properties of mitochondria are coded by the nuclear DNA (outer membrane, soluble respiratory enzymes, replication, recombination-damage repair, transcription and protein synthesis elements as aminoacyl-tRNA synthetases, initiation, elongation and termination factors, etc.). Most of the mitochondrial proteins are also synthesized on the 80S cytoplasmic ribosomes. However, it has been shown that a few of the nuclear-coded mitochondrial proteins are synthesized on the 70S organelle ribosomes. In the first case, the synthesized proteins must be transported into the organelle, and in the second, the nuclear messenger RNAs must cross the membrane to enter the organelle.

If one relates the number of endogenous mitochondrial genetic products to the minimal RNA and protein complement for a procyte, one sees that mitochondria code for about 35% at most, and thus are 65% dependent on the cell nucleus. Nevertheless, the mitochondria have at their disposal an endogenous genetic system with specific protein biosynthesis and their own metabolism. However, the larger part of the elements of this apparatus are coded by the nucleus and synthesized in the cytoplasm. The mitochondria are thus not autonomous, but at best semiautonomous or semiautoreproductive components of eucytes.

3. Chloroplast,

Careful analysis has shown that chloroplasts also contain circular, mostly histone-free, membrane-associated DNA molecules (Figure 5). Its molecular weight of 10^8 daltons (10^{-11} to 10^{-12} mg per molecule) corresponds to that of the simpler procytes such as mycoplasms and rickettsiae. The plastids account for about 6% of the total DNA in a plant cell. The

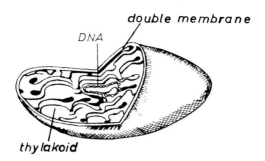

FIGURE 5.   Schematic representation of a chloroplast, showing stacked thylakoids with layered membrane bands. (From Strassburger[80a].) Relations of the thylakoids to the inner membrane, see Figure 1.

differences between the size of the DNA molecules in the chloroplasts of lower and higher plants is not as great as the differences in mitochondrial DNAs. For example, the chloroplast DNA from certain protozoa is 55 μm long, while chloroplast DNA from seed plants is 40 μm long. This is equivalent to 250,000 to 300,000 base pairs, which suffices for the synthesis of about 200 proteins with an average molecular weight of 40,000. However, activity measurements have shown that at most 20% of the genome is active, which would mean that the synthesis of at most 50 different proteins and nucleic acids is possible. The rest of the genetic information might be redundant genetic ballast (also exous) which arose through previous duplication, or inactivated genes.

Plastid DNA contains only the information for the ribosomal RNAs of the 70S ribosomes (5S, 16S, 23S), the 20 transfer RNAs, for about 5 to 10% of the proteins of the inner organelle membrane, including the thylakoids, probably for a few ribosome proteins and for a few enzymes (e.g., photosystem I, proteins of the electron transport chain) and their subunits (e.g., the 50,000-daltons subunit of ribulose 1,5-bisphosphate carboxylase). All the other properties of the organelle appear to be coded by the nuclear genome (Figure 6).* These include almost all the enzymes for chlorphyll synthesis and for photosynthesis.

Again, if one compares the plastid-coded synthesis products to the RNA or protein minimal complement, then the organelle controls at most 50%. The plastids thus would have a genetic autonomy of 50% and a genetic dependence on the nucleus of 50%. Like the mitochondria, the plastids have an endogenous genetic apparatus with the accompanying machinery for protein biosynthesis and metabolism, but here too, essential parts are coded by the nucleus and synthesized by the cytoplasm. Thus the chloroplasts are also only half-autonomous, or semiautoreproductive cell organelles, although they are more independent than the mitochondria.

A comparison of the degree of genetic autonomy of the DNA-containing organelles yields an order which runs exactly counter to their presumed phylogenetic age (Table 5). This fact supports the endosymbiont hypothesis, which would suggest that the most recent, and therefore least integrated or reduced organelles of endosymbiotic origin, namely the chloroplasts, would also have the greatest genetic autonomy. The phylogenetically oldest organelles, the flagella, would be the most reduced, or the best integrated in the cell. The organelle's high degree of genetic dependence on the nucleus has been used as an argument for the compartment hypothesis and against the endosymbiont hypothesis. However, this is not a convincing argument, because the phylogenetic development of the endocytobiont**

---

*   Figure 6I is located after page 134.
** For detailed discussion of endocytobiosis, see Chapter 8. For definition, see glossary.

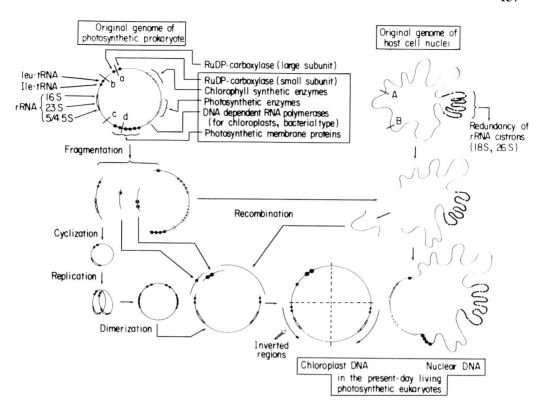

FIGURE 6.II. A model on the evolutionary origin of the chloroplast genome on the basis of organization of certain genes on its circular DNA molecule seen in present-day living plants. After immigration of a photosynthetic prokaryote as a symbiont into the host cell, the behavior of the gene group with DNA segments is depicted.[26] (Figure 6.I. is located after page 134.)

Table 5

COMPARISON OF THE POSTULATED OR PROVEN GENETIC AUTONOMY OF FLAGELLA, MITOCHONDRIA, AND CHLOROPLASTS WITH THEIR PRESUMED PHYLOGENETIC AGE.[69]

Estimated data	Flagella	Mitochondria	Chloroplasts
Genetic autonomy (% essential genes)	20%(?) >	35% >	50%
Phylogenetic age (years)	2 billion <	1.5 billion <	1 billion

into a cell organelle would require an increasing integration of the former symbiont into the cell. For example, the necessary coordination of division of the two partners is better guaranteed when the DNA-polymerase required for the replication of the organelle DNA is coded in the nucleus.

The question thus arises, whether the low coding capacity of the DNA-containing organelles is a result of a transfer of the DNA units from the endosymbiotic organelle to the nucleus, in the course of evolution, so that the DNA remaining in the organelle is a relict of a necessary endocytobiotic adaptation. It would now be interesting to determine whether there are modern groups of endocytobionts which are in the process of integration into the cell, and for which the intermediate stages are known. Do they demonstrate a genetic transfer to the cell nucleus comparable to that observed in cell organelles?

C. Analogies with Endocytobionts
1. Flagellum
Many protozoa live in more or less intensive symbiosis with motile bacteria.[38] The zooflagellate *Pyrosonympha,* for example, harbors episymbiotic spirochetes. One end of each spirochete is firmly anchored to the surface of the flagellate, so that one has the impression of a ciliated surface or additional flagellation. The symbionts are "rooted" on the surface of the host cell in a manner quite similar to that of cilia. However, dissolution of the cell wall and cell membrane in the contact zone cannot be seen. These protozoa were first incorrectly assigned to the group of hypermastigina, on account of their symbionts. The episymbiosis with spirochetes is functionally more advanced in the zooflagellata *Mixotricha paradoxa,* which live in the intestines of termites. Electron micrographs have shown that another type of symbiotic bacteria of unknown nature is arranged in regular patterns on the surface of the flagellate (Figure 7,A). Symbiotic spirochetes are "inserted" into the bacteria, across an extension of the host cytoplasm, like a kind of flagellar basal body (Figure 7,B). The spirochetes are about the size of flagella. They move by means of an axial rod, which in this case is composed of about 17 long fibrils (Fi) which are capable of contraction (Figure 7,C). The bacteria-spirochete-host complex is also capable of coordinated beating, which serves only the locomotion of the flagellate. Its four polar flagella only act as rudders to steer the organism. The episymbiotic complex therefore was formerly (and incorrectly) thought to be a eucyte flagellum.

However, the symbiosis is not essential for either the host or the symbionts; it is a facultative arrangement. Each partner is viable on its own. Aside from the small extension of the host cytoplasm, the symbionts do not live intracellularly, but only episymbiotically. They also display no recognizable reduction or integration characteristics. Their membrane is always completely intact, and their genetic information is not reduced. In contrast, the flagellum is not completely surrounded by a membrane, at least in the region of the basal body, and its own DNA is either not present or completely or largely integrated in the nucleus. *For lack of final stages, it is therefore not possible, with the present knowledge, to draw a direct analogy between the eucyte flagellum and recent spirochete-like endocytobionts.*

2. Mitochondrion
The cicadas as a group, display a number of increasingly integrated or reduced endocytobionts (for details, see Chapter 8, and Schwemmler[66-68]). The phylogenetically more recent rickettsia- or eubacteria-like *accessory symbionts* are usually still incorporated into nonspecific insect cells. They are largely independent of their host cells. There is no existential reduction, in particular of the cell wall, which is induced by the host. The host cell and the symbiont are both viable on their own; their symbiosis is thus facultative. The phylogenetically old protoplastoid* *primary and auxiliary symbionts* of the cicadas, in contrast, are always maintained in special host cells (Figure 8,A). When they are taken up in these cells, they are surrounded by a membrane from the host cell. Only traces of the symbionts' cell

---

* *Protoplastoids* are a group of intracellular procytes, newly described as such, which are similar in many ways to the mycoplasms and rickettsiae.[65,68]

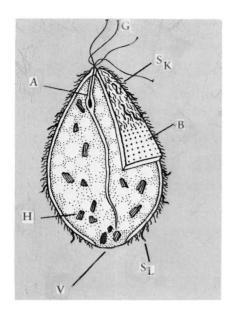

A

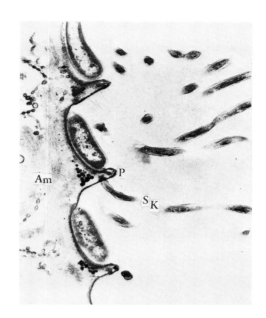

B

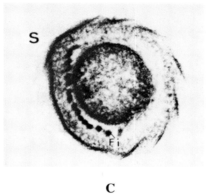

C

FIGURE 7. (A) Organization of the symbiosis of the host flagellate *Mixotricha paradoxa* with episymbiotic bacteria and spirochetes.

A	= axostyle	P	= plasma projection
B	= bacterium	S_L	= longer spirochete
Fi	= filament	S_K	= shorter spirochete
G	= (bacterium) flagellum	V	= digestion zone
H	= wood		

(B) Section through the surface of the host flagellate (Am) showing the regular arrangement of the spirochetes (S_K) "inserted" in the episymbiotic bacteria (B). Each bacterium-spirochete complex is tightly bound to the surface of the flagellate by a plasma projection (P).

(C) Cross section of an episymbiotic spirochete (S) showing the typical spirochete filament pattern (Fi) in the protoplasmatic cylinder (A, B and C from Grimstone[22]).

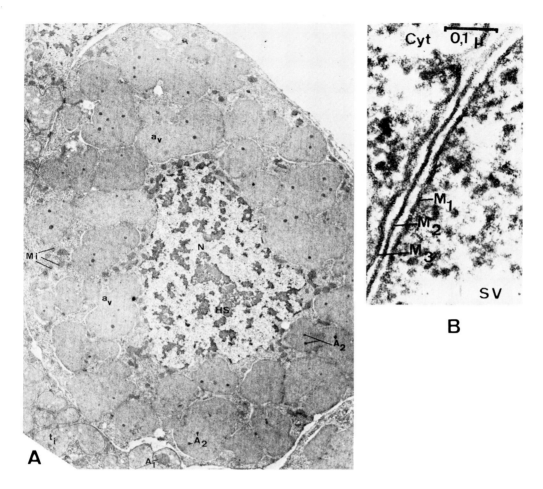

FIGURE 8. (A) Embryonic host cell or bacteriocyte of the cicada *Euscelis incisus* filled with protoplastoid primary type *a* symbionts. The symbionts contain many optically dark inclusions; the nucleus of the bacteriocyte is polyploid.[69]

a_v	=	primary symbiont *a* in the vegetative form
$A_{1,2}$	=	various types of inclusions
HS	=	heterochromatin structure of the cell nucleus
Mi	=	mitochondrion
N	=	bacteriocyte nucleus
t_i	=	auxiliary symbiont *t* in infectious form

(B) Section through the membrane of the vegetative form (a_v) of the primary symbiont *a*, showing the layer formed by the host cell cytoplasm (M_3), the layer formed by the symbiont (its cell membrane) (M_1), and between M_3 and M_1 another membrane layer M_2, presumably derived from the symbiont, which contains isolated remnants of cell wall.[37]

walls can be detected (Figure 8,B; Louis and Laporte[37]). Their metabolism is also influenced by the host cell. Outside the host, they are not even able to reproduce, and soon die. The symbionts have thus lost their autonomy in part, but the symbiosis is also obligatory for the host. If the symbiotic union is destroyed, the host dies within a few days. The endocytobionts are therefore passed on to the next generation via the host egg, in a kind of continuous *"symbiont chain"*. If the transfer of symbionts with the egg is spontaneously or artificially interrupted, the eggs develop non-viable partial embryos, consisting of head and thorax, but

Table 6
ANALOGIES BETWEEN MITOCHONDRIA AND PROTOPLASTOID ENDOSYMBIONTS OF CICADAS[68,69]

Protoplastoid (endocytobiont)	*Mitochondrion* (organelle)
– *Outer symbiont membrane of cytoplasmic origin*	– *Outer mitochondrial membrane possibly of cytoplasmic origin (e.g. enzyme components)*
– *Symbiont plasma membrane*	– *Inner mitochondrial membrane (cardiolipin)*
– *Space between outer symbiont membrane and symbiont plasma membrane with traces of cell wall*	– *Space between outer and inner mitochondrial membrane without traces of cell wall*
– *Invagination of the inner symbiont membrane (e.g. membrane bodies, mesosomes, division membrane)*	– *Invagination of the inner mitochondrial membrane (e.g. cristae / tubuli, division membrane)*
– *Symbiont matrix (circular chromosome (?): maximal 10^8 daltons, 70S ribosomes, etc.)*	– *Mitochondrial matrix (circular chromosome: ~10^7 daltons, 70S ribosomes, etc.)*
– *Inclusions: chelate ($Ca^{2\oplus}, PO_4^{3\ominus}$, proteins)*	– *Granula: hydroxyapatite ($Ca^{2\oplus}, PO_4^{3\ominus}$, amino acids?)*
– *Regulation of pH, endorhythms and osmotic pressure of the host cell*	– *Regulation of pH, endorhythms and osmotic pressure of the eucyte*
– *Infectious and vegetative forms*	– *Condensation and orthodox forms*
– *Reproduction dependent on the host cell (semiautoreproductive)*	– *Eucyte-dependent reproduction (semiautoreproductive)*

no abdomen. The endocytobionts also synthesize building materials for the host from the host's waste products. In doing so, they probably regulate the pH and osmotic pressure in connection with the formation of cell inclusions. Beyond this, they also influence certain endogenous rhythms in the energy metabolism of the host cell.[71,72] The endocytobionts thus exercise functions which are comparable with those of mitochondria. The molecular weight of the DNA in the endocytobionts is a little higher than that of mitochondria, and lower than that of primitive procytes, which indicates a reduction caused by the host cell. Table 6 lists the most important analogies between protoplastoid cicada symbionts and mitochondria.

In the giant amoeba *Pelomyxa palustris*, endocytobiotic bacteria which take the place of mitochondria can be observed (compare also the mitochondria-less flagellate *Mixotricha paradoxa*). These bacteria, like the mitochondria, appear in two substructure types: a small, optically dense form (condensation form) and a larger, optically more transparent form (orthodox form). The larger forms are "woven into" a net of tubes comparable to the endoplasmic reticulum. The bacteria appear to carry out respiration, thus taking over the function of the absent mitochondria.[86] In addition, the symbiosis appears to be obligatory for both partners. The symbionts have lost part of their autonomy, but the host cell has also become essentially dependent on the endocytobionts. This requires a measure of gene transfer between the two systems. *The integration or reduction between the host eucyte and the procytic symbiont found here has a degree comparable to that of mitochondria.*

3. Chloroplast

Syncyanoses[52] are symbioses between blue-green algae and zooflagellates, fungi or plastid-free phytoflagellates. There is a continuous spectrum from loose associations which may be dissolved at any time to indissoluble endocytobioses, in which the functions of the symbiont are vital for the host cell (review, see Schenk[60]). The phylogenetically recent endocytobiosis between the primitive fungus *Geosiphon pyriforme* and the blue-green alga of the genus *Nostoc* stands at the beginning of this spectrum. The fungus continually takes up procytic algae from the soil and also releases them again. It surrounds them with its own plasma membrane in so-called phagocytosis vacuoles. The blue-green algae are not visibly changed in the host cell. There is no reduction of their cell wall, and they can be isolated at any time from the host cell and cultivated independently. The host is also viable without the blue-green algae. It follows that the endocytobiosis is not vital for either partner, but rather facultative.

The endocytobiosis between the unicellular, plastid-free *Glaucocystis nostochinearum*, which resembles green algae, and blue-green algae of unknown nature (compare *Cyanophora paradoxa*) stands at the other end of the scale. In this case, too, the blue-green algae are maintained in phagocytotic vacuoles (Figure 9,A). The host-derived membrane of the phagocytosis vacuole has, with its large amount of cholesterol, the same composition as the outer plastid membrane. The plasma membrane of the symbiont exhibits a typical procytic enzyme complement and thus resembles the inner plastid membrane. The blue-green algae, like plastids, form pigment-carrying thylakoids, the sites of photosynthesis by invagination of the inner membrane. In this symbiosis, however, the cell wall of the blue-green algae has been reduced until only traces of it remain (Figure 9,B). There is next to no murein in the space between the phagocytosis vacuole membrane and the symbiont plasma membrane. Thus the space is analogous to the space between the inner and outer plastid membranes. Other important analogies between endocytobiotic blue-green algae and plastids are summarized in Table 7. The blue-green algae in this syncyanosis are so firmly integrated, or so greatly reduced (e.g. genomes of plastid size), that they are no longer viable outside the host cell. The host, too, cannot live without the symbionts, which photosynthetically fix CO_2 for it. Thus the symbiosis is obligatory for both partners, and they have reached a high degree of mutual dependence. In the unicellular eucaryotic alga *Glaucosphaera vacuolata*, the host cell division is even synchronized with that of the symbiotic blue-green algae. The high degree of integration or reduction suggests genetic transfer between the two systems. It therefore seems justifiable to place such endocytobiotic blue-green algae at the level resembling the chloroplasts. *In order to emphasize their organelle-like character, these endocytobiotic blue-green algae are also called cyanelles.*

Observation of recent endocytobioses has thus shown that there are endosymbionts with levels of reduction or integration comparable to those of chloroplasts and mitochondria, and also, to a lesser degree, to that of flagella. In any case, it is in effect only a question of definition whether one labels them endocytobionts or cell organelles. The mutual relations are genetically controlled and regulated, as is also the case with the DNA-containing organelles, by the nucleus. In another case, the nuclear genes of the host cell *Paramecium aurelia* even control the reproduction of the endocytobiotic bacteria of the Kappa type[29,56]).

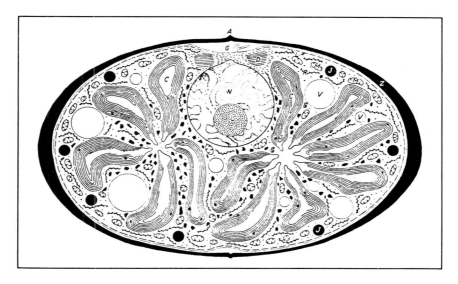

A	= equator	N	= nucleus
C	= cyanelle	R	= endoplasmic reticulum
G	= flagellum	S	= starch grains
J	= osmiophilic globuli	V	= vacuole
M	= mitochondrion	Z	= host cell wall

FIGURE 9A.   Organization of the endocytobiosis between the plastid-free host green alga *Glaucocystis nostochinearum* and a blue-green alga or cyanelle of unknown nature.

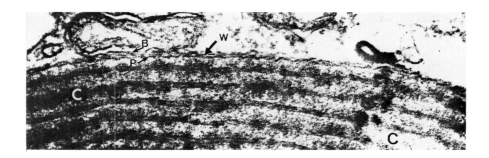

FIGURE 9B.   Section through the membrane of the cyanelle (C): a symbiont-derived membrane (W) with traces of cell wall is found between the host-derived membrane (B) and the symbiont plasma membrane (P). (A and B according to Schnepf and Brown[62]).

The ciliate *Euplotes aediculatus* even loses the ability to divide if its endocytobiotic bacteria are removed.[25] All these phenomena may require a gene exchange between host and symbiont. However, the genetic correlation between host eucytes and endocytobiotic procytes is still poorly understood. Intensive studies in this area are therefore not only desirable, but essential, if a better understanding of the total complex is to be obtained.

   All of the facts discussed above support the endosymbiont hypothesis, rather than the compartment hypothesis. The fact that in spite of intensive search no examples of a compartmental intracellular development of a cell-organelle containing DNA have been discov-

Table 7

ANALOGIES BETWEEN PLASTIDS AND ENDOSYMBIOTIC BLUE-GREEN ALGAE

Blue-green alga (cyanelle)	Plastid (organelle)
– Phagocytosis vacuole membrane of cytoplasmic origin	– Outer plastid membrane possibly of cytoplasmic origin (enyzme components)
– Symbiont plasma membrane	– Inner plastid membrane
– Space between phagocytosis vacuole and symbiont plasma membrane with traces of cell wall	– Space between outer and inner plastid membrane without traces of cell wall
– Invagination of the inner symbiont membrane (e.g. thylakoid, division membranes)	– Invagination of the inner plastid membrane (e.g. thylakoid, division membranes)
– Symbiont matrix (circular chromosome 5×10^8 daltons:10–25 copies, 70 S ribosomes, etc.)	– Plastid matrix (circular chromosome of 10^8 daltons:2–6 copies, 70 S ribosomes, etc.)
– Reproduction dependent on the host cell (semiautoreproductive)	– Reproduction dependent on the eucyte cell (semiautoreproductive)
– Photosynthetic CO_2 fixation	– Photosynthetic CO_2 fixation
– Cyanelle grains (starch)	– Plastid-pyrenoids (starch)
– Phosphate inclusions	– Phosphate granula
– Photopigments	– Photopigments
– Metabolic pathways	– Metabolic pathways
– Antibiotic sensibilities	– Antibiotic sensibilities

Adapted from Schnepf and Brown.[62]

ered also speaks against the compartment hypothesis. On the contrary, there is a complete lack of intermediate steps between procytes and eucytes, with the possible exception of the cell nucleus of the dinoflagellates. However, both hypotheses hold that this organelle arose endogenously (compartimentally). Since these organisms have few if any chromosomal proteins, and since the DNA is attached to the nuclear membrane, and also on account of their primitive cell division mechanism, they may be assigned an intermediate position (e.g., Haapala and Sorsa;[23] for details, see Section IV.B.). The lack of intermediates between procytes and eucytes again supports the endosymbiont hypothesis. In an evolution of the eucytes via endocytobiosis formation, transitionless phylogenetic jumps are more to be expected than in evolution through mutation and selection.

As review of recent endocytobioses shows, the phylogenetic derivation of the DNA-containing organelles from prokaryotic endocytobionts is, in principle, possible. The question now arises whether the organelles are in fact homologous to recent procytes (compare also Chapter 6,II.E).

D. Homologies with Procytes

The relatedness of different DNA or RNA molecules can be determined through comparison of their sequences (compare also Chapter 6, II. E). This is accomplished by base-pairing or hybridization experiments. Under certain conditions, the double-stranded DNA can be converted into a single-stranded form (melted) which can be paired with other single-stranded DNA or RNA molecules. RNA can be paired directly with other RNA molecules. The number of base pairs formed, which can be estimated from the temperature at which the complex dissociates (melts), is then an expression of the degree of relatedness. The comparison of the amino acid sequences of proteins can also be used for this purpose. Here we will, for the most part, make use of comparative sequence analyses as homology criteria.

1. Flagellum

DNA has not as yet been isolated from basal bodies or centrioles. More recent results even deny the existence of such DNA.[24,53] Specific RNA has also not yet been quantitatively isolated from the motile organelles, so that comparative sequence analyses are not possible, at least at present. Margulis postulates and has demonstrated homologies between certain proteins of the eucyte motile apparatus and spirochete proteins (Margulis et al.[46,47]). However, there is also evidence that the flagellar apparatus may be derived from photergic bacteria like the halobacteria with an auxiliary (9 + 2) axial skeleton pattern.[68] Until there is some experimental confirmation, however, such proposals, according to which the eucytic flagellum with its (9 + 2) or (9 + 0) homologs (flagella, cilia, basal bodies, centrioles, centromeres(?), and other microtubuli systems) arose from primitive spirochetes or photergic, spirochete-like bacteria with an axial skeleton pattern (9 + 2), remain unproved working hypotheses.

2. Mitochondrion

Mitochondrial ribosomes, like bacterial ribosomes, have a sedimentation constant of 70S (for review, see Arnold[5]). In contrast, only 80S ribosomes are found in the cytoplasm of eucytes. Both mitochondrial and bacterial 70S ribosomes, not however cytoplasmic ribosomes, consist of 50S and 30S subunits. In bacteria and plant mitochondria, the 50S subunit usually contains a 23S and a 5S subunit (animal or fungal mitochondria do not contain 5S RNA), whereas the 30S subunit contains a 16S RNA. The sequences of these ribosomal RNAs are nearly identical in mitochondria and bacteria, but there is no correlation with the corresponding RNAs from cytoplasmic ribosomes (see also Table 8).

For protein biosynthesis, mitochondrial ribosomes can be replaced functionally by bacterial ribosomes, but not by cytoplasmic ribosomes. Mitochondrial and bacterial protein biosynthesis is started by the same N-formylmethionine tRNA, whereas eucyte protein biosynthesis is started by another tRNA, methionine tRNA. The messenger RNA required for protein biosynthesis is transcribed from DNA by the enzyme RNA polymerase. In an in vitro system, the function of mitochondrial RNA polymerase from the fungus *Neurospora* can be exercised by RNA polymerase from the bacteria *Escherichia coli,* but not by cytoplasmic RNA polymerase. Furthermore, protein biosynthesis in mitochondria and bacteria can be blocked by the same antibiotics, chloramphenicol, streptomycin, erythromycin, and tetracycline, but these antibiotics do not affect synthesis on cytoplasmic 80S ribosomes. And conversely, antibiotics like cycloheximide, which inhibit cytoplasmic protein biosynthesis, do not affect that of mitochondria and bacteria.

In the cases listed above, mitochondria and bacteria are more similar to each other than to the cytoplasm. Mitochondria appear to be especially closely related to certain types of bacteria. The amino acid sequence of the mitochondrial respiratory enzyme cytochrome *c* is nearly identical to cytochrome *c* of respiring bacteria. There are other far-reaching similarities between the respiratory chain of the mitochondria and that of the bacteria. Mito-

Table 8
HOMOLOGIES BETWEEN THE CHLOROPLAST/MITOCHONDRION AND
THE PROCYTE SYSTEMS, AND ANALOGIES WITH THE EUCYTE SYSTEM[a]

Procyte/Chloroplast/Mitochondrion	*Eucyte (Cytoplasm)*
Genetic apparatus	
Circular chromosome: $<10^{12}$ daltons	Chain chromosome: $> 10^{12}$ daltons
No histones (or low content)	Associated with histones
Associated with the membrane	Not bound to the membrane
Specific base composition: $G-C > A-T$	Specific base composition: $G-C < A-T$
One starting point for replication	Several starting points for replication
Non-Mendelian genetics	Mendelian inheritance
Rapid renaturation	Slow renaturation
Protein biosynthesis	
Specific mRNAs	Specific mRNAs
rRNAs (23S, 16S, 5S)	rRNAs (28S, 18S, 5S?)
tRNAs	tRNAs
70S ribosomes (50S, 30S)	80S ribosomes (60S, 40S)
specific ribosomal proteins	specific ribosomal proteins
Specific aminoacyl tRNA synthetases	Specific aminoacyl tRNA synthetases
Protein synthesis begins with	Protein synthesis starts with
tRNA $^{N-F-Met}$	tRNA^{Met}
Rifamycin sensitivity (transcription)	α-Amanitin sensitivity (transcription)
Chloramphenicol sensitivity	Cycloheximide sensitivity
(translation)	(translation)
Membrane components	
Special enzyme pattern	Special enzyme pattern
Specific glycolipids and phospholipids	Specific glycolipids and phospholipids
(for example cardiolipin: not in	(for example cholesterol)
plastids)	

[a]   Exception: dinoflagellates with nearly histone-free, circular, membrane-associated DNA molecule[69]

chondria also occasionally fuse and subsequently exchange genetic material, as has been observed in the flagellate *Polytoma*. This behavior is also observed in the corresponding bacteria. These and other criteria, which have not been listed here,[38,46] substantiate the procyte nature of the mitochondria, so that the descent of mitochondria from aerobically respiring bacteria appears possible (compare Figures 10 and 11 and Chapter 6, Figure 4,B).

Beyond the more conservative, procytic features in structure, function and information, the mitochondria also exhibit eucytic features. Some of their genes are split into smaller pieces (introns), separated by noncoding genetic material (exons), as are the genes of the eukaryotic nucleus and those of the plastids.[15] Following transcription of the entire gene, the introns are removed by special RNA-splicing enzymes and reunited to a functioning mRNA molecule. Because of the exons, mitochondrial DNA is rich in adenine and thymine and poor in guanine and cytosine (~18% in yeast), as is typical for the eukaryotic system. The same is true of the G/C content of mitochondrial rRNA and tRNA, although we find here great inter- and intraspecific diversity (see Table 4). On the other hand, the mitochondrial biosynthesis of DNA, RNA, protein, and their aggregates resembles both the eucytic and procytic models. The structure and evolution of tRNA and mRNA of the mitochondria have, however, other features which do not correspond to either model; that is, they are unique.

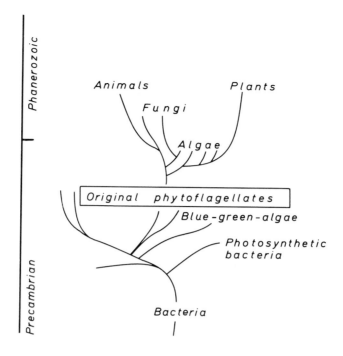

FIGURE 10.   The classical view of the evolution of bacteria, protozoan, plant, fungal and animal cells (adapted from Margulis[38-41]).

The mitochondria possess a different heterogeneous code in which UGA codes for tryptophan and is not, as usual, a stop codon. Six of the usual codes are not used at all by the mitochondria (GUG: Val; UCG: Ser; ACG: Thr; CGG: Arg; AGG: Arg; ACG: Ser). Some are used only by a few mitochondrial species, so that they have fewer than the usual 32 tRNAs, i.e. 20 to 25.

Mahler (in Schwemmler and Schenk[74]) has called these differences and heterogeneities of the mitochondrial replication, transcription, and translation apparatus the "mitochondrial dilemma". The question of the exogenous or endogenous origin of mitochondria becomes a problem which is difficult to resolve. A possible solution of this dilemma has been suggested by Doolittle and Bonen (in Fredrick[21]). They regard the heterogeneity of the mitochondrial system (code) and the splicing phenomenon (exon-intron) as conservative features, whereas the diversities (inter- and intraspecific genetic polymorphism) are regarded as a result of the diverging evolution of individual cell compartments, i.e. as derived features (mosaic evolution[15]). Thus the mitochondria were formed endosymbiotically and all organisms could be traced back to three great developmental lines: the archaebacteria with mycoplasms (term introduced in 1977 by Woese), halobacteria and methane bacteria; the nucleocytoplasm system; and the eubacteria with the mitochondria and the plastids (see Figure 11,B). According to this assumption, mitochondria and chloroplasts have conserved in contrast to eubacteria a number of primitive features by means of endocytobiosis formation (Table 12). These lines of descent are supported by the homology studies of 16 and 18S rRNA, as well as by such studies on initiator tRNA of these 4 groups (e.g. Woese and Fox,[88] Woese,[89] Schwartz and Dayhoff,[64] as well as Doolittle and Bonen in Fredrick[21]). Prerequisite for the validity of such a developmental scheme would be the exchange of genes between nuclear and mitochondrial genomes in the course of millions of years of common evolution. The case of homologous mitochondrial ATPase-proteolipids, which is coded in the mitochondrial

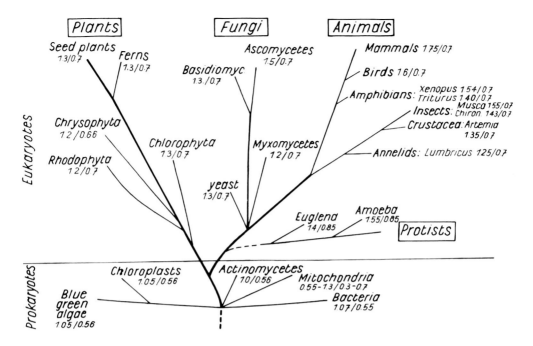

FIGURE 11A.   Phylogenetic tree of the five kingdoms of organisms and the molecular weight (in 10^6 daltons) of the two high molecular ribosomal RNAs of selected species (after Parthier[51]). Recent data on cytochrome and ferredoxin structure, on gene expression and gene product structure, especially on 5S rRNA and tRNA structure and nucleotide sequences also favor the endosymbiotic origin of eukaryotic cell.[10,18,20,21,27,51,64,74,78]

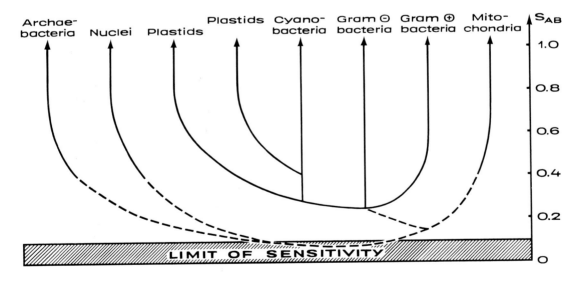

FIGURE 11B.   A schematic phylogenetic tree, based on 16S and 18S rRNA catalogs relating cyanobacteria, plastids, bacteria (including the archaebacteria: halobacteria, methane bacteria, mycoplasms), mitochondria, and the nucleocytoplasmic components of eukaryotic cells (data from Woese and Fox,[88,89] adapted from Doolittle and Bonen in Fredrick[21]). $S_{AB} = 2 N_{AB} (N_A + N_B)$, where N_A = total number of nucleotides (bases) represented by oligonucleotides present in catalog A (table of all determined sequences of the oligonucleotides of the species A), N_B = total number of nucleotides represented by oligonucleotides present in catalog B, and N_{AB} = total number of residues represented by oligonucleotides common to both catalog A and B.

DNA in yeast and in nuclear DNA in *Neurospora*, shows that gene transfer between cell compartments is possible. Börner and Hagemann (in Schwemmler and Schenk[74]) have discussed mechanisms which explain the passage of whole genes through the mitochondrial membrane. This seems to be possible, at least in the case of nuclear-coded organelle proteins which are synthesized within the organelles, since the required RNAs must pass through the nuclear and organelle membrane. A radical alternative to this concept would be to regard mitochondrial heterogeneity and the splicing phenomenon as derived features (degeneration of the genetic apparatus, mosaic evolution) and certain diversities as primitive features. This could support the idea of endogenous origin of mitochondria and coincide with the finding that mitochondrial DNA is, with regards to physical aspects (e.g. G/C content), very similar to nuclear DNA (Mahler and Cavalier-Smith in Schwemmler and Schenk[74]). If such a non-symbiotic origin of mitochondria is assumed, a number of nucleotide sequence homologies between bacteria and this organelle would remain unexplained, as Taylor points out (in Schwemmler and Schenk,[74] see also Bonen et al.[12]).

3. Chloroplast,

Hybridization experiments have shown that chloroplast DNA from *Euglena viridis* is at most about 1% homologous to cytoplasmic RNA from the same cell, up to 4.5% homologous to bacterial RNA, and up to 47% homologous to blue-green algal RNA,[55] the sequence analyses of 16S rRNA also indicate a far-reaching homology between chloroplasts and blue-green algae (Dayhoff and Schwartz in Schwemmler and Schenk;[74] see also Figure 11). This indicates little similarity between the chloroplast and the cytoplasm. The conservative differences are basically the same as those already discussed in detail for the mitochondria (Table 8). On the other hand, the same general agreement with the procytic system can be observed (e.g. behavior in the presence of antibiotics, genetic recombination through plastid fusion, as in *Chlamydomonas* and *Polytoma*[7] (Table 8). The chloroplasts are especially closely related to the blue-green algae. Table 9 summarizes some of the more general homologies between the transcription and translation systems of chloroplasts and blue-green algae (for review, Parthier;[50] Bonen and Doolittle,[10] and Zablen et al.[91]). It should also be mentioned that the base sequence of the individual ribosomal RNAs from chloroplasts are nearly homologous with the corresponding RNAs from blue-green algae. They differ no more than those from different forms of bacteria. Furthermore, various components of the transcription and translation apparatus from chloroplasts and blue-green algae, such as messenger RNAs, transfer RNAs, and the associated amino-acid coupling enzymes (aminoacyl-tRNA synthetases) can be substituted for each other in cell-free systems. The physical and chemical properties of the four protein subunits from the chlorophyll complex of photosystem I are equivalent to those from blue-green algae. The two subunits of the chromoprotein phycoerythrin from chloroplasts can even be exchanged with the corresponding subunits from blue-green algae without any loss of activity. These and other criteria[38,46] are evidence that chloroplasts and blue-green algae have descended from a common procyte ancestor, which today no researcher seriously doubts (reviewed in Schwemmler and Schenk[74]).

Thus there are basic differences or analogies between DNA-containing organelles and the cytoplasmic system with respect to the transcription and translation apparatus, and to metabolism; on the other hand, there are demonstrable homologies or relatedness to recent procytes (Table 8). This is also unequivocally supported by the most recent phylogenetic analyses of the sequences of two proteins, ferredoxin and cytochrome c, and two nucleic acids (5S RNA, 16S RNA) by Schwartz and Dayhoff in Schwemmler and Schenk[74]). Their results indicate that the mitochondria are much more closely related to aerobic and photosynthetic bacteria and that plastids are closer to the blue-green algae; both cell organelles are phylogenetically clearly distant from the cytoplasm system of eukaryotic protists, plants, fungi, and animals (compare Figure 11,B and C). This supports the endosymbiont hypothesis

Table 9

COMPARISON OF THE TRANSCRIPTION AND TRANSLATION SYSTEMS IN CHLOROPLASTS AND BLUE-GREEN ALGAE

Components	Chloroplasts in		Blue-green algae
	Green algae	Seed plants	
Ribosomes: S-values	70	70	70
subunits	50.30	50.30	50.30
Ribosomal RNA: mol.wt. x 10^6			
a. 23 S RNA	1.10	1.10	1.10
b. 16 S RNA	0.56	0.56	0.56
c. 5 S RNA	present	present	present
RNA synthesis: inhibited by			
rifamycin (antibiotic)	yes	yes	yes
Transfer RNA: capable of			
acylation by cytoplasmic			
enzyme?	no	no	no
Y-bases in tRNA^{Phe}	yes	yes	yes
Heat stability of amino-			
acid-activating enzyme	high	high	high
Protein synthesis: inhibi-			
tion by chloramphenicol			
(antibiotic)	yes	yes	yes

From Parthier.[50]

and contradicts the compartment hypothesis, according to which the organelles ought to resemble the eucytic and not the procytic system. To be sure, both hypotheses make the assumption that at the beginning of eucyte evolution the conditions were procytic, i.e. that all ribosomes were 70S. While the organelle was very conservative, outside it, due to the different evolutionary velocity and direction, the eukaryotic system with its 80S ribosomes evolved (mosaic evolution).

Evidence in favor of the compartment hypothesis and against the endosymbiont hypothesis has been presented, in which there is no agreement between the DNA-containing organelle and the corresponding procyte. Differences in the DNA, ribosomes, and metabolism of bacteria and mitochondria, on the one hand (Table 4) and blue-green algae and chloroplasts on the other are presented (Table 10). According to Arnold,[5] such examples can hardly weaken the endosymbiont hypothesis, let alone disprove it: "For plastids are naturally not blue-green algae, and mitochondria are not bacteria. The hypothesis only means that they have a common origin. This goes back to a stage in the development of life that lies before the existence of eukaryotic unicellular organisms! Therefore one can not expect to find complete identity in comparative studies of prokaryotic organisms and eukaryotic cell organelles. The point is to detect relicts which indicate a common descent."

Other arguments against the endosymbiont hypothesis are based on differences between the DNAs and RNAs found in various mitochondria or plastids (Tables 4 and 10). For

Table 10

DIFFERENCES IN THE PROPERTIES OF DNA FROM CHLOROPLASTS AND FROM BLUE-GREEN ALGAE

Properties	Chloroplasts from		Blue-green algae
	Green algae	Seed plants	
Circular molecule	yes	yes	no
Length in μm	44	40	100 - 360
Mol. wt. x 10^7	9	9	20 - 30
Buoyant density (g/cm^3)	1.696	1.698	1.715
% Guanine-cytosine content	38	37	56
T_m value in °C	80	84	92
Speed of renaturation	rapid	rapid	slow
5'-Methylcytosine present	no	no	yes

Adapted from Parthier.[50]

example, the mitochondria from plants contain ribosomes with a sedimentation constant of about 78 S, and circular DNA with a length of approximately 12 to 30 μm. The mitochondria of higher animals, in contrast, have DNA which is only 5 μm long and miniribosomes with sedimentation constants of about 55 S. In addition, only the plastids of the red algae have phycoerythrin and phycocyanin in common with the blue-green algae, while the chloroplast of the eukaryotic green algae do not contain these biliproteins. The plastid genome has a size between 85×10^6 daltons in bryophytes (cf. mycoplasm) and 15×10^8 daltons in *Acetabularia* (cf. blue-green algae). In other respects we also find genetic polymorphism in the nature, organization, expression, and regulation of the mitochondrial and plastid genome between different species and within one species. However, these facts are hardly an argument against the endosymbiont theory, but rather one in favor of the polyphyletic origin of the mitochondrial and plastid endocytobioses, although the possibility of their monophyletic development cannot be completely disregarded. If, however, one assumes endosymbiotic origin for chloroplasts but not for mitochondria, then it is difficult, according to Taylor (in Schwemmler and Schenk[74]) to explain their close structural, functional, and genetic analogies, as well as their similar topological relationships.

III. EUCYTE AS ENDOCYTOBIOSIS

In conclusion, we wish to emphasize the variety of arguments for and against the compartmental and the endosymbiont hypothesis, *which are, nevertheless, only evidence, not proof.* However, direct proof is rare in questions of evolution. In general, it is only possible to determine to what degree the data can be fit into the total biological picture without contradiction.

In this sense, the facts and conclusions discussed above all unequivocally favor an endocytobiotic rather than a compartmental intracellular development of the chloroplasts. An endocytobiotic, rather than a compartmental, origin of the mitochondria would appear to be very probable. The question of the origin of the flagellum and its (9 + 2) or (9 + 0) homologues remains open. Margulis has summarized evidence indicating their endocytobiotic origin. Even if it should prove impossible to detect endogenous DNA in the flagellum, this would not disprove its endocytobiotic nature. In light of other criteria,[38,44,46] it would still be necessary to consider the possibility that the DNA of this presumably oldest eucytic endocytobiont has, in the course of billions of years of cellular organization, gradually been integrated into the nuclear genome. The disolution of the cell boundaries of the motile apparatus may have been coupled to the complete integration of a possible flagellar DNA. The compartmental development of the motile apparatus remains a radical alternative. The essential arguments relevant to this problem are presented in Table 11, which compares the endocytobiotic and compartmental points of view. Sequence analyses and hybridization experiments could, in future, shed light on these questions.[47]

It is thus quite probable that the eucytes emerged not from one, but from several procytes, through the formation of endosymbioses. The exact number of procytes involved in this process must remain open for the time being. On the other hand, it is clear that organelles such as the nucleus, Golgi-apparatus, etc. arose by compartmental intracellular development. Thus the endosymbiont hypothesis has in principle proved correct, but the compartment hypothesis has been found to possess a certain validity, as is often the case for alternative hypotheses. As it has been confirmed in many areas, the endosymbiont hypothesis has been adopted in many textbooks as the endosymbiont theory *or better endocytobiotic cell theory*. Its importance for the understanding of unsolved cell problems, e.g. cancer, polarity, pattern, endorhythm, etc., should not be underestimated, since the solution of all these problems depends on the clarification of the historically developed relationships between the structure, function, information, and evolution of the individual cell organelles.

IV. HYPOTHETICAL RECONSTRUCTION OF EUCYTE EVOLUTION

Two basically different evolutionary trees can be constructed, one based on the compartment hypothesis (Figure 10), the other on the endocytobiotic cell theory (Figures 11 and 12). As we have shown above, the evidence favors the endocytobiotic point of view, without, of course, proving it. In the following, therefore, we present a short outline of eucyte evolution as seen from the endocytobiotic point of view. This summary is based on the unmodified endocytobiotic cell theory, even though the nature of the motile apparatus has not yet been settled (Figure 12; see also Section II.C).

The development of the most important procyte metabolic types must have been completed at least 2 billion years ago. The evolution of the procytes had reached a dead end, partly because their complement of genetic information could not be increased (see Chapter 6, IV), and partly because the various metabolic modes, fermentation, photergy, respiration, and photosynthesis could not be fully active at the same time. The lack of separate areas in which different metabolic pathways could run simultaneously in different directions (compartmentation) meant that the different pathways inhibited each other. This effect was discovered by Pasteur, who observed that anaerobic fermentation and aerobic respiration in bacteria could not occur simultaneously (Pasteur effect). The transition from one type of metabolism to another usually involves a substructural and chemical redifferentiation of the procyte. For example, the formation of thylakoids by invagination of the plasma membrane and incorporation of photopigments accompanies the transition from respiration to photosynthesis. Therefore, a certain amount of time is needed before a newly activated metabolic pathway is completely functional, and this created a selection pressure for compartmentation of metabolic pathways, in order to make them simultaneously and instantaneously available. Procytes which achieved this goal by endosymbiosis had a selective advantage over those which "attempted" to achieve it by intracellular differentiation, since the prefabricated parts were available for the formation of endocytobioses. They could be more quickly incorporated than adequate compartments could evolve intracellularly. Thus a series of interrelated en-

Table 11

COMPARISON OF THE TWO OPPOSING POINTS OF VIEW ON THE ORIGIN OF THE MICROTUBULE SYSTEM OF THE EUCYTE

Exogenous View (Margulis, 1970, 1978)	Endogenous View (Pickett-Heaps, 1974, 1975)
1. Complex (9+2) flagella preceded the evolution of the microtubule system in eukaryotes and were ultimately derived from ectosymbiotic surface microbes.	1. Microtubules and their organizing centers preceded the origin of flagella and were ultimately derived endogenously, perhaps from nuclear membrane sites.
2. Microtubules will be found in spirochetes, eventually a spirochete with a (9+2) array of intracellular tubules may be found.	2. Microtubules will be found in the cytoplasm of the prokaryotes that were ancestral to the eukaryotes. There is no reason to expect them in spirochetes.
3. MTOC's (microtubule organizing centers) are replicative, RNA replication will accompany "reproduction" of centrioles, basal bodies, kinetochores and so forth.	3. MTOS's are protein self-assembly systems.
4. All eumitotic organisms have flagellated ancestors; (9+2) organelles may be inducible in them.	4. Some eumitotic organisms are primitively nonflagellated (red algae, higher fungi), thus (9+2) organelles could never be induced in them.
5. The genes coding for the tubulin and microtubule associated proteins will be homologous to the genes coding for such protein in selected spirochetes.	5. The genes coding for tubulin and associated proteins are in the nucleus where they originated, such genes would only be expected among ancestral prokaryotes on the direct line to eukaryotes.
6. The flagellum is monophyletic, mitosis and meiosis are polyphyletic in several lines of protists.	6. Mitosis is monophyletic; the flagellum subsequently evolved in a mitotic organism and was retained for its selective advantage.
7. Once evolved, mitosis was not lost; flagellates and amoeboids lacking mitosis are primitive.	7. Mitosis was lost in several lines of flagellated protists.

From Margulis.[44,45]

docytobioses arose, more or less of necessity, although precipitated by chance. The result was endocytobiotic complexes which united the most important procytic metabolic types, fermentation, photergy, respiration, and photosynthesis. These complexes were capable of adapting rapidly and simultaneously to all four of the elementary ecological niches which had been individually occupied by procytes (see Chapter 6, Table 2).

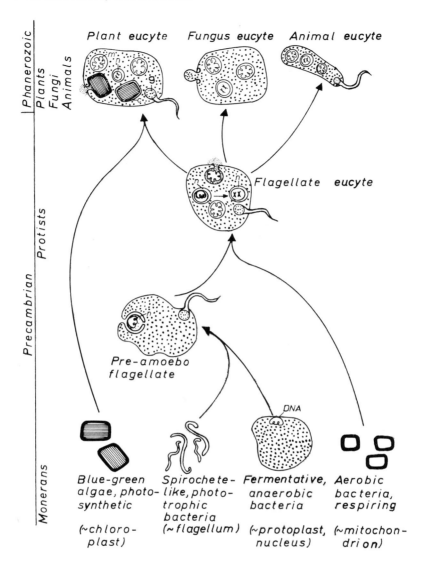

FIGURE 12.   Hypothetical model of the development of the most important types of eucytes, or their organelles, by formation of endocytobioses (adapted from Margulis;[44] details, see text). c = chloroplast; g = flagellum; n = nucleus; m = mitochondrion.

A. The Protozoan Cell (see Table 12)

Before the formation of free oxygen more than 2 billion years ago, selection probably favored endocytobioses between amoeboid procytes and those with an axial skeleton with a (9 + 2) pattern. Analogously to recent symbioses, the mobile procytes were probably at first only episymbiotically associated with the surface of the ameboid host. Later their head portions were assimilated into the host cytoplasm. This would not have occurred by *phagocytosis,* since phagocytosis is unknown among procytes. It would seem more likely that the symbionts penetrated the host actively, in the manner of the *Bdellovibrio* bacteria. The motile symbionts would then have dissolved the host plasma membrane enzymatically at the point of penetration. The membranes of the invading heads would then in turn be lysed by the host, releasing the membrane-bound DNA of the symbiont. After a time, the symbiont

Table 12

CHARACTERISTICS OF ARCHAEBACTERIA, EUBACTERIA, AND EUKARYOTES[a]

	Archaebacteria	Eubacteria	Eukaryotes
Cell length	Approx. 1 μm	Approx. 1 μm	Approx. 10 μm
Subcellular organelles	Absent	Absent	Present
Nucleus membrane	Absent	Absent	Present
Cell wall	Many types, none containing muramic acid	Many varieties of one type, all containing muramic acid	No cell wall in animal cell, otherwise many types
Membrane lipids	Ester with branched chains	Ester with straight chains	Ester with straight chains
Transfer RNA molecules: thymine in the same arm	Absent	Present in most tRNAs of most species	Present in all species in most tRNAs
Dihydrouracil	Present in one genus only	Present in all species in most tRNAs	Present in all species in most tRNAs
Amino acid of the starter tRNA	Methionine	Formylmethionine	Methionine
Ribosomes: size of subunits	30S, 50S	30S, 50S	40S, 60S
Approximate length of 16S (18S) RNA	1500 nucleotides	1500 nucleotides	1800 nucleotides
Approximate length of 23S (25S, 28S) RNA	2900 nucleotides	2900 nucleotides	3500 nucleotides or more
Elongation factor	Reacts with diphtheria toxin	Does not react with diphtheria toxin	Reacts with diphtheria toxin
Sensitivity to			
chloramphenicol	Insensitive	Sensitive	Insensitive
anisomycin	Sensitive	Insensitive	Sensitive
kanamycin	Insensitive	Sensitive	Insensitive
Messenger RNA binding sites AUCACCUCC at the 3' end of 16S (18S) RNA	Present	Present	Absent

[a] The three large groups of *archaebacteria, eubacteria* and *eukaryotes* are differentated on the basis of characteristic features in molecular structure. Certain features are found only in archaebacteria; others are also found in a similar form in eubacteria or eukaryotes. Thymine and dihydrouracil can replace uracil in RNA-transfer molecules. The elongation factor is part of the translation apparatus of the cell (according to Woese[89]).

DNA would finally be integrated in the host genome, in the same way as episomes. The cytoplasm and ribosomes of the symbiont mingled with those of the host. The membrane of the symbiont tail fused with the host membrane. The symbionts with their mobile appendages had come under the genetic control of the host, forming the *flagellum*, the first typical eucyte organelle.

A possible candidate for a living descendant of the amoeboid host cell is the mycoplasm *Thermoplasma acidophilum*.[21,76] This microaerobic bacterium is without a cell wall, resistant to acid (optimal pH 1 to 2) and to heat (optimal temperature 59°C) and is considered to be an Archaebacterium on the basis of the nucleotide sequence of its 16S RNA[88] (cf. also Section II.D). *Thermoplasma* has a number of similarities with the nucleo-cytoplasm system, i.e. similar DNA and RNA sequences; histone-like arginine-lysine-rich proteins which surround the DNA as a heat and acid shield (nucleosomes, Kornberg and Klug[30]); mobile, actomyosine-like proteins (cytoskeleton: inhibition by cytochalasin B); Ca-modulating proteins; one b-type cytochrome; peroxisomal activity; superoxide-dismutase, which seems to show both pro- and eukaryotic features (possessing two zinc atoms but not inhibited by cyanide); AA-sequence homologies in glycolytic active proteins; phospholipid-protein composition of the plasma membrane.

Certain spirochetes or better spiroplasms are regarded by Margulis[46] as possible candidates for living descendants of the hypothetical mobile symbionts. The anaerobic or microaerobic bacteria have in their plasma microtubuli, which are 20 ± 5 nm in diameter, similar to those in eukaryotes. Preliminary immunofluorescence studies suggest that their protein may be eukaryotic tubulin and not flagellin as in bacterial flagella. There are some indications that the flagellum symbiont had a photergic metabolism (see Figure 10; however, modern representatives of such photergic, spirochete-like forms are unknown). Under certain conditions, the photergic genes conserved in the host genome can apparently still be activated. Certain eukaryotic algae form membrane structures which contain carotenoids at the bases of their flagella. These carotenoid-containing structures, together with the adjacent eye spot, are the basis for the alga's light-oriented movements. The flagellar basal bodies of certain cells in the retina of vertebrates also form carotenoid-containing membrane structures. They are involved in light-induced cell contraction, and thus have an important role in the visual process.

By incorporating the flagellar symbiont, the host apparently obtained both protection from light through pigment integration and greater mobility in connection with light orientation. It was now able to survive anaerobically both in the dark depths and in the higher, illuminated water layers. In the dark, it lived chemotrophically, through fermentation, and in the light, phototrophically, by photergy. Active mobility also enabled the organism to search for ever diminishing organic substrate over a greater radius. This symbiosis, once it arose, had such a selective advantage that it multiplied rapidly and thus superseded other, similar associations. A pre-eucyte in the form of a pre-protozoan had emerged and, through the improvement in its metabolism, had taken an important hurdle on the way to the conquest of land by the eucytes. Today, such hypothetical pre-eucytes are long extinct. They were probably not able to hold their own against further-developed endocytobioses.

B. The Animal and Fungal Cell

During the development of the present oxidizing atmosphere, between 2 and 1.5 billion years ago, the second series of endocytobiotic forms could have arisen, probably by the incorporation of whole procytes through phagocytosis. The pre-eucytes incorporated aerobically respiring bacteria at first as food and then as symbionts in their cytoplasm. This development may have occurred several times, that is polyphyletically. Incorporation was quite analogous to recent symbioses. The host membrane invaginated and phagocytized the symbionts along with the aqueous phase surrounding them (Figure 12). The cell wall between the double membrane formed in this way was degraded in time. Symbiont and host began to exchange genetic material. However, a regulated gene transfer made a physiological barrier around the host genome necessary. Such a barrier was formed when the host membrane invaginated at the attachment site for DNA and other sites and enclosed the host genome with a double membrane. This formed the cell nucleus. The second stage of eucyte evolution was generally marked by a great deal of membrane folding activity. *The endosymbiont phase was thus followed by a period of intracellular, compartmental differentiation.*

The membrane motions were apparently favored, or perhaps first made possible, by the contractile microtubule system "imported" with the flagellar symbionts. The most important eucyte organelles developed in this period.[14,85] Not only did aerobic symbionts and host nuclei with double membranes emerge, membrane folds also formed a system of canals through the entire host cell (the endoplasmic reticulum). Further invaginations led finally to the Golgi apparatus, lysosomes, peroxisomes, and the entire system of endo- and exocytosis.* The result was the separation of the various cell functions into their own *reaction spaces,* or compartments, which could readily communicate with each other by membrane fusion. At the same time, the increase in the surface area of the membrane resulted in an increase in the enzyme and metabolic activities associated with it. This also applies to the membrane folds which were developed by the aerobic symbionts, leading to the formation of cristae and tubuli. The endocytobionts had become mitochondria. A primitive eucyte had emerged. It can be regarded as the predecessor of animal and, possibly, fungal cells.

The facultatively anaerobic, Gram-positive bacterium *Paracoccus denitrificans* (i.e. Bdellovibrio) seems to be a suitable candidate for a leftover relative of the mitochondria (Whatley in Schwemmler and Schenk[74]). *Paracoccus* is presumed to be a nonphotosynthetic, aerobic descendant *of the purple, sulfuric* bacteria. It has a cell wall but no ATP transport system and reduces nitrate or nitrite, as an alternative electron acceptor to N_2 when the cell is grown anaerobically in the presence of nitrate (compare Chapter 6, Figure 4B). *Paracoccus* contains in its plasma membrane an electron transport chain (cytochromes, flavoproteins, quinons, Fe/S proteins) and a phosphorylating mechanism (F_1 ATPase more similar to mitochondria than other bacteria) which very closely resemble these systems in the inner membrane of the modern mitochondria (compare Chapter 6, Figure 17C). Similarities include also the respiratory inhibitors, details of respiratory control, and P/O and H^+/O ratios.

The large amoeba *Pelomyxa palustris* lacks mitosis, a Golgi apparatus, and mitochondria.[74,86] *Pelomyxa* contains instead mitochondria, a population of aerobic bacteria-like endocytobionts. These bacteria are not found as free-living forms and cannot be cultured separately: they appear to be well integrated with their host. *Pelomyxa* with its bacteria represents a modern parallel to an intermediate stage in the evolution of a eukaryotic cell with mitochondria.

The mitochondria gave the primitive eucyte the ability to carry out oxidative phosphorylation, a capacity which the pre-eucyte did not possess. This meant, aside from the enormous increase in ATP synthesis (Chapter 6, III.E), that an inorganic acceptor of H^+ had been found, and thus a step toward independence from organic substrates had been taken. The eucyte made use of inorganic substrates for energy. In addition, its ability to live aerobically made it possible for it to live in the oxygen-rich upper layers of the water, thus coming one step closer to the colonization of the land. However, it still lacked a mechanism for cell division which would, among other things, exactly distribute the genetic information obtained from the symbionts among its progeny. Such a distribution was necessary, however, in order for the eucyte to replicate itself precisely and thus to prevail over the competing procytes. At this point a further intracellular, successive differentiation must have taken place.

The process of differentiation of a cell division mechanism must have begun even before the formation of the cell nucleus. One possible mechanism would be through binding of microtubuli of the flagellar symbiont to the membrane attachment sites of the host chromosome just after it had replicated.[14] By growing in length they separated the two DNA molecules. Then a system of microtubuli perpendicular to the first formed and provided for an equatorial invagination of the membrane, leading to division at this point, and guaranteeing that each half obtained a copy of the chromosomes. This was a primitive form of mitosis, or *ur-mitosis.* The formation of the nuclear membrane by invagination of the cell membrane with its associated DNA must have occurred after this point (see above). After a while, the extra copies of the host genome disappeared. A eucyte had now emerged of a type which

---

* Endocytosis denotes the uptake of foreign particles in small phagocytosis vacuoles; exocytosis is the process by which membrane vesicles containing cell substance (or waste) are secreted.

is now represented only by dinoflagellates[23]* and some fungi.[32] Soon after the formation of the nuclear membrane, the chromosomes were released from the inner membrane. The differentiation of special binding structures for the division microtubuli on the chromosomes themselves eliminated the selective advantage of the membrane attachment. The *centromeres* of the chromosomes later developed from the attachment sites for the microtubuli. The "membrane-free" host chromosome thereafter developed differently and more rapidly than the membrane-bound mitochondrial chromosomes (see 80S, 70S ribosomes). The host chromosome rapidly grew to a multiple of its original length by repeated duplications.[23] The enormously increased amount of genetic information still had to be read, however, so the DNA "wrapped itself" in proteins. The protein coat served to regulate the genetic activity. The giant ring chromosome eventually fell apart into a number of protein-wrapped DNA chains. The only living representatives of such eucytes with chain chromosomes and ur-mitosis is the amoebo-flagellate *Tetramitus*. The growing number of linear chromosomes made another improvement in the division mechanism necessary. Homologies of the flagellar basal body with the (9 + 0) pattern differentiated to centrioles, which became the organizers of the division microtubuli. At first, the centrioles penetrated the nucleus and organized the nuclear division without dissolving the nuclear membrane (amoeba, ciliates, sporozoa, certain fungi). Later, two centriole pairs moved extranuclearly to opposite poles of the cell. They formed a division spindle, and caused the nuclear membrane to dissolve before separating the replicated chromosomes (zooflagellates, animal cells). Nuclear division was followed by cytoplasmic division. The process of *mitosis* or *eumitosis* had developed. The sexual fusion of two eucytes, presumably as a response to starvation, could have led to the polyphyletic evolution of *meiosis*. A mechanism which permits an exact pairing of homologous chromosomes is a prerequisite for meiosis.

C. The Plant Cell

The microfossil evidence supports the hypothesis that the third and last endocytobiosis leading to eucyte cells developed about 1 billion years ago. Several eucyte lines probably took up blue-green algae or related forms in phagocytosis vacuoles.[62] Analogously to recent symbioses, the cell wall of the symbiont was gradually degraded, so that it was finally surrounded only by a double membrane. Like the mitochondria, the symbiont could then exchange genetic materials with the nucleus. This led to a reduction of the symbiont genome. The plastid-like cell organelles of the phytoflagellata and higher plant cells are the descendents of these endocytobionts.

The discovery of the oxygenic-photosynthetic prokaryote *Prochloron didemni* by Lewin[33,34] again supports the idea, which was first presented in a complete form by Raven,[58] that plastids have arisen at least twice by endosymbiosis (see also Figure 11 and Chapter 6, Figure 4 B).

Prochloron lacks the phycobilisomes and phycobilins characteristic (phycocyanin, phycoerythrin) of the blue-green algae and has instead chlorophyll b as accessory pigment. With the loss of phycobilins and the acquisition of chlorophyll b, stacking can take place and a thylakoid pattern is achieved which resembles more that in chloroplasts of green algae, of

---

*   Just as ontogenetic sequences and atavistic effects have been used to demonstrate early phylogenetic stages of organisms, Giesbrecht and Drews have tried to extend these principles to organelles.[210] The nucleus of eukaryotic cells has been considered not to be a homogenous structure like mitochondria or plastids, but rather a composite system, formed by an assembly of more or less independent "karyomers" (as is still the case during early stages of oogenesis in many organisms). Each karyomer is said to correspond to one single chromosomal apparatus of a prokaryotic cell. Furthermore, on the basis of a so-called cellular atavism, dinoflagellates have been considered to reflect early phylogenetic stages in nuclear development ("elementary karyokinesis"), in spite of the fact that an analysis of certain RNA sequences has demonstrated that, in fact, recent dinoflagellates are not very early organisms.[250]

other protists (e.g., euglenoids) and plants, than that of blue-green algae. *Prochloron* is thus possibly a *missing link* between photosynthetic prokaryots on the one hand and the plastids of green algae, certain protists and plants on the other and has therefore been classified in an independent group, the prochlorophytes.[46] *Prochloron* is a likely candidate for a plastide ancestor form of green algae, other protists and plants, whereas the blue-green algae represent possibly a living descendant of the plastid ancestor form of the other cell systems with plastids (e.g., red algae, dinoflagellates, cryptomonades). The endocytobiosis of *Cyanophora* or *Glaucocystis* with the blue-green algae, the cyanelles, appears to represent modern parallels to intermediates between the free-living and chloroplast state.[74]

Whatley et al.[86] (see also Gibbs and Gillott, both in Schwemmler and Schenk[74]) have researched the possible influence of the acquisition of organelles of one eucyte from another with regard to the phytogenesis of the algae. This aspect was first introduced by Taylor[74] in connection with euglenoids. According to Whatley, the presence in chloroplasts of more than two surrounding membranes (two membranes: green algae, red algae, modern symbiotic blue-green algae) implies an origin involving a second act of endocytobiosis of eukaryotic rather than prokaryotic photosynthetic organisms. The suggested mode of formation of these more complex chloroplasts is by the uptake of "naked" chloroplasts from a eukaryotic donor (three membranes: euglenoids with chlorophyll a + b, dinoflagellates with chlorophyll a + c, chloroplast endocytobionts in opistobranches) or by the uptake of a complete eukaryotic alga (four membranes: chromophyts with chlorophyll a + c, *Zoochlorella* in *Hydra/ Platymonas/Convoluta*). That is consistent with patterns of change in the photosynthetic pigments, in the thylakoid system, in the location of the glucose polymer synthesizing system and in the progression from predominantly α-1,4 to β-1,3-glucan storage products.

The plastids brought the eucyte complete independence from an organic substrate and fulfilled the most rigorous criterion for the conquest of the land. However, the colonization of land was first achieved by multicellular systems.

V. MECHANISMS OF EUCYTE EVOLUTION

The eucytes appear to have developed in three stages: pre-protozoa $\rightarrow$ pre-animals, pre-fungi $\rightarrow$ pre-plants. According to the endocytobiotic cell theory, this was possible because a fermenting procyte host of physiochemical type I successively incorporated three groups of procyte endocytobionts (Figure 13). The photergic flagellar symbionts with physiochemical type I/II were first incorporated in the form of obligatory *primary symbiosis*. Then the mitochondrial respiratory endocytobionts with physiochemical type II/III were incorporated in an obligatory *auxiliary symbiosis*. Finally, plastid photosynthetic endocytobionts of physiochemical type III were integrated as a facultative *accessory symbiosis* (physiochemical types, see Chapter 6, II.D).

In this way, eucytes were able to occupy new ecological niches and adapt to the changes in milieu and substrate involved. The incorporation of photergers made possible the shift from dark to light, or from chemotrophy to phototrophy. The incorporation of the respirers facilitated the shift from anaerobiosis to aerobiosis, or from organotrophy to partial inorganotrophy (inorganic H^+ acceptors), and the incorporation of photosynthesizers made the change from water to land, or from C-heterotrophy to C-autotrophy possible.

The presumed endocytobionts of the eucyte thus seem to have mediated, chemically and physiologically, between the physiochemically lower type of the host and the higher type of its new substrate. Mitochondria and plastids have this capacity, too. They regulate pH and osmotic pressure, as well as the type and amount of inorganic ions within the eucyte and thus mediate between the inner and outer milieu of the eucyte.

Both procytes and eucytes are in principle capable of fermentation, photergy, respiration and photosynthesis. Their synthetic capabilities are entirely comparable. What, then, was

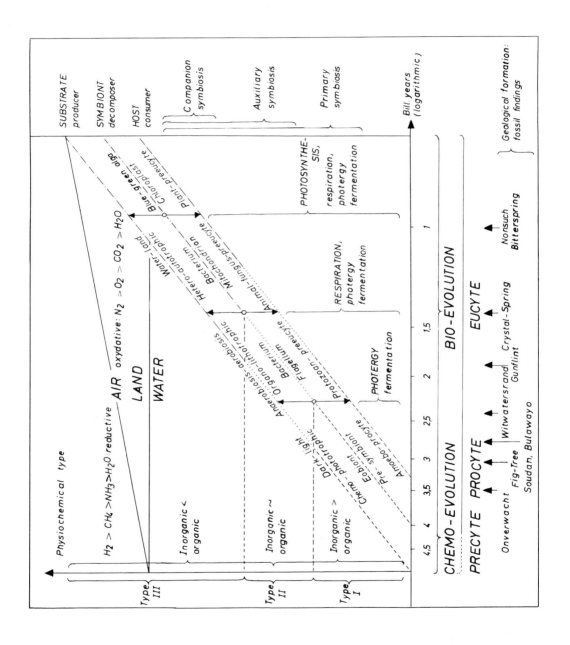

FIGURE 13. Probable evolutionary relationship between the host consumer, substrate producer, and the hypothetical endocytobiont decomposers in the eucyte. According to this scheme, the eukaryotic cell developed from fermenting bacteria via the protozoan, then the animal or fungal cell, and finally the plant cell by integration first of bacteria-like photergic symbionts (flagella/centriole/centromere), then bacteria-like respiratory symbionts (mitochondria), and finally blue-green-alga-like photoassimilating symbionts (chloroplasts), in order to adapt to the constantly changing substrate. They thus developed from chemotrophy to phototrophy, organotrophy to lithotrophy, C-heterotrophy to C-autotrophy. The presumed endocytobionts appear to have mediated, thanks to their intermediate physiochemical type, between the lower type of the host and the higher type of the substrate.[68]

the selective advantage of the eucyte over the procyte, which resulted in the further development only of the former? There can be no doubt that the explanation is to be found in the compartmentation of metabolism in the eucyte, which permits different pathways to operate simultaneously, rather than one at a time, as in the procytes. In the procyte, countercurrent metabolic pathways inhibit each other mutually (e.g., Pasteur effect). The procytes remained dependent on an aqueous, or at least moist medium for the complete development of their vital processes and for replication. Some eucytes, or at least their multicellular representatives, on the other hand, succeeded in colonizing the land. There were many difficulties to overcome, including the danger of drying out, the large changes in temperature, support against gravity, and the absorption of nutrients and reproduction. The improved supply of energy and the capacity to rapidly adapt to a new ecological niche provided by metabolic compartmentation helped the eucytes to gradually overcome these problems.

REFERENCES

1. **Allsopp, A.,** Phylogenetic relationships of the procaryotes and the origin of the eucaryotic cell, *New Phytol.,* 68, 591, 1969.
2. **Altmann, R.,** *Die Elementarorganismen und ihre Beziehung zu den Zellen,* Veit. u. Comp., Leipzig, 1890.
3. **Arnold, C. G.,** Gene ausserhalb des Zellkerns. Die genetischen und molekularen Grundlagen der extra-karyotischen Vererbung, *Biologie in unserer Zeit,* 1, 111, 1971.
4. **Arnold, C. G.,** III. Extrakaryotische Vererbung, *Fortschr. Bot.,* 35, 217, 1973.
5. **Arnold, C. G.,** IV. Extranuclear Heredity: The Phylogenetic origin of extranuclear Heredity, *Fortschr. Bot.,* 39, 182, 1977.
6. **Arnold, C. G.,** Die Entstehung der eukaryontischen Zelle (Euzyte), In *Evolution,* Sieving, R., Ed., Fischer Verlag (UTB-Taschenbuch), Stuttgart, 155, 1978.
7. **Arnold, C. G. and Gaffal, K. P.,** Die räumliche Struktur von Mitochondrien und Plastiden, *Biologie in unserer Zeit,* 9 (2), 45, 1979.
7a. **Barckhausen, R., Heger, W., Hollihn, K.-U., Lootz, J., Lüdcke, J., Pulvermacher, C., Schwemmler, W., Seipel, S. and Timner, K.,** *Kompendium Biologie für Mediziner.* 2. Auflage, Gustav Fischer Verlag, Stuttgart, New York, 1981.
8. **Berthold, M.,** *Analyse der Interaktionen zwischen Kern- und Plastengenom* (Review), Staatsexamensarbeit, Freie Universität, Berlin, 1979.
9. **Bogorad, L.,** Evolution of organelles and eukaryotic genomes, *Science,* 188, 891, 1975.
10. **Bonen, L. and Doolittle, W. F.,** On the prokaryotic nature of red algae chloroplasts, *Proc. Nat. Acad. Sci.,* 72 (6), 2310, 1975.
11. **Bonen, L. and Doolittle, W. F.,** Partial sequences of 16S rRNA and the phylogeny of blue-green algae and chloroplasts, *Nature,* 261, 669, 1976.
12. **Bonen, L., Cunningham, R. S., Gray, M. W., and Doolittle, W. F.,** Wheat embryo mitochondrial 18S ribosomal RNA: evidence for its prokaryotic nature, *Nucleic Acids Res.,* 4, 663, 1977.
13. **Buetow, D. E.,** Phylogenetic origin of the chloroplasts, *J. Protozool.,* 23 (1), 41, 1976.
14. **Cavalier-Smith, T.,** The origin of nuclei and of eukaryotic cells (a review), *Nature,* 256, 463, 1975.
15. **Chambon, P.,** Gestückelte Gene — ein Informationsmosaik, *Spektrum der Wissenschaft,* 7, 104, 1981.
16. **Cohen, S. S.,** Are/were mitochondria and chloroplasts microorganisms? *Am. Scientist,* 58, 281, 1970.
17. **Cook, C. B., Pappas, P. W., and Rudolph, E. D.,** Cellular interactions in symbiosis and parasitism, *5th Biosciences Colloquium,* Ohio State University, Ohio State University Press, Columbus, 1980.
18. **Dickerson, R. E.,** Cytochrom c und die Entwicklung des Stoffwechsels, *Spektrum der Wissenschaft,* 5, 47, 1980.
19. **Dodson, E. O.,** Crossing the prokaryote-eukaryote border: endosymbiosis or continuous development? *Can. J. Microbiol.,* 25, 652, 1979.
20. **Eigen, M. and Winkler-Oswatitsch, R.,** Transfer-RNA: The early adaptor, *Naturwissenschaften,* 68, 217, 1981.
21. **Fredrick, J. F.,** Origins and evolution of eukaryotic intracellular organelles, *Ann. N.Y. Acad. Sci.,* 361, 1981.
21a. **Giesbrecht and Drews,** in *Die Zelle: Struktur und Funktion,* Metzner, H., Ed., Wissensch. Verlagsgesellschaft, 3 Aufl., Stuttgart, 1981 (and personal communication).

22. **Grimstone, A. V.**, The structure of *Mixotricha* and its associated microorganisms, *Proc. Br. Royal Soc.*, (London) 159, 668, 1964.
23. **Haapala, O. and Sorsa, V.**, Evolution of eukaryotic chromosome organization, *Biol. Zbl.*, 95, 317, 1976.
24. **Hartmann, H.**, The centriole and the cell, *J. Theor. Biol.*, 51, 501, 1975.
25. **Heckmann, K.**, Omikron, ein essentieller Endosymbiont von *Euplotes aediculatus*, *J. Protozool.*, 22, 97, 1975.
25a. **Hinnebusch et al.**, *J. Mol. Evol.*, 17, 334, 1981.
26. **Ishida, M. R.**, The evolutionary origin of chloroplast DNAs based on the organization of their genes, *Viva Origins*, 9(2), 37, 1981.
27. **Kandler, O.**, (1981): Archaebakterien und Phylogenie der Organismen, *Naturwissenschaften*, 68, 183, 1981.
28. **Kaplan, W.**, *Der Ursprung des Lebens. Biogenetik, ein Forschungsgebiet heutiger Naturwissenschaft*, 2. Aufl., Georg Thieme Verlag, Stuttgart, 1978.
29. **Koizumi, S. and Kobayashi, S.**, A study on the mate-killer toxin by micro-injection in *Paramecium*, *Genet. Res.*, 27, 179, 1976.
30. **Kornberg, R. D. and Klug, A.**, Das Nucleosom, *Spektrum der Wissenschaft*, 4, 29, 1981.
31. **Lee, R. E.**, Origin of plastids and the phylogeny of algae, *Nature*, 237, 44, 1972.
32. **Leighton, T. J., Dill, B. C., Stock, J. J., and Philips, C.**, Absence of histones from the chromosomal proteins of fungi, *Proc. Nat. Acad. Sci.*, 68, 677, 1971.
33. **Lewin, R. A.**, Prochlorophyta as a proposed new division of algae, *Nature*, 261, 697, 1976.
34. **Lewin, R. A.**, *Prochloron*, type genus of the prochlorophyta, *Phycologia*, 16, 217, 1977.
35. **Loeblich, A. R.**, Protistan phylogeny as indicated by the fossil record, *Taxon*, 23, (2/3), 227, 1974.
36. **Lorenzen, H. and Wiessner, W.**, Intracellular and intercellular regulation and recognition in algae and symbionts, *Proc. Int. Coll. Göttingen, 1980*, Ber. Deutsch. Bot. Ges., in press, 1981.
37. **Louis, C. and Laporte, M.**, Caractères ultrastructureaux et différenciation des formes migratrices des symbiotes chez *Euscelis plebejus* (Homoptera, Jassidae), *Ann. Soc. Ent.*, 5, 799, 1969.
38. **Margulis, L.**, *Origin of Eukaryotic Cells*, Yale University Press, New Haven, London, 1970.
39. **Margulis, L.**, Symbiosis and evolution, *Am. Scientist*, 225 (2), 48, 1971.
40. **Margulis, L.**, The origin of plant and animal cells, *Am. Scientist*, 59 (2), 230, 1971.
41. **Margulis, L.**, Whittaker's five kingdoms of organisms: Minor revisions suggested by considerations of the origin of mitosis, *Evolution*, 25 (1), 242, 1971.
42. **Margulis, L.**, The classification and evolution of prokaryotes and eukaryotes, in *Handbook of Genetics*, King, R. C., Ed., Vol. 1, Plenum Press, New York, 1974.
43. **Margulis, L.**, On the evolutionary origin and possible mechanism of colchicin-sensitive mitotic movements, *BioSystems*, 6, 16, 1974.
44. **Margulis, L.**, The microbes contribution to evolution, *BioSystems*, 7, 266, 1975.
45. **Margulis, L.**, Microtubules and evolution, in *Microtubules and Microtubule Inhibitors*, Borgers, M. and de Brabander, M., Eds., North Holland Publishing, Amsterdam, 1975.
46. **Margulis, L.**, *Symbiosis in Cell Evolution*, W. H. Freeman, San Francisco, 1981.
47. **Margulis, L., Leleng, T., and Chase, D.**, Microtubulus in prokaryotes, *Science*, 200, 1118, 1978.
48. **Mereschkowsky, C.**, Über Natur und Ursprung der Chromatophoren im Pflanzenreich, *Biol. Centralbl*, 25, 593, 1905.
49. **Nass, S.**, The significance of the structural and functional similarities of bacteria and mitochondria, *Int. Rev. Cytol.*, 25, 55, 1969.
50. **Parthier, B.**, Zur Evolution von Chloroplasten und Mitochondrien, *Nova Acta Leopoldina NF*, 42 (218), 223, 1975.
51. **Parthier, B.**, Evolutionary aspects of gene expression-organization in macro-compartments, in *Leopold. Symp. Cell Compartmentation and Metabolic Channelling*, Nover, L., Lyen, F., and Mothes, K., Eds., Elsevier Company/Fischer, Amsterdam, Jena, 1979.
52. **Pascher, A.**, Ueber Flagellaten und Algen, *Ber. Deutsch. Bot. Ges.*, 32, 136, 1914.
53. **Pickett-Heaps, J.**, The evolution of mitosis and the eukaryotic condition, *BioSystems*, 6, 37, 1974.
54. **Pickett-Heaps, J.**, Aspects of spindle evolution, *N.Y. Acad. Sci.*, 253, 352, 1975.
55. **Pigott, G. H. and Carr, N. G.**, Homology between nucleic acids of blue-green algae and chloroplasts of *Euglena gracilis*, *Science*, 175, 1259, 1972.
56. **Preer, J. R., Preer, L. B., and Jurand, A.**, Kappa and other symbionts in *Paramecium aurelia*, *Bact. Rev.*, 38, 113, 1974.
57. **Raff, R. A. and Mahler, H. R.**, The non-symbiotic origin of mitochondria, *Science* 177, 575, 1972.
58. **Raven, P. H.**, A multiple origin for plastids and mitochondria, *Science*, 169, 641, 1970.
59. **Richmond, M. H. and Smith, D. C.**, The cell as a habitat, *Proc. Royal Soc. London*, B 204 (no. 1155), p. 113, Royal Society, London, 1979.

60. **Schenk, H. E. A.,** Inwieweit können biochemische Untersuchungen der Endocyanosen zur Klärung der Plastiden-Entstehung beitragen? *Arch. Protistenk.,* 119, 274, 1977.

61. **Schimper, A. F. W.,** Über die Entwicklung der Chlorophyllkörner und Farbkörper, *Bot. Z.,* 41, 105, 1883.

62. **Schnepf, E. and Brown, R. M.,** On the relationships between endosymbiosis and the origin of plastids and mitochondria, in *Origin and Continuity of Cell Organelles,* Reinert, J. and Ursprung, H., Eds., Vol. 2, Springer Verlag, Berlin, 1971.

63. **Schopf, J. W. and Zeller-Oehler, D.,** How old are the eukaryotes? *Science,* 193, 47, 1976.

64. **Schwartz, R. M. and Dayhoff, M. O.,** Origins of prokaryotes, eukaryotes, mitochondria, and chloroplasts, *Science,* 199, 395, 1978.

65. **Schwemmler, W.,** Intracellular symbionts; a new type of primitive prokaryotes, *Cytobiology,* 3, 427, 1971.

66. **Schwemmler, W.,** Beitrag zur Analyse des Endosymbiosezyklus von *Euscelis plebejus* F. (Hemiptera, Homoptera, Cicadina) mittels in vitro-Beobachtung, *Biol. Zbl.,* 92, 749, 1973.

67. **Schwemmler, W.,** Zikadenendosymbiose: ein Modell für die Evolution höherer Zellen. (Zur Verifikation der Endosymbiontentheorie der Eukaryontenzelle), *Acta Biotheoretica,* 23, 132, 1974.

68. **Schwemmler, W.,** Allgemeiner Mechanismus der Zellevolution, *Naturwiss. Rdschau.,* 28 (10), 351, 1975.

69. **Schwemmler, W.,** Die Zelle: Elementarorganismus oder Endosymbiose? *Biologie in Unserer Zeit,* 7 (1), 7, 1977.

70. **Schwemmler, W.,** *Mechanismen der Zellevolution. Grundriss einer modernen Zelltheorie,* W. de Gruyter, New York, 1978.

71. **Schwemmler, W. and Herrmann, M.,** Oszillationen im Energiestoffwechsel von Wirt und Symbiont eines Zikadeneies. I. Analyse möglicher stoffwechselphysiologischer Korrelationen beider Systeme, *Cytobios,* 25, 45, 1979.

72. **Schwemmler, W. and Herrmann, M.,** Oszillationen im Energiestoffwechsel von Wirt und Symbiont eines Zikadeneies. II. Analyse möglicher endogener Rhythmen beider Systeme, *Cytobios,* 27, 193, 1980.

73. **Schwemmler, W., Hobom, G., and Egel-Mitani, M.,** Isolation and characterization of leafhopper endosymbiont DNA, *Cytobiologie,* 10 (2), 249, 1975.

74. **Schwemmler, W. and Schenk, H.,** Endocytobiology. Endosymbiosis and cell biology. Synthesis of recent research, *Proc. Int. Coll. Tübingen, 1980,* Vol. I, W. de Gruyter, New York, 1980.

75. **Schwemmler, W.,** The endocytobiotic cell theory and the periodic system of cells, *Acta Biotheoretica,* 31, 45, 1982.

76. **Searcy, D. G., Stein, D. B., and Green, G. R,** Phylogenetic affinities between eukaryotic cells and a thermoplasmic mycoplasm, *BioSystems,* 10, 19, 1978.

77. **Searcy, D. G. and Stein, D. B.,** Nucleoprotein subunit structure in an unusual prokaryotic organism: *Thermoplasma acidophilum, Biophys. Acta,* 609, 180, 1980.

78. **Stackebrandt, E. and Woese, C. R.,** Primärstruktur der ribosomalen 16S RNS — ein Marker der Evolution der Prokaryonten, *Forum Mikrobiologie,* 4, 183, 1979.

79. **Stanier, R. Y.,** Some aspects of the biology of cells and their possible evolutionary significance, *Symp. Soc. Gen. Microbiol.,* 20, 1, 1970.

80. **Starr, M.,** The *Bdellovibrio* association as model, in Society of Experimental Biology Symposium 29, Jennings, D., Ed., University Press, Cambridge, 1975.

80a. **Strassburger, E.,** *Lehrbuch der Botanik.,* 30. Neubearbeitete Auflage. Gustav Fischer-Verlag, Stuttgart, 1971.

81. **Tappan, H.,** Protistan phylogeny: multiple working hypothesis, *Taxon,* 23 (2/3), 271, 1974.

82. **Taylor, F. J. R.,** Implications and extensions of the serial endosymbiosis theory of the origin of eukaryotes, *Taxon,* 23, 229, 1974.

83. **Taylor, F. J. R.,** Autogenous theories for the origin of eukaryotes, *Taxon,* 25, 377, 1976.

84. **Tiedtke, A.,** Enthalten Kinetosomen Nukleinsäuren? (Cytochemische Untersuchungen an reinen Pellikula-Präparationen von *Tetrahymena pyriformis), Zool. Anz. Suppl.,* 41, 1977, *Verh. d. Dtsch. Zool. Ges. Erlangen* 306, 1977.

85. **Uzzell, T. and Spolsky, C.,** Mitochondria and plastids as endosymbionts: a revival of special creation? *Am. Scientist,* 62, 334, 1974.

86. **Whatley, J. M.,** Bacteria and nuclei in *Pelomyxa palustris:* comments on the theory of serial endosymbiosis, *New Phytol.,* 76, 111, 1976.

87. **Whittaker, R. H.,** New concept of kingdoms of organisms, *Science,* 163, 150, 1969.

88. **Woese, C. R. and Fox, G. E.,** The concept of cellular evolution, *J. Mol. Evol.,* 10, 1, 1977.

89. **Woese, C. R.,** Archäbakterien — Zeugen aus der Urzeit des Lebens, *Spektrum der Wissenschaft,* 8, 74, 1981.

90. **Wolf, K.,** Mitochondriale Genetik der Hefen, *Biologie in unserer Zeit,* 9 (3), 65, 1979.

91. **Zablen, L. B., Kissel, M. S., Woese, C. R., and Buetow, D. E.,** Phylogenetic origin of the chloroplast and prokaryotic nature of its ribosomal RNA, *Proc. Nat. Acad. Sci.,* 72 (6), 2418, 1975.

Chapter 8

EVOLUTION OF POSTCYTES

The term *symbiosis* was introduced in 1879 by De Bary to describe a partnership between different species of organisms which, in contrast to parasitism, is useful for both partners. In analogy to parasitic relationships, the larger partner is called the *host,* but the smaller is called the *symbiont.* A special form is intracellular symbiosis or endocytobiosis in which the symbionts — referred to here as endocytobionts — live in special cells of their host. Such host cells are called especially *mycetocytes* (yeast as endocytobiont) or *bacteriocytes* (bacterium as endocytobiont) and in general *"symbiocytes"* or *"postcytes"*. In extreme cases endocytobiosis is essential for host and symbiont and must be passed on from generation to generation. In such systems, the partners are not viable when separated.

Endocytobioses are to be found in nearly all major plant and animal phyla.[2,11,18,20,48] Nitrogen-assimilating bacteria which are located in special cells, the nodules, of the rootstock allow certain legumes such as beans and peas to live on nitrogen-poor soils. Insects which are nutritional specialists, in so far as they consume only mammalian blood, plant sap, or hard to digest plant substances such as cellulose and wood, often require endocytobionts. In these cases, the host and endocytobiont show the highest degree of mutual dependence. But, what is the significance of endocytobionts for the host cells?

I. WORKING HYPOTHESES

One concept could be called the *nutritional endocytobiosis hypothesis.* The nutritional endocytobiosis hypothesis was conceived by the pioneer of symbiosis research, Paul Buchner,[2] and then developed further by Koch,[13] Schwartz and Koch,[30] Malke and Schwartz,[17] and Ehrhardt.[5,6] According to this hypothesis, the endocytobionts only digest or supplement the one-sided diet of the host. They are chiefly responsible for the breakdown of hard to digest food (cellulose, blood, etc.) or poisonous metabolic wastes from the host (urea, uric acid, etc.). They convert products of host metabolism into essential substances such as vitamins, amino acids, and enzymes for themselves and their host. The host and symbiont are thought to exchange only metabolites and gene products, in the form of enzymes; an exchange of the genes themselves is thought not to have occurred (Figure 1A).

More recently, however, it has become apparent that many endocytobionts have structure, function, and information comparable to those of the DNA-containing organelles of eukaryotes such as mitochondria and chloroplasts. These endocytobionts seem to be essentially integrated into the host's development and metabolism. Such an integration should require a genetic exchange between host genome and symbiont genome in the course of their common evolution. This concept will be called the *organelle endocytobiosis hypothesis* (see Figure 1B). Examples of this phenomenon were reviewed by Margulis[18-20] and Schwemmler.[43,46,47]

II. EXPERIMENTAL RESULTS

The endocytobiosis of the Hemiptera (lice, bugs, cicadas) is one of the few for which important aspects of the structural, functional, and genetic interdependence between eukaryotic host cell (bacteriocyte) and prokaryotic endocytobionts has been studied. It serves here as a better example than a review-like presentation. On the basis of the presented facts, it should be possible then to decide whether the hemipteran endocytobionts only provide a nutritional supplement without gene transfer, or are instead similar to DNA-containing cell organelles with gene transfer. Only those developmental, metabolic, genetic and physiochemical findings which are relevant to this problem will be discussed here.

A NUTRITIONAL ENDOCYTOBIOSIS HYPOTHESIS

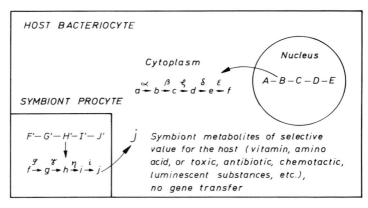

B ORGANELLE ENDOCYTOBIOSIS HYPOTHESIS

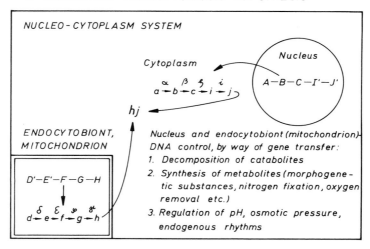

$A \cdots Z$ = Genes of the nucleo-cytoplasm
$A' \cdots Z'$ = Genes of the endocytobiont / mitochondrion
$a \cdots z$ = Substrates or metabolites
$\alpha \cdots \omega$ = Gene-controlled enzymes

FIGURE 1. Scheme of the integration of procyte endocytobionts in eucyte host cells and bacteriocytes (adapted from Margulis[19] and Schwemmler[47]). According to the *nutritional endocytobiosis hypothesis* (A), the symbionts serve only to supplement the host's nutrition; there is no exchange of genes between host and symbiont genome. According to the *organelle endocytobiosis hypothesis* (B), many endocytobionts have, in addition, the function of DNA-containing cell organelles, based on a transfer of genes in the course of their common evolution with the host. They build anabolites using the host's catabolites and thereby regulate the pH, osmolarity, and certain endogenous rhythms of the host.

$$\text{bacteriocyte} \xleftarrow[\;?\;]{\text{gene transfer}} \text{endocytobiont}$$
$$\text{(eukaryotic)} \qquad\qquad\qquad \text{(prokaryotic)}$$

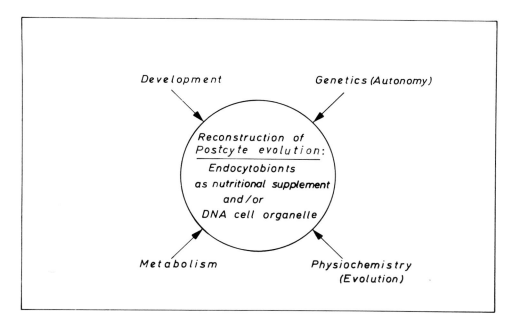

A. Development

The small, plant phloem-sucking cicada *Euscelis incisus* K.,* which belongs to the sister group of the lice (homoptera), harbors two types of essential endocytobionts. These were labelled by Müller[21] the *primary symbiont a* and the *auxiliary symbiont t,* which are considered to represent an intracellular group of primitive prokaryotic microorgansms, the *Protoplastoids*[31,40] (see also Figure 6). Type *a*, which is also found in lice and in primitive forms of lice and cicadas, can occur alone. The type *t* appears, when present, only together with a type *a* symbiont. A third, rickettsia-like endosymbiont is not essential for the cicada. It can be referred to as the *accessory symbiont KR_E*. The three different types of endocytobionts are passed on to the next host generation by way of egg infection (Figure 2). This occurs with the help of a special extracellular *infectious stage* enclosed in a double membrane.

The host inserts the infectious stage of the primary and auxiliary endocytobionts in the form of a so-called symbiont ball into the area between the egg cell and the egg coat at the posterior pole. The fate of the endocytobionts then varies.[14] The *a*-endocytobionts are soon taken up by polyploid host cells of the egg plasma, the so-called *a_1-bacteriocytes,* These, however, disintegrate during a later stage. The released *a*-endocytobionts then enter the final *a_2-bacteriocytes,* which are cells of gonadic origin with two cell nuclei. Only during this stage the *t*-endocytobionts are phagocytized by polyploidic *t-bacteriocytes* which have previously wandered down from certain relatively far-removed abdominal segments of the embryo. Within these bacteriocytes the *a*- and *t*-endocytobionts then differentiate into their reproductive forms, the so-called *vegetative stages,* which are surrounded by a triple membrane.

In the further course of development, the *a*- and *t*-bacteriocytes merge in a strictly predetermined fashion to form a common symbiont organ, the *bacteriome*. This organ then divides to build two lateral bacteriomes which lie on the left and right side of the host abdomen. Each consists of a central *t*-bacteriome surrounded by an *a*-bacteriome. The lateral

---

*   The earlier name *Euscelis plebejus* is no longer accepted. According to Metcalf's general catalog, 1967, the species must now be called *Euscelis incisus* Kbm. 1858. The name *E. plebejus,* introduced in 1806 by Fallén, was based on the rules of priority, but must be dropped because of primary homonomes (see Strübing, H. and Hasse, A., *Zool. Beiträge Berlin,* 20 (3), 1974).

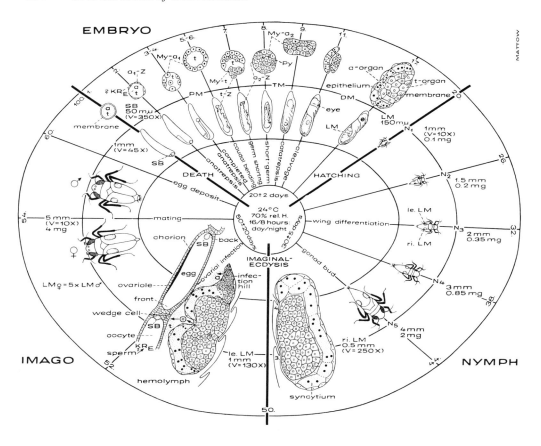

FIGURE 2.   Analysis of the developmental cycle of the host, the cicada *Euscelis incisus,* and its symbionts (details see text; embryonic stage according to Körner[14]). The developmental cycle consists of the embryonic or egg stage, the larval or nymphal stage, and finally the imaginal or adult stage. The inner circle of the figure represents the stages of the host, the outer, the corresponding endocytobiotic structures. The time span refers to standardized breeding conditions. The data of the figure was mainly obtained by means of cell culture methods. Subsequent histological treatment of material in vivo and in vitro was carried out by means of normal and ultrastructural sections and by live observation and preparation according to standard methods.

a	= primary endocytobiont a	My-t	= t - bacteriocyte
a_1-Z	= a_1 - cell	N	= nymphal stage
a_2-Z	= a_2 - cell	PM	= primary bacteriome
DM	= double bacteriome	Py	= pycnosis
I	= imaginal stage	SB	= symbiont ball
KR_E	= rickettsia-like accessory endocytobiont	TM	= transitory bacteriome
LM	= lateral bacteriome (ri = right, le = left)	t	= auxiliary endocytobiont t
My-a_1	= a_1 - bacteriocyte	t-Z	= t - cell
My-a_2	= a_2 - bacteriocyte	V	= enlargement

bacteriomes are still in this state when the embryo hatches and undergo little change in the course of larval development. Further important differentiation occurs only in the bacteriomes of adult females. Ovary cells migrate into the a-bacteriome and form the so-called infectious cell mound (see ''Infektionshügel'', Buchner[2], in which the a-infectious stages mature and from which they are released into the hemolymph. The t-infectious stages are formed at places of contact between the ovarioles and the t-organ in the motile t-bacteriocytes. These migrate into the hemolymph and release the symbionts after bursting. Both infectious stages

Table 1
ANALYSIS OF THE CONTROL MECHANISMS OF LEAFHOPPER
ENDOCYTOBIOSIS[37,38]

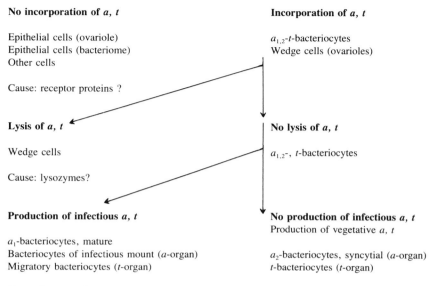

No incorporation of *a, t*

Epithelial cells (ovariole)
Epithelial cells (bacteriome)
Other cells

Cause: receptor proteins ?

Lysis of *a, t*

Wedge cells

Cause: lysozymes?

Production of infectious *a, t*

a_1-bacteriocytes, mature
Bacteriocytes of infectious mount (*a*-organ)
Migratory bacteriocytes (*t*-organ)

Cause: hormones?

Incorporation of *a, t*

$a_{1,2}$-*t*-bacteriocytes
Wedge cells (ovarioles)

No lysis of *a, t*

$a_{1,2}$-, *t*-bacteriocytes

No production of infectious *a, t*
Production of vegetative *a, t*

a_2-bacteriocytes, syncytial (*a*-organ)
t-bacteriocytes (*t*-organ)

Note: Classification of the different cells in the endocytobiotic system, corresponding to the incorporation, lysis, and production of endocytobionts and the possible distribution of lysosomes, hormones, and of recognition-specific receptor proteins of the cell membrane (compare with Figure 2).

reach the ovarioles by way of the hemolymph. Here they infect the egg, before it is enclosed in the chorion by entering special wedge cells. The developmental cycle is thus completed.

The leafhopper host seems to control the cellular specificity of incorporation of its endocytobionts by *a*- or/and *t*-specific systems of membrane proteins and the lysis specificity of the endocytobionts by the amount of antibacterial lysozyme it produces differently in its cells; furthermore, the host regulates multiplication of the endocytobiotic infectious form by direct coupling to its sexual hormone system (Table 1; Schwemmler[37-40]).

The question which arises at this point, is whether this extensive structural interdependence between the cicada host and its symbiont is based on a reciprocal genetic integration of the two systems. In a first step towards answering this question, the DNA molecular weight of the endocytobionts was determined.

B. Genetics
1. DNA Molecular Weight

Since the symbiont ball of a single egg contains only approximately 0.2% of the symbiont content of a single bacteriome, it was not possible to prepare enough symbiont ball material to perform a density and sedimentation analysis. This material was only used for electron microscopic analysis. For the density and sedimentation analyses, approximately 400 bacteriomes per test were used. These bacteriomes had previously been marked with thymidin-6-H^3 and prepared from adult cicada females.[40]

The density analysis of such bacteriome preparations in CsCl gradients produces two bands: one large band, and a smaller, less dense one (Figure 3, IA). Since the smaller band is totally absent in control preparations of absolutely symbiont-free host cells (Figure 3, IB),

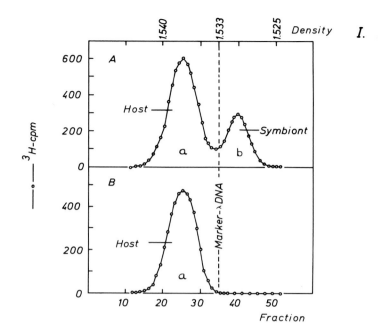

FIGURE 3 (I).   Analysis of host and endocytobiont DNA.[40] Density of thoracic and bacteriomal *Euscelis* DNA. The thorax and bacteriome of cicadas labeled with radioactive (^{3}H)thymidine were lysed with sodium dodecylsulphate (SDS) and centrifuged in cesium chloride (CsCl) gradients containing ethidium bromide. (^{14}C)Thymidine-labeled bacteriophage λDNA was added as a marker. The gradients were then collected in 3-drop fractions, and the DNA was precipitated with TCA. The radioactivity in each fraction was then measured by liquid scintillation counting. The density of the DNA (ρ = 1.533) was used as a basis for the calculation of the density of the thoracic and bacteriomal DNA from the cicadas.

   A: Endocytobiont (b) and host DNA (a) from the bacteriome.

   B: Symbiont-free host DNA from thorax cell (a). cpm = counts per minute.

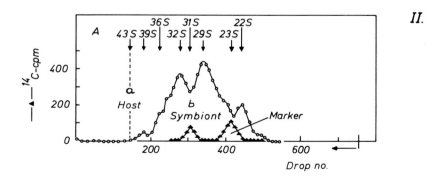

FIGURE 3 (II).   Sedimentation of *Euscelis* enocytobiont DNA. The purified endocytobiont and host DNA from (I) was extracted and dialyzed, then sedimented in a sucrose density gradient. The marker used here was (^{14}C)thymidine-labeled circular DNA (tightly coiled λdv DNA = 31.5 S; uncoiled DNA = 23.2 S). 10-drop fractions were again precipitated with TCA and the radioactivity determined by scintillation counting.

   (a) Host DNA (density ρ = 1.700).

   (b) Endocytobiont DNA (density ρ = 1.707).

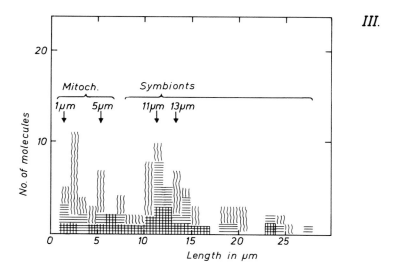

FIGURE 3 (III).   Electron microscopic conformational analysis of the endocytobiont DNA molecules. The various fragments were divided into length classes with 1 μm intervals, and the frequency distribution was determined. The molecules came from three different purified endocytobiont DNA fractions:
⊞: Endocytobiont DNA from the symbiont ball.
≡: Endocytobiont DNA from the male bacteriomes (density ρ = 1.711).
}}} : Endocytobiont DNA from female bacteriomes (density ρ = 1.700).

it was assumed that it represents the endocytobiont DNA. By comparison with a marker a buoyant density of ρ = 1.707 for the host cell DNA, and of ρ = 1.700 for the endocytobiont DNA, was determined.

The DNA fractions of the host and symbiont were then purified several times in CsCl gradients until visual fluorescence tests at 360 nm showed no further noticable impurities. The fractions were then subjected to sedimentation analysis in neutral sucrose gradients. The host cell DNA was shown to be a homogenous population of fragments of about 6×10^7 daltons. The fraction of endocytobiont DNA, on the other hand, was determined to consist of two dominating, likewise linear molecular populations, i.e. one of 22×10^6 and one of 26×10^6 daltons, as well as a number of small components; 11×10^6, 43×10^6, 50×10^6, and 32×10^6 daltons (Figure 3, II).

Finally, electron microscopic or conformational analyses were carried out with the purified bacteriome DNA, as well as with the unpurified, but also radioactively marked symbiont ball DNA. The preparations of host DNA contain randomly broken DNA fragments, which are however all significantly larger than the individual species of endocytobiont DNA molecules. The DNA histogram of the endocytobiont fragments shows, in contrast, two series of oligomers: 11, 22, 32, 43×10^6 and 26, 50×10^6 daltons, whereby the species of 22×10^6 and 26×10^6 dominate (Figure 3, III). These two series are in good agreement with the sedimentation analyses, since DNA molecules of a double size have double the radio-activity. Control preparations with mitochondrial DNA (ρ = 1.707) contain significantly shorter fragments and a totally different frequency distribution. The most attractive interpretation of this data is that the two dominating types of molecules represent the a- and t-endocytobiont DNA. Accordingly, the cicada endocytobionts have a DNA molecular weight of maximally 10^8 daltons. This corresponds to approximately 150,000 bases with an informational capacity for coding 150 proteins with a mean molecular weight of approximately 40,000.

In comparison, mycoplasms and rickettsiae, which are among the prokaryotes with the smallest known genome, have a DNA molecular weight which is ten times greater at about 10^9, corresponding to a coding capacity sufficient for approximately 2000 proteins (see Figure 6). It would thus appear to be interesting to study how great the autonomy of the endocytobionts is from their hosts, in view of their low DNA molecular weight. To this end, in vitro culture experiments with the endocytobionts outside of their host cell were undertaken.

2. Endocytobiont Autonomy

Extracellular, in vitro culture experiments with the endocytobionts were performed in a tissue culture medium which was otherwise successfully employed for the cultivation of host cells[33,35,36] (Figure 4). However, neither reproduction nor growth of the endocytobionts occurs in such cell-free cultures. Only the addition of certain cell metabolites, e.g., ATP, into the medium, or the presence of host cells in the cultivation vessel leads to a few solitary divisions. However, these division stages also show no growth.

If, however, in vitro cultures of symbiont-free host cells, so-called *pseudobacteriocytes*, are infected with the endocytobionts formerly held in vitro, the latter are phagocytized by the pseudobacteriocytes and immediately begin to divide and grow intracellularly. This corresponds to in vivo conditions, where proliferation of the endocytobionts also has been observed only intracellularly, never extracellularly. Apparently, the endocytobionts are only capable of identical reproduction within the host cell. The endocytobionts are, genetically speaking, only semiautonomous, a feature which they share with the DNA-containing cell organelles, for example mitochondria and plastids. Thus, the in vitro behavior of plastids is comparable with that of the endocytobionts[49] (Figure 5). The supposed symbiont's DNA molecular weight of 2.2 to 2.6 × 10^7 daltons lies exactly between that of the mitochondria, with approximately 10^7 daltons, and that of the plastids with about 10^8 daltons[48] (Figure 6).

In the further course of work, we attempted to investigate whether the host is dependent in turn on the endocytobiont for its identical reproduction.

3. Host Autonomy

By feeding antibiotica to larval or adult hosts, one can totally eliminate their endocytobionts[32-37] (Figure 7). Hosts which have lost their symbionts in this manner die within a short period of time. The elimination of the symbionts during egg development also has devastating consequences. If egg infection is prevented by applying lysozyme to the cicada female, symbiont-free eggs are laid. These develop into *cephalothorax embryos* without an abdomen[34,38,39,46,47] (Figure 8). Spontaneously symbiont-free eggs also develop into cephalothorax embryos (Sander 1959). If, however, the symbiotic infectious mass is separated from the egg after egg laying, normal embryos with symbiont-free bacteriomes, so-called *pseudobacteriomes*[25] develop (the symbiont-free host cell cultures are won from these pseudobacteriomes). The provisions for endocytobiosis are thus pre-programmed before the egg is laid: the endocytobionts have no more influence on this process. Before the egg is laid, however, the endocytobionts seem to directly or indirectly influence the activities of the factors which control the formation of the abdomen.[46,47] The endocytobionts thus represent a kind of complement to the host genome which must be passed on from generation to generation, thereby enabling the host to reproduce identically. If this process is interrupted, the host dies directly or cannot develop into a viable organism. Thus the host is, from a genetic standpoint, also only a semiautonomous system. What are the metabolic consequences of this extensive reciprocal genetic interdependence?

D. Metabolism

A comparison of the minimal diet (= completely synthetic food meeting minimal re-

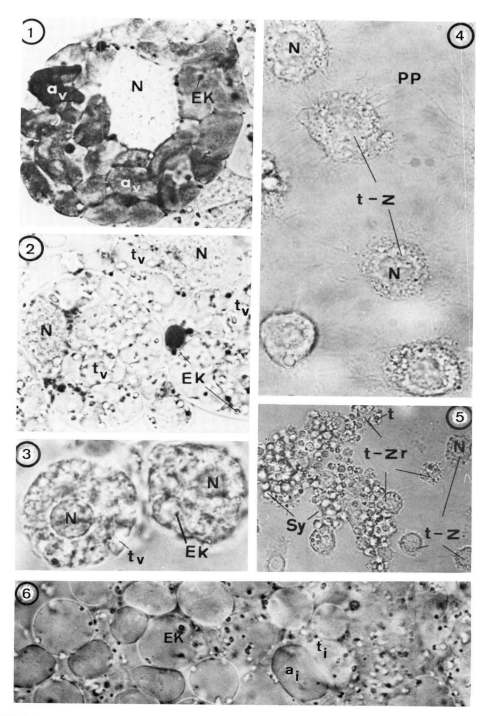

FIGURE 4. In vitro reinfection tests with cicada endocytobionts.[33,49] (1) Vitally stained a bacteriocyte with *a*-endocytobionts in vitro (× 786). (2) Vitally stained *t*-bacteriocyte with *t*-endocytobionts in vitro (× 786). (3) Two-week old in vitro culture of *t*-bacteriocytes carrying symbionts (× 786). (4) Two-week old culture of noninfected *t*-cells (prospective *t*-bacteriocytes), (× 748). (5) Reinfected *t*-cell cultures with in vitro conserved endocytobionts (see Figure 5), (× 262). (6) In vitro conservation of infectious endocytobionts of the symbiont ball for one week (× 786). $a_{i,v}$ — primary symbiont *a* (infectious, vegetative); EK — symbiotic inclusion body; HS — heterochromatin; N — nucleus; PP — pseudopodia; Sy — symbiont; $t_{i,v}$ — auxiliary symbiont *t* (infectious, vegetative); and *t*-Zr — *t*-cell (reinfected).

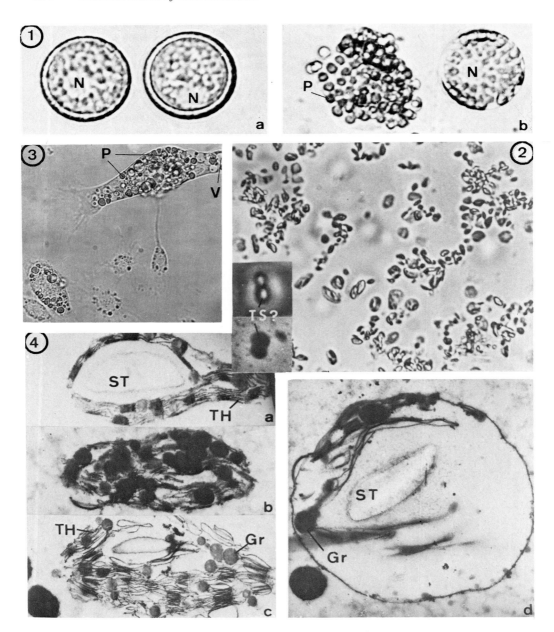

FIGURE 5. In vitro reinfection tests with sterile plastids.[133] (1a) In vitro fraction of free, etiolized protoplasts from wheat plants kept in the dark ($\times$ 659). (1b) In vitro fraction of free plastids (protochloroplasts) after rupture of the protoplast cell membrane ($\times$ 659). (2) In vitro conservation of wheat plastids (chloroplasts) in a synthetic tissue culture medium for one week with possible division stages ($\times$ 489, detail $\times$ 848). (3) "Infection" of animal cultures with the plastids kept in vitro ($\times$ 323). (4) Analysis of the fine structure plastids kept in vitro for different long periods (chloroplasts): all transitions from intact (a) and poor conservation (b) to total degeneration (c,d) are shown ($\times$ 8,075). Gr — granula, N — nucleus, TH — thylakoid, TS — division stages, ST — starch particle, and P — plastid.

quirements of quality and quantity) of normal cicadas with that of symbiont-free or semi-symbiont-free cicadas shows that the endocytobionts synthesize, in normal cases, choline, almost all essential amino acids, and approximately half of the necessary vitamins for their

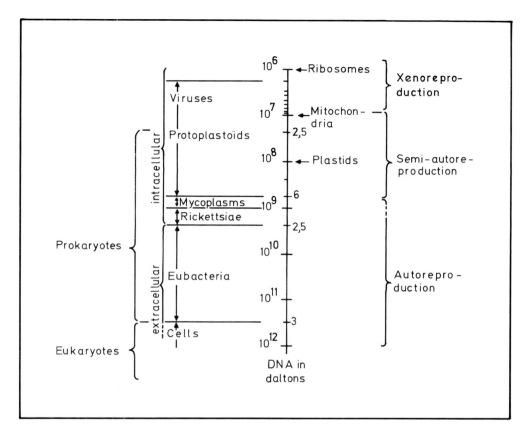

FIGURE 6. Classification of eukaryotes, prokaryotes, and viruses according to their DNA quantity (in daltons) in comparison to the DNA-molecular weight of the DNA-containing or RNA-containing cell organelles (ribosomes, mitochondria, plastids) and the degree of their ability for self-reproduction (details see text, Schwemmler[31]).

host.[34,47] Feeding experiments with radioactively marked substances show furthermore that the endocytobionts are capable of producing cholesterol as well as those amino acids which contain sulfur, using only sulfate as a sulfur source.[5,6,12] The endocytobionts are therefore essential factors for the host's anabolism. The endocytobionts probably use host catabolites to synthesize metabolites for themselves and their hosts. This is inferred from the observation that animals freed from symbionts suffer an abnormal accumulation of wastes such as uric acid, particularly in their fat bodies.[34]

During the developmental cycle, the host and symbiont furthermore undergo characteristically correlated changes in their respective pH values. The host goes from alkaline during the early embryonal phase to acid during the late embryonal and the entire larval phase and finally back to alkaline in the adult phase, whereas the pH changes in the endocytobiont run parallel in the other direction, changing from acid to alkaline and back to acid (Figure 9A). The osmotic values of host and symbiont also fluctuate several times during the developmental cycle in a similar, antagonistic manner. Finally, the pH and osmotic pressure values in the hemolymph and in the egg plasma of normal animals are lower than the corresponding values of symbiont-free cicadas. This evidence appears to indicate a regulation of pH and osmotic pressure between host and symbiont (Figure 9B); however, such a regulation has not yet been proven with absolute certainty. The energy metabolism and endogenous rhythms of host and symbiont also appear to be correlated. Whenever the host cell exhibits a high energy level, that of the symbiont is low, and vice versa[42,44,45] (Figure 10).

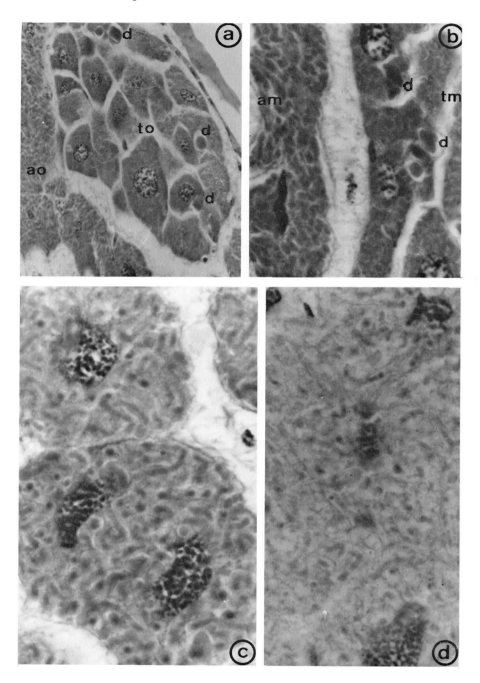

FIGURE 7. Elimination of endocytobionts *a* and *t* by means of antibiotica (tetracycline[37,38]). (a,b) Artificially induced double infection of *t*-bacteriocytes of the t-bacteriome by *t*- and *a*-endocytobionts (a: × 420; b: × 1050). (c) *a*-bacteriocytes of *a*-bacteriome in which *a*-endocytobionts are partially lysed (× 1050). (d) *a*-bacteriocytes, completely lacking *a*-endocytobionts (× 1050).

ao and to = *a*-, *t*-bacteriome; am and tm = *a*-, *t*-bacteriome; d = double infection.

In conclusion, it can be determined that the endocytobionts appear to be so firmly integrated into their cicada hosts, not only structurally and functionally, but, in particular, genetically, that they cannot be removed without causing extensive mutual disturbance of the system.

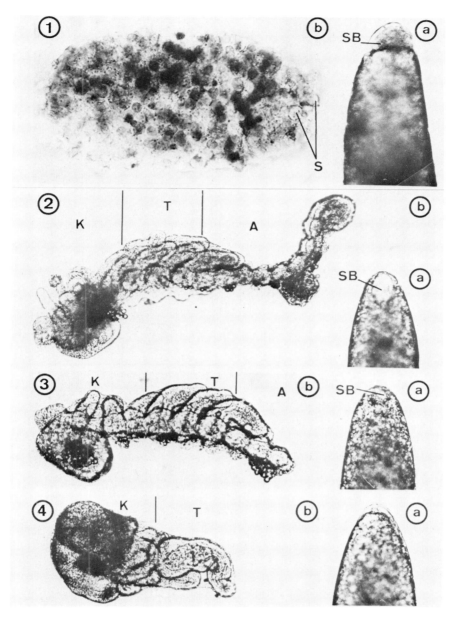

FIGURE 8. Prevention of the infection of Euscelis incisus eggs.[33,35-39]

(1a) Posterior pole of an x-rayed egg, about 2 to 3 days old, with diffuse symbiont ball. Such eggs can develop into symbiont-free cephalothorax embryos (× 100).

 (b) Non-invaginated, darkened SB of a three-day old egg, x-ray-treated (× 420).

(2a) Posterior pole of a normal egg with normal SB (∼ 90,000 μm³) (× 100).

 (b) 6-day-old normal embryo with normal primary bacteriome from the above egg (× 100).

(3a) Posterior pole of an egg with reduced SB (∼ 20,000 μm³) from a female treated with tetracycline before oviposition. Lysozyme-treated females produce similar eggs, with reduced SB (× 100).

 (b) 6-day-old embryo with reduced SB from an egg laid by a lysozyme-treated female (× 100).

(4a) Posterior pole of an egg without a symbiont ball, laid by a tetracycline-treated female. The results are similar after lysozyme treatment (× 100).

 (b) 6-day-old embryo cephalothorax from an x-ray-treated egg, similar to embryos from symbiont-free eggs, without an abdomen, from females treated with tetracycline or lysozyme (× 100).

A	= abdomen	K	= head	SB(r)	= symbiont ball (reduced)
E	= symbiont inclusion body	Pd	= denatured polar plasma	S	= symbionts (a, t)
Ex	= extremity	PM	= primary bacteriome	T	= thorax

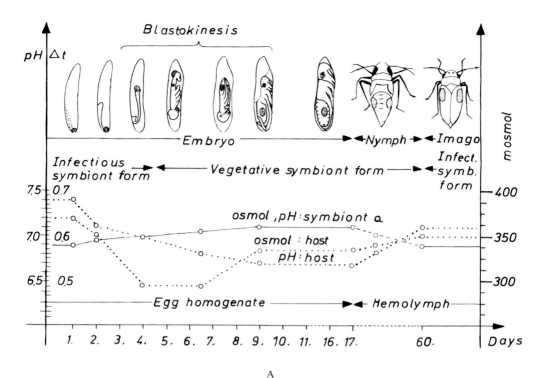

A

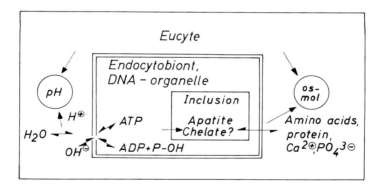

B

FIGURE 9.   (A) pH and osmol values of the host cicada (eggs, larvae, imagos) and its symbionts (infectious and vegetative forms). Data points represent the averages of three measurements. Scatter for the pH and osmol values were at most 0.2 units. Host pH values are always on the acid side when symbiont values are alkaline, and vice versa. Osmol values of the host and symbiont are also antagonistical;[39,41b,47,53] sketches of the embryonic stages from Körner.[14]

(B) Postulated mechanism of pH and osmotic regulation of the eucyte by the endosymbiotic procyte or mitochondrion. pH values might be reversibly regulated by withdrawal of OH⁻ due to ATP hydrolysis whereas osmotic pressure could be reversibly regulated by withdrawal of ions and molecules during the formation of inclusion bodies.[35]

Note: the vegetative forms of the cicada symbionts have three membranes.

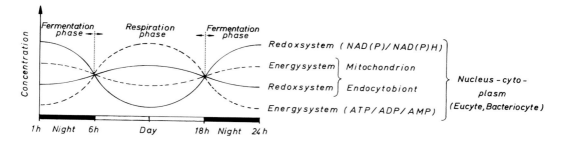

FIGURE 10.   Idealized scheme of the working hypothesis that the endogenous rhythms in the energy metabolism and redox system of the cicada egg are counterbalanced by the rhythms in its endocytobionts. The hypothesis implies mutual regulation between the host cell and its symbionts. We suspect that the relationship is analogous to that in the mitochondria.

E. Probable Gene Transfer

The low DNA molecular weight and the genetic semiautonomy of the endocytobionts indicate that, in the course of the common evolution with their hosts (see Section III), a transfer of genes from the symbiont genome into the cicada genome probably took place. Accordingly, part of the metabolites which the endocytobionts need for reproduction are coded by the nucleo-cytoplasm system of the host and are also synthesized there. The semiautonomy of the host also indicates a possible gene transfer in the opposite direction, that is from the host genome into the symbiont genome, resulting in the coding and synthesizing of essential host metabolites by the endocytobiont DNA.

In this context it is interesting to note that the endocytobiont of the single cell organism *Euplotes aediculatus,* called *omikron,* has a DNA molecular weight of only 0.5×10^9 daltons, corresponding to about one-fourth of non-parasitic bacteria.[9] This endocytobiont is structurally and functionally so extensively integrated into the host system that it has been considered by many researchers to be a DNA-containing cell organelle. Observations of metabolic products and gene products of the unicellular organism *Paramecium aurelia* give indirect evidence of gene transfer between the host and symbiont genome.[22]

The only direct evidence to date of gene transfer has been provided by Schell and his co-workers[28] for the endocytobiont DNA of the bacteria *Agrobacterium tumefaciens* and the genome of their plant cells. The DNA information of the endocytobionts is integrated into the host genome in the form of plasmids. Following transfer of the plasmid, the plant cell develops into a crown gall; the information for this morphogenetic change is provided by the plasmid.

The mutual structural, functional, and genetic integration of such endocytobionts in the nucleo-cytoplasm system calls to mind an analogous situation in the case of the mitochondria. Both endocytobionts and mitochondria synthesize anabolites for their system by using its catabolites. In doing so, both apparently regulate the pH and osmotic pressure values of their nucleo-cytoplasm system. The various inclusion bodies found in the endocytobiont and mitochondria, among them Ca^{2+}, HPO_4^{2-}, amino acids, and/or proteins, are the most visible manifestation of this regulative activity. Furthermore, both appear to have a regulative influence on certain endogenic rhythms of their particular systems. Thus, each would seem to originate from a kind of intracellular ecosystem, in which the endocytobionts and mitochondria mediate as decomposers between the consuming nucleo-cytoplasm system and its nutrition-producing environment, on the chemical, as well as on the physiological level (see also Figure 10; Chapter 7, III; and Chapter 8, IV).

According to the endocytobiosis hypothesis, such endocytobionts are comparable to DNA-containing organelles. One possibility of falsifying or verifying this hypothesis would be to analyze the genetic system of the cicada endocytobionts. In order to determine possible homologies to the mitochondrial system, endocytobiont DNA or RNA could be cloned and subjected to sequencing or hybridization. Cicada endocytobionts should prove to be particularly suited for such studies since, as a very old phylogenetic group, they possibly represent a kind of living fossil in the form of a missing link that has survived for millions of years under the protection of its cicada host (see Section III).

Since a prerequisite for gene transfer is a long common evolution of host and symbiont, we should examine the necessary ecological conditions for such an evolution. For this purpose we must now consider the physiochemical correlations between insect host (consumer), its plant food (producer) and the procyte endocytobionts (decomposers) in more detail.

F. Physiochemistry

The qualitative and quantitative compositions of the physiochemical types and their correlation with the metabolic types of the procyte have already been discussed in Chapter 6, II. D. The reader is advised at this point to briefly review that section. The extra- and intracellular fluids of plants and animals also have similar characteristic physio-chemical values, including those for pH, osmol, inorganic ions, and organic molecules. The *concentration relationships* of the ions and molecules are constant for a given organism. Duchâteau[4] was the first to evaluate systematically such relationships in different insect hemolymphs, in order to establish hemograms. Sutcliffe[54] grouped the different hemograms and those of other animal fluids into three hypothetical basic phylogenetic types (Figure 11). These were later expanded to include the physiological values of pH and osmol, and their application was extended to plants.[32,33,39] These physiochemical types can be qualitatively and quantitatively characterized (Figure 12, Table 2). Among the inorganic ions (left part of the diagram), going from type I to type III, Na^+ is reduced in favor of K^+, and Ca^{2+} or Mg^{2+} and Cl^- in favor of HPO_4^{2-}. Among the organic molecules (right side of the diagram), sugar is reduced in favor of organic acids, including the amino acids.[41b,47,53] Thus an intra- or extra-cellular animal or plant fluid can be roughly assigned to one of the physiochemical types on the basis of its pH and osmol values, or the ion and molecule concentration relationships. To be sure, one and the same animal or plant may belong to several physiochemical types. The crab *Astacus*, for example, has hemolymph type I, but its muscle cells have cell type II/III. There are also characteristic differences between animals and plants. The sugar concentration of the plant phloem type III, for example, is higher than the amino acid concentration, in contrast to the corresponding animal hemolymph type. The physiochemical type of extracellular animal and plant fluids is determined by the sum of the various physiochemical types of all the cells comprising the entire system. Of course, superimposed regulatory mechanisms such as anal pumps, Malphigian glands, liver, etc. must be taken into account. Lifestyle and nutrition also have a certain influence on the physiochemical type. Thus the division into three physiochemical types is a very handy, but highly simplified representation of the actual facts. However, this concept can lead to concrete and useful results, as can again be demonstrated with the hemiptera as an example[32,33,41,52]

In the course of their evolution, the *hemiptera* have developed various physiochemical types in connection with their habitats and nutrition. This can be clearly seen in the cation ratios of some of their typical representatives[8] (Figure 12). The water bug *Corixa* feeds on algae of cell sap type I/II and fish of hemotype I. The bug's cation ratio places it in hemotype I. The water bug *Neotrephes* also leads an aquatic life, but feeds preferentially on salamander eggs of cell sap type II and salamander larvae of hemotype II and only occasionally on crustaceans of hemotype I. This bug has developed hemotype II, in agreement with its nutrition. The leaf bugs *Palomena* and *Graphosoma* have hemotype III, corresponding to their terrestrial habitat and nutrition, the fruits of angiosperms, with fruit sap type III. The predatory bugs *Triatoma* and *Rhodnius* occupy terrestrial, but relatively moist habitats and

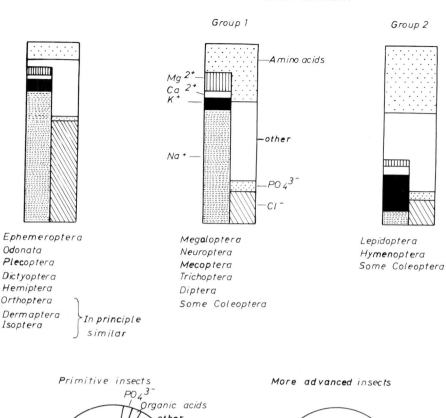

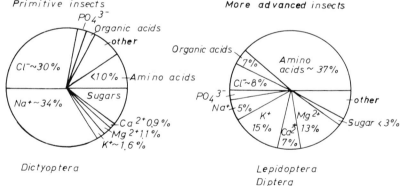

FIGURE 11. Sutcliffe[54] summarized the various data from analyses of insect hemolymph into three phylogenetic types of hemograms. Landureau[15] proposed two such types. The amounts of the hemolymph substituents are expressed as a percent of the total in meq/ℓ.[41b]

feed on mammalian blood of type I. On the basis of their cation ratios, they belong to hemotype II. The scale insect *Rastrococcus* and the cicada *Euscelis* occupy a moist terrestrial transition zone, but they feed on angiosperms of phloemtype III and have hemotype II. *Euscelis* stages which have been artificially freed of their symbionts even have the pH and osmotic characteristics of hemotype I.

It is now interesting to compare the cation ratio of the hemolymphs of the above-mentioned hemipterans, the consumers, with the cation ratios of their food, the *producers* (Table 3). This comparison will show whether the type of hempteran symbiosis is dependent on the

Table 2
QUALITATIVE SCHEME OF THE THREE POSTULATED BASIC PHYSIOCHEMICAL TYPES

Physiochemical type	I	II	III
Physiological characteristic	pH ∿ 8 ∿ 450 mosmol	pH ∿ 7 ∿ 350 mosmol	pH ∿ 6 ∿ 250 mosmol
Chemical characteristic	Inorg. > Org. Na > Mg Na > Ca Na > K K > Ca K > Mg Mg > Ca Cl > PO₄ Sugar > Org. acids	Inorg. ∿ Org. Na > Mg Na > Ca Na > K K < Ca K < Mg Mg < Ca Cl ∿ PO₄ Sugar ∿ Org. acids	Inorg. < Org. Na < Mg Na < Ca Na < K K < Ca K < Mg Mg < Ca Cl < PO₄ Sugar < Org. acids
Examples — Animals	Hemiptera (hemolymph) *Euscelis*, asymbiotic *Corixa* Crayfish (hemolymph) Intestinal cells (soluble part)	Hemiptera (hemolymph) *Euscelis* *Triatoma* Amphibians (blood) Muscle cells (soluble part)	Hemiptera (hemolymph) *Palomena* *Graphosoma* Cyanocytes, some (soluble part)
Examples — Plants	Algae, anaerobic I/II*(soluble part) Root cells (soluble part)	Mosses (cell sap) II/ III* Stem cells (soluble part)	Seed plants (cell sap) Leaf cells (soluble part)
Examples — Microbes	Clostridia (soluble part) Primary symbionts I/II*(soluble part)	Eubacteria (soluble part) Auxiliary symbionts II/III*(soluble part)	Blue-green algae, aerobic Companion symbiont (soluble part)
Examples — Organelles		Mitochondria (II/III)* (soluble part)	Chloroplasts (soluble part)

Note: Qualitative scheme of the three postulated basic physiochemical types of (extra- and intracellular) fluids of animals, plants, microbes and cell organelles, characterized by their pH and osmol values and by the relative concentrations of their inorganic molecules.[39,41] (For the data from cation determinations, see Table 3.)

*   In intermediate types the physiochemical characteristics represent transitions between the main types.

relationship between its hemolymph type and the physiochemical type of its nutritional source.[8] The cation ratios of *Corixa* and *Neotrephes* correspond in principle to those of the producers on which they feed (Figure 13). They also harbor no symbionts, aside from the usual extracellular intestinal bacteria (intestinal flora). *Graphosoma* and *Palomena,* on the other hand, differ from their producers with respect to one cation ratio. They live in a loose symbiosis with extracellular bacteria, which are housed in special folds or crypts of the intestine. The bacteria are passed on to the next generation by smearing the eggs with feces or by coprophagia.*

*   Kopros, Greek = feces, and phagein, Greek = to eat.

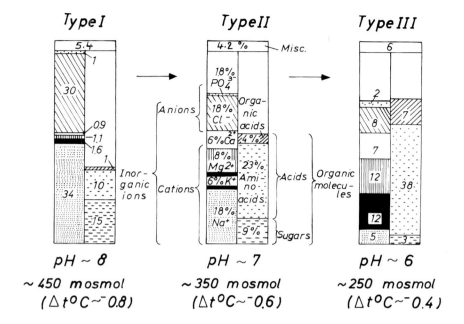

FIGURE 12. The hypothetical three basic physiochemical types of extra- or intracellular animal or plant fluids, as distinguished by the percent of the total osmolarity (meq/ℓ) represented by the various inorganic ions (left side of each bar) and organic molecules (right side of each bar) and by their pH and osmol values.[39]

Note: AA = amino acid.

Triatoma and *Rhodnius* differ from their producers with respect to two cation ratios, and harbor vitally necessary bacterial symbionts intracellularly in the intestinal epithelium and extracellularly in intestinal folds. These symbionts are also passed on by smearing the eggs. *Rastrococcus* differs from its producers with respect to about four cation ratios. It also harbors essential bacterial and fungal endocytobionts, which are present in the fat body of the insect. They are passed on to the next generation by insertion between the egg coat and the egg cells. *Euscelis* differs from its nutrition with respect to at least four cation ratios. The symbiont-free cicadas probably differ with respect to all six ratios. Here we also find the strictest form of obligate endocytobiosis (see above).

It is evident that a symbiosis is always involved when the hemipteran differs from its producer with respect to cation ratios. The greater the difference, the more intense the symbiosis. Thus the degree of endosymbiosis seems to stand in direct relationship to the difference in cation ratios of consumer and producer, on the one hand, and to the level of symbiosis on the other (Figure 14):

- in the case of complete agreement between the cation ratios, i.e., agreement between the physiochemical types of consumer and producer, there is no special symbiosis (endosymbiosis degree 0);
- when there are differences in up to three cation ratios, equivalent to the difference between type I and III, or between type II and III, an exosymbiosis is present (endosymbiosis degree < 3);
- when more than three cation ratios are different, equivalent to the difference between type I and III, an intracellular symbiosis = endocytobiosis is present (endosymbiosis degree >3).

Apparently, the symbionts, as extracellular or endocytobiotic decomposers, mediate chem-

Table 3
SELECTED DATA ON THE CATION CONTENTS OF TYPICAL ANIMALS, PLANTS, MICROBES, CELLS AND DNA-ORGANELLES[41b]

Organisms	meq/ℓ (kg)				Σ	% of the Σ				Physio-chem. type	Author
	Na	K	Mg	Ca		Na	K	Mg	Ca		
1. *Crustacea*											
Astacus (hemolymph)	218	6	1	5	230	94	3	1	2	I	Scholles (1933), Rockstein (1964)
2. *Hemiptera*											
a) *Heteroptera*											
Water-bugs:											
Corixa (hemolymph)	112_S	31_S	4_C	8_C	155	73	20	2	5	I	Sutcliffe (1962), Clark/Craig (1953)
Notonecta (hemolymph)	155_S	21_S	18_C	31_C	225	68	10	8	14	II	Clark/Craig (1953) Sutcliffe (1962)
Land bugs:											
Triatoma (hemolymph)	132_B	10_B	2_C	41_C	185	71	6	2	22	II	Boné (1944), Clark/Craig (1953)
Palomena (hemolymph)	$4^?_M$	42_M	80_M	54_M	180	3	23	44	30	III	Boné (1944), Mullen (1957)
b) *Homoptera*											
Jassidae, genus											
(f.i. *Euscelis*)											
Cinara (hemolymph)	59_S	20_S	30_C	21_C	130	46	15	23	16	II	Jass.:Sutcliffe (1963) Cin.:Clark/Craig (1953)
3. *Dictyoptera*											
Periplaneta (hemolymph)	157	8	4	6	175	90	4	3	3	I	van Asperen/Esch (1956)
4. *Coleoptera*											
Hydrophilus (hemolymph)	119	14	44	23	200	59	7	22	12	II	Duchâteau et al. (1953)
5. *Lepidoptera*											
Phlogophora (hemolymph)	12	35	67	36	150	8	23	45	24	III	Duchâteau et al. (1953)
6. *Pisces*											
Lophius (blood)	189	9	4	8	210	90	4	2	4	I	Forster/Berglund (1953)
7. *Anura*											
Rana (blood)	96	3	5	6	110	88	3	4	5	II	Fenn (1936)
8. *Mammalia*											
Equus (blood)	84	31	3	2	120	70	26	3	1	I	Duchâteau et al. (1953)

	1	2	3	4	5	6	7	8	9	Type	Reference
9. Algae											
Nitella (cell sap)	79	58	18	20	175	46	34	9	11	I/II	Hoagland/Davis (1930)
10. Bryophyta											
Sphagnum (cell sap)	11	33	5	11	60	18	56	8	18	II/III	Zailer/Wilk (1907)
11. Angiospermae											
Abies (press sap)	0,5	4,5	1	4	10	5	45	10	40	III	Councler (1886)
12. Gymnospermae											
Poa (press sap)	1	14	1	2	18	5,5	78	5,5	11	III	Strigel (1912)
Vicia (press sap)	1	24	3	12	40	2	60	8	30	III	Strigel (1912)
Pyrus (press sap)	<3?	14	3	10	<30	<10	47	10	33	III	Kenworthy (1950)
Daucus (press sap)	27	175	36	212	450	6	39	8	47	III	Duchâteau et al. (1953)
Yucca (phloem sap)	20	9320	280	80	9700	0,5	95,5	3	1	III	Tammes (1966) in:Crafts/Crisp (1971)
13. Enterobacteria											
Escherichia coli (soluble part)	484	143	99	374	1100	44	13	9	34	II/III	Senez (1968)
14. Athiorhodaceae											
Rhodospirillum (soluble part)	31	130	>130	<130	±421	7	31	>31	<31	II/III	Stenn (1968)
15. Eukaryotic cells											
Crayfish-muscle cell (soluble part)	—	—	—	—	—	30	60	4	6	II	Rockstein (1964)
Human muscle cell (soluble part)	10	145	40	5	200	5	72	20	3	II/III	Lehninger (1970)
16. Cell organelles (DNA) — μeq/g											
Mitochondrion (soluble part)	46	81	200	<81	<408	12	19	50	<19	II/III	Lehninger (1964)
	7	70	11	6	94	7,5	74,5	11,5	6,5	III	Jacobus/Brierley (1969)
Chloroplast (soluble part)	36	340	>340	>340	>1056	4	32	>32	>32	III	Larkum (1968)
	65	620	103	39	827	7	78	11	4	III	

Note: The numbers given in some cases are rounded off or converted. In some cases, the data are for different representatives of the same species and are taken from different authors. They serve roughly to categorize the individual species in one of the three physiochemical types.[38,39,42,43]

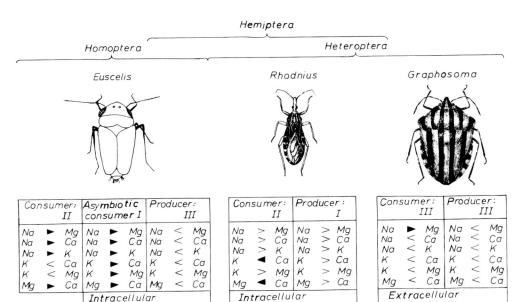

	Hemiptera	
Homoptera	Heteroptera	
Euscelis	Rhodnius	Graphosoma

Consumer: II	Asymbiotic consumer I	Producer: III
Na ► Mg	Na ► Mg	Na < Mg
Na ► Ca	Na ► Ca	Na < Ca
Na ► K	Na ► K	Na < K
K < Ca	K ► Ca	K < Ca
K < Mg	K ► Mg	K < Mg
Mg ► Ca	Mg ► Ca	Mg < Ca
LAND	Intracellular endosymbionts, intraovarial infection	

Consumer: II	Producer: I
Na > Mg	Na > Mg
Na > Ca	Na > Ca
Na > K	Na > K
K ◄ Ca	K > Ca
K > Mg	K > Mg
Mg ◄ Ca	Mg > Ca
Intracellular endosymbionts, coprophagy	

Consumer: III	Producer: III
Na ► Mg	Na < Mg
Na < Ca	Na < Ca
Na < K	Na < K
K < Ca	K < Ca
K < Mg	K < Mg
Mg < Ca	Mg < Ca
Extracellular endosymbionts, coprophagy	

WATER

Neotrephes

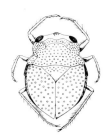

Consumer: II	Producer 1: II	Producer 2: I
Na > Mg	Na > Mg	Na > Mg
Na > Ca	Na > Ca	Na > Ca
Na > K	Na > K	Na > K
K < Ca	K < Ca	K < Ca
K > Mg	K > Mg	K > Mg
Mg < Ca	Mg < Ca	Mg > Ca
Intestinal flora		

Corixa

Consumer: I	Producer: I
Na > Mg	Na > Mg
Na > Ca	Na > Ca
Na > K	Na > K
K > Ca	K > Ca
K > Mg	K > Mg
Mg < Ca	Mg < Ca
Intestinal flora	

FIGURE 13. Comparison of the cation ratios (see physiochemical types) in the hemolymph of insect hosts (consumers) with those in their food (producers) with reference to the degree of symbiosis in typical repesentatives of the homoptera and the heteroptera, which belong to the superorder hemiptera. A solid triangle (►) indicates a difference in the cation ratios of consumer and producer. The greater the number of differences in the cation ratios, the higher the degree of symbiosis (Schwemmler;[39] insect sketches, except for *Euscelis*, from Grassé[8]).

Coprophagy = transmission of symbionts by smearing the eggs with the feces; the larvae become infected after hatching by eating the feces.

The Roman numerals after consumer and producer indicate the physiochemical type.

ically and physiologically between the host and its nutrition. The same is presumably true of the DNA-containing cell organelles (see Chapter 7, V). As has been shown in the case of *Euscelis*, endocytobionts actually are capable of maintaining such a nutritional-physio-

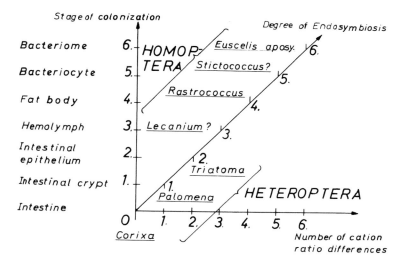

FIGURE 14. Diagram of the degree of endosymbiosis as a plot of the integration level of the symbionts against the difference in cation ratios between the insect host and its food for typical homoptera and heteroptera. With the difference in cation ratios, the level of integration rises and thus, with it, the degree of endosymbiosis. The data for *Lecanium* and *Stictococcus* are incomplete (Schwemmler;[39] data see Table 2, 3).

logical relationship to the consumer and producer. They are able to use the host's poisonous waste materials as starting materials for the synthesis of vital substances which it lacks. They also regulate its intra- and extracellular pH and osmol values to correspond to those of its nutrition. In addition, they seem to be involved in the control of certain endogenous rhythms of the host cell.[44,45]

This physiochemical correlation between host and endocytobiont could only have evolved in the course of millions of years of common development. Therefore, the conditions under which far-reaching gene transfer could have occurred must have existed. This would again be evidence in favor of the organelle endocytobiosis hypothesis. It has previously been assumed that, in accordance with the nutritional endocytobiosis hypothesis, the symbionts served only to compensate one-sided nutritional sources and to remove host waste products. This assumption was in need of revision, if only because the omnivorous *cockroaches* are known to support a highly developed endocytobiosis, although their nutrition is hardly one-sided. Furthermore, it has been found that several substrates, such as the *phloem* of angiosperms, are not at all nutritionally poor media.[55] *Phloem* contains all necessary nutritional elements, although their concentrations may be low.

III. HYPOTHETICAL RECONSTRUCTION OF BACTERIOCYTE EVOLUTION

A short hypothetical outline of bacteriocyte (postcyte) evolution in the hemipteran group, based on the organelle endocytobiosis hypothesis, is presented below.[41] It should be taken as no more than a first approximation, since far too many detailed facts and comparisons with other systems are missing for an exact description.

A. Primary Bacteriocytes

The predecessors of the *Mandibulate* were the last completely aquatic ancestors of the insects.[10] About 600 million years ago (Pre-Cambrium = Archaeozoic), this group split into the aquatic *Crustacea* and the *Tracheata*, which were partly aquatic and partly terrestrial.

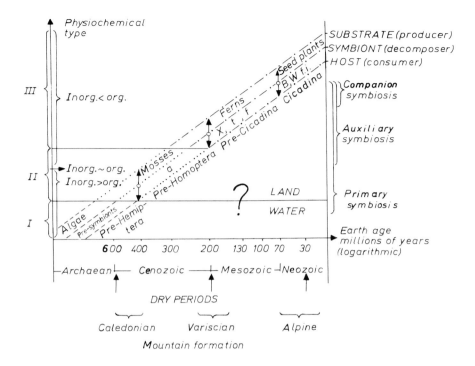

FIGURE 15.   Probable relationships in the evolution of the cicada as host-consumer and its food or substrate-producer and symbiont decomposer. The aquatic hemiptera evolved into moisture-loving, peloridiid-like forms. Either the aquatic or the semiaquatic forms then developed into the modern, purely terrestrial cicada species through integration of primary, auxilary, and companion symbionts, thereby adapting to new producers; i.e., mosses, then ferns, and finally angiosperms. The symbionts must in each case have mediated between the more primitive physiochemical type of the host and the higher physiochemical type of its food (Schwemmler;[39,41] data see Table 2, 3).

About 500 million years ago, (Cambrium/Paleozoic) the *Myriapoda* and the *Insecta* emerged from the latter group. The water bugs *(Hydrocorixae,* subgroup of the Hemiptera) are still or again aquatic today. The *Corixa* of physiochemical type I as well as possibly the hemipteran ancestors, feed almost exclusively on algae which are also type I/II[29] (Table 2). No particular symbiosis has been found in this group (Figure 15). The *Peloridiidae,* which are about 400 million years old, are thought to be the ancestral form of cicadas, or of the homoptera, to which the cicadas belong.[2] They feed on moist forest mosses, physiochemical type II/III, which evolved about the same time that they did. *Hemiodoecus fidelis* (physiochemical type I/II?*), a representative of the Peloridiidae, harbors only the essential *a*-endocytobionts. These bacteria should be part of the basic complement of endocytobionts common to the entire hemipteran group, and are thus called primary symbionts.[21] The primary symbionts are harbored in the host's primary bacteriocytes, where they should be integrated structurally, functionally, and genetically by gene transfer during their long common evolution (see Section 2).

B. Auxiliary Bacteriocytes

The oldest fossil cicadas date from the Permian, about 200 million years ago (Mesozoic). They are the direct ancestors of the forms living today and may have fed on the ferns which developed at that time. These ferns probably had the physiochemical type II/III (?*). The

---

* There are no exact experimental data for this system; the physiochemical classification is therefore hypothetical.

ancestral forms of the cicada group *(Procicadoidae, Profulgoroidae)* are thought to have harbored the essential auxiliary symbionts *t*, *f* and *X* (Figure 15), in the so-called *auxilary bacteriocytes*. There, they should be integrated by gene transfer, too.

C. Companion Bacteriocytes and Accessory Bacteriocytes

Modern cicadas feed primarily on the phloem of the angiosperms, whose evolution began about 130 million years ago (Mesozoic). They have taken up the facultative companion symbionts, including *B* and *W*. This is thought to have occurred in the sister group of the cicadas, the scale insects *(Coccidae)*, in the Tertiary, about 60 million years ago (Cenozoic; Figure 15). The companion symbionts were in each case taken up into *companion bacteriocytes* of the host. For the sake of completeness, it should be mentioned that a fourth group of accessory endocytobionts, such as the rickettsia-like KR_E-symbiont, has very recently been taken up in *accessory bacteriocytes* of the host. This process is still going on among the cicadas. Companion and accessory endocytobionts are integrated only structurally and functionally in their host cell system.

IV. ENDOCYTOBIOSIS AS INTRACELLULAR ECOSYSTEM

An ecosystem is generally understood to be a well-defined biotop with a species variety of organisms which together make up a living community or biocenose; such communities are characterized by the interdependent relationships of these organisms.

According to this definition, one can apply the term intracellular ecosystem whenever the intracellular space is a biotop in which at least one representative each of two different species develops and multiplies in mutual interdependence (in collaboration with Schenk[53]). As in the case of extracellular ecosystems, the individual organisms involved in an intracellular ecosystem must be genetically independent of one another and self-reproductive. Thus, we can refer to endocytobiotic relationships between self-reproductive endocytobionts as oligogenetic, intracellular ecosystems[52] (Figure 16B). Therefore, endocytobiosis represents in general *one of the oldest, smallest and most intensive intracellular ecosystems*. The interdependent relationship between the individuals of an intracellular ecosystem must not necessarily involve a total division of functions according to producers, consumers, and decomposers. On the contrary, one individual can assume more than one of these functions.

The eucyte cannot be classified as an intracellular ecosystem according to this definition (Figure 16B; cf. Chapter 7). The DNA-containing organelles of the eucyte are only semi-autonomous; i.e., their reproduction is neither independent of the nuclear genome, nor does it proceed according to organelle-specific laws (cf. the biogenesis of the chloroplasts, which is regulated by phytochrome). The eucyte is a functional unit in which a variety of regulatory mechanisms are combined to form a central regulatory system.

A partial integration of the endocytobiont's genome into the genome of the host cell goes hand in hand with the transition from endocytobiosis as an intracellular ecosystem, to the eucyte as a central, cybernetic, regulatory system. This is a gradual transition. Leafhopper endocytobiotic forms, which are characterized by semi-autonomous, organelle-like endocytobionts, represent such a missing link between an intracellular ecosystem and a central regulatory system (Figure 16B). Formerly free-living procytes (decomposers) break down intracellularly the food (from producers) for the leafhopper host cells (consumers). Furthermore, they synthesize the essential anabolites for the host cell by utilizing their catabolites or waste products. The endocytobionts thereby appear to regulate the pH, osmotic pressure, and certain internal rhythms of the host cell and so form various cell inclusions (see Section II.B). Consequently, the leafhopper endocytobionts are essential factors in the circulation of materials and in the energy flow of the complex net of food chains.

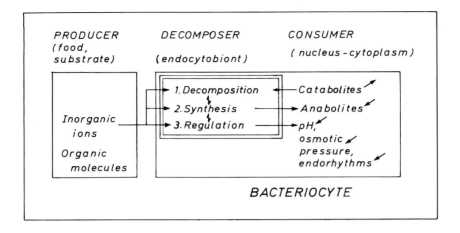

A

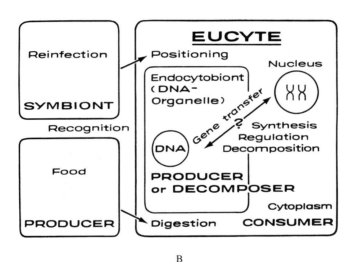

B

FIGURE 16.   (A) Bacteriocyte: representation of endocytobiosis as an intracellular ecosystem: the endocytobionts function as: decomposers of catabolites and waste products of the host cell or of the producer, synthesizers of anabolites for the nucleocytoplasm or consumer, and regulators of the pH, osmotic pressure, and probably of certain endorhythms of the host cell plasm (Schwemmler [42,50]).

(B) Eucyte: representation of eukaryotic-cell with its semiautonomous DNA-containing organelles as a central regulatory, cybernetic system, which is more highly structured than the most complex intracellular ecosystem of endocytobiosis with its diverse components. According to the endocytobiotic cell theory the eucyte has developed phylogenetically of an endocytobiotic oligogenetic ecosystem (Schwemmler [52,53]).

V. MECHANISM OF POSTCYTE EVOLUTION

During the evolution of the primary, auxiliary, and companion bacteriocytes, the *global climate* was arid, due to geological folding processes. The facts and conclusions discussed here suggest that the bacteriocytes developed from aquatic hemipteran predecessors of physiochemical type I, via (or from) moisture-loving, peloridiidean forms of the physiochemical type I/II, to the modern purely terrestrial forms of physiochemical type II/III. This development was probably only possible through the uptake and incorporation of intracellular primary, auxiliary, and companion symbionts. Only in this way were the cicadas able to adapt to the changing plant nutrition, i.e. mosses of physiochemical type I/II, ferns of possible type II/III and seed plants of type III. The phylogenetic function of the edocytobionts seems to have been to mediate chemically and physiologically between the lower physiochemical type of the host and the higher physiochemical type of the food source. This, however, is precisely the function which the endocytobionts had already fulfilled ontogenetically.

In summary, we can observe that there are structural, functional, genetic, and perhaps evolutionary analogies between the endocytobionts of the bacteriocytes and the DNA-containing organelles of the eucytes. Both appear to be surrounded with a host-derived membrane and are also firmly integrated into the developmental cycle of the host. They cannot be removed without serious disturbance. Outside the host cell, the endocytobionts are for their part not viable and the DNA-organelles are not capable of division. Symbionts and organelles have their own systems of DNA, RNA, protein, redox potential, and energy charging, which are different from those of the nucleo-cytoplasm. They are not attacked by the host cell's defense mechanisms (Table 1). Both endocytobionts and the cell DNA-organelles are capable of regulating the pH and osmotic pressure of the host cells. [42,43] The various inclusion bodies in the symbiont's cytoplasm, consisting of Ca^{2+}, HPO_4^{2-}, amino acids, proteins, etc. (see Figure 9B) are a visible manifestation of this regulatory activity. In addition, they synthesize materials needed by the host from its waste products. They mediate between the external and internal milieus of the host, and regulate certain endogenous rhythms in the host cell.

In the course of their evolution, the eukaryotic cell and the cicada bacteriocytes appear to have integrated three groups of comparable endocytobionts: first the vitally necessary primary symbionts of physiochemical type I/II, which are part of the endocytobiotic basic equipment, and which are now in some cases present only in rudimentary form; then the auxiliary symbionts of physiochemical type II/III, which are also vitally necessary; and finally, the companion or accessory symbionts of physiochemical type III, which are not absolutely necessary, and which are not present in all groups (Figure 17). The incorporation of each class of endocytobiont was associated with a change in the host's source of nutrition and milieu. It appears that the endocytobionts mediated chemically and physiologically between the host which was being forced to adapt and the changed substrate/milieu.

Thus postcytes and eucytes both originate from intracellular ecosystems (see Figure 3). Accordingly, both are essential factors in the flow of matter or energy in the complicated network of food chains. The evolution of the postcytes also shows that the process of integration of endocytobionts in the form of DNA-containing cell organelles is far from being completed. Rather, the evolution of the eucyte has been continued by that of the postcyte, and it is still in full swing.

The analogies between structure, function, information, and evolution of DNA-containing cell organelles and of certain endocytobionts raises, on the other hand, two questions which are coupled to one another: Are these cell organelles of endocytobiotic origin and have certain endocytobionts acquired the status of DNA cell organelles? A molecular explanation of the entire complex can be obtained by separate analysis of the nucleo-cytoplasm and endocytobiont parts or DNA-organelle parts of one eukaryotic cell system in separate but parallel experiments. [46,47] Such experiments with the DNA-containing organelles are possible only in exceptional cases, such as with the alga *Acetabularia*. However, they are often possible with the components of an endocytobiotic system, which are more easily separated. *A new synthetic research field, endocytobiology, is thus arising between endosymbiosis and cell biology.* [3,7,16,23,48,53] In endocytobiology combined methods from cell and symbiosis research are employed. Thus, this new field of research can possibly contribute to the analysis

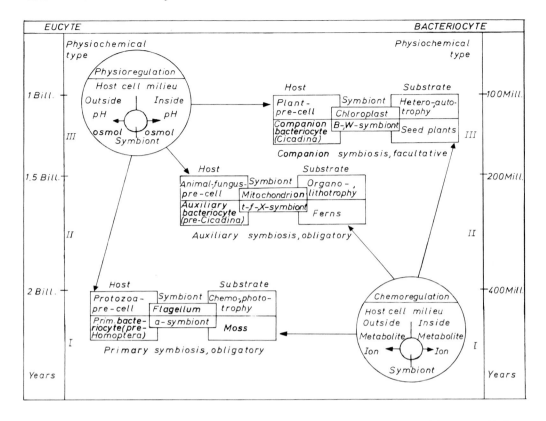

FIGURE 17.   Comparison of the postulated primary, auxiliary, and companion symbioses of the eucyte with those of the hemiptera bacteriocytes. The two different successive symbioses arose with a phase difference of about 1.5 billion years. The symbionts occupying each of the three levels are functionally comparable in the two systems. They regulate chemically the levels of metabolites and ions, and physiologically the pH and osmol, and thus mediate physiochemically between the lower level of the host and the higher level of its nutritional substrate (Schwemmler;[41,41b] details see in text; data see Table 2,3).

of the eukaryotic system. It appears that the basic characteristic of every eukaryotic cell is the interplay between cytoplasmatic fermentation and mitochondrial respiration. This interplay is perhaps the common cause of cell polarity and endogenous cellular rhythms; when it is disturbed, the result may be a kind of cancer-like growth of the cell.[46,47]

REFERENCES

1. **Barton, I.,** *Der Einfluss von Tetracyclin auf Mortalität und Fortpflanzung der Kleinzikade Euscelis plebejus Fall,* Diplomarbeit, Zoologisches Institut Freiburg, 1976.
2. **Buchner, P.,** *Endosymbiosis of Animals With Plant Microorganisms,* Wiley & Sons, New York, 1965.
3. **Cook, C. B., Pappas, P. W., and Rudolph, E. D.,** *Cellular interactions in symbiosis and parasitism,* Ohio State University Press, Columbus, 1980.
4. **Duchâteau, G., Florkin, M., and Leclercq, Y.,** Concentration des bases fixes et types de composition de la base totale de l'hémolymphe des insectes, *Arch. Intern. Physiol.,* 61, 518, 1953.

5. **Ehrhardt, P.,** Nachweis einer durch symbiontische Mikroorganismen bewirkten Sterinsynthese in künstlich ernährten Aphiden (Homoptera, Rhynchota Insecta), *Experientia,* 24, 82, 1968.
6. **Ehrhardt, P.,** Die Rolle von Methionin, Cystein, Cystin und Sulfat bei der künstlichen Ernährung von *Neomyzus (Aulacorthum) circumflexus* Buckt. (Aphidae, Homoptera, Insecta), *Biol. Zbl.,* 88, 335, 1969.
7. **Fredrick, J.,** Origins and evolution of eukaryotic intracellular organelles, *Ann. N.Y. Acad. Sci.,* 361, 1981.
8. **Grassé, P. P.,** *Traité de Zoologie (Anatomie, Systématique, Biologie),* Vol. IX, Masson & Cie, Paris, 1949.
9. **Heckmann, K.,** Omikron, an essential endosymbiont of *Euplotes aediculatus,* in *Endocytobiology. Endosymbiosis and Cell Biology.* Intern. Cell. Proc., Vol. I, Schwemmler, W. and Schenk, H., Eds., Walter de Gruyter, New York, 1980.
10. **Henning, W.,** *Die Stammesgeschichte der Insekten,* Kramer Verlag, Frankfurt, 1969.
11. **Henry, S. M.,** *Symbiosis,* Vol. II, Academic Press, New York, 1967.
12. **Houk, E. J.** Lipids of the primary intracellular symbiote of the pea aphid, *Acyrthosiphon pisum, J. Insect Physiol.,* 20, 471, 1974.
13. **Koch, A.,** Intracellular symbiosis of insects, *Ann. Rev. Microbiol.,* 14, 121, 1960.
14. **Körner, H.,** Die embryonale Entwicklung der symbionten-führenden Organe von *Euscelis plebejus* (Homoptera, Cicadina), *Oecologia* (Berlin) 2, 319, 1969.
15. **Landureau, J. C.,** Obtention de plusieures lignées de cellules embryonnaires de blattes. 2. Colloquio internazionale sulle culture di tessuto degli invertebrati, 1967, Milano, 3–9, 1968.
16. **Lorenzen, H. and Wiessner, W.,** Intracellular and intercellular regulation and recognition in algae and symbionts, Proceedings in: *Ber. Deutsch. Bot. Ges.,* G. Fischer, Stuttgart, in press.
17. **Malke, H. and Schwarz, W.,** Untersuchungen über die Symbiose von Tieren mit Pilzen und Bakterien. XII. Die Bedeutung der Blattiden-Symbiose, *Z. allgem. Mikrobiol.,* 6, 34, 1966.
18. **Margulis, L.,** *Origin of Eukaryotic Cells,* Yale University Press, New Haven, 1970.
19. **Margulis, L.,** Genetic and evolutionary consequences of symbiosis (a review), *Experim. Parasitol.,* 39, 277, 1976.
20. **Margulis, L.,** Symbiosis and Cell Evolution, W. H. Freeman, San Francisco, 1981.
21. **Müller, H. J.,** Zur Systematik und Phylogenie der Zikaden-Endosymbiosen, *Biol. Zbl.,* 68, 343, 1949.
22. **Preer, J., Preer, L., and Jurand, A.,** Kappa and other symbionts in *Paramecium aurelia, Bact. Rev.,* 38, 113, 1974.
23. **Richmond, M. H. and Smith, D. C.,** *The Cell as a Habitat,* Royal Society, University Press, Cambridge, 1979.
24. **Sander, K.,** Analysen des ooplasmatischen Reaktions-systems von *Euscelis plebejus* Fall. (Cicadina) durch Isolieren und Kombinieren von Keimteilen, *Roux Archiv.,* 151, 430, 1959.
25. **Sander, K.,** Entwicklungsphysiologische Untersuchungen am embryonalen Mycetom von *Euscelis plebejus* Fall. (Homoptera, Cicadina). I. Ausschaltung und abnorme Kombinationen des symbiontischen Systems, *Develop. Biol.,* 17, 16, 1968.
26. **Sander, K.,** Pattern specification in the insect embryo, in *Cell patterning,* (Ciba Foundation Symposium 29), Amsterdam, 1975.
27. **Sander, K.,** Anomalien der Oogenese und Embryogenese symbiontenführender Kleinzikaden unter Tetracyclin-Einfluss, *Zool. Anz. Suppl.,* 41, (70. Verh. d. Dtsch. Zool. Ges. Erlangen 1977), 257, 1977.
28. **Schell, J. et. al,** Interactions and DNA transfer between soil bacteria and host plant cells. in *Endocytobiology. Endosymbiosis and Cell Biology. Int. Col. Proc.,* Vol. I. Schwemmler, W. and Schenk, H., Eds., Walter de Gruyter, New York, 1980.
29. **Schlee, D.,** Morphologie und Symbiose: ihre Beweiskraft für die Verwandtschaftsbeziehungen der Coleorrhyncha. Phylogenetische Studien . . . IV: Heteropteroidea (Heteroptera, Coleorrhyncha) als monophyletische Gruppe, Stuttgarter Beitr., *Naturkunde,* 210, 1, 1969.
30. **Schwarz, J. and Koch, A.,** Der Tribolium-Test als quantitativer Test für 8 B-Vitamine (I), *Z. Vitam. Horm. Fermentforschg.,* 12 (4), 291, 1962.
31. **Schwemmler, W.,** Interacellular symbionts: a new type of primitive prokaryotes, *Cytobiologie,* 3, 427, 1971.
32. **Schwemmler, W.,** Physikochemische Korrelation zwischen Lebewesen, Lebensraum und Symbiose am Beispiel der Hemipteren, *Zool. Anz. Suppl.,* 35 (Verh. d. Dtsch. Zool. Ges. 1971), 144, 1972.
33. **Schwemmler, W.,** In vitro Vermehrung intrazellulärer Zikadensymbionten und Reinfektion asymbiontischer Myzetozytenkulturen, *Cytobios* 8, 63, 1973.
34. **Schwemmler, W., Duthoit, J. L., Kuhl, G., and Vago, C.,** Sprengung der Endosymbiose von *Euscelis plebejus* Fall und Ernährung asymbiontischer Tiere mit synthetischer Diät (Hemiptera, Cicadina), *Z. Morph. Tiere,* 74, 297, 1973.
35. **Schwemmler, W.,** Zikaden leben mit dem Erbgut ihrer Symbionten, *Umschau,* 14, 438, 1973.
36. **Schwemmler, W.,** Beitrag zur Analyse des Endosymbiose-Zyklus von *Euscelis plebejus F.,* (Hemiptera, Homoptera, Cicadina) mittels in vitro-Beobachtung, *Biol. Zbl.,* 92, 749, 1973.

37. **Schwemmler W.,** Control mechanisms of leafhopper endosymbiosis, in *Contemporary Topics in Immunobiology,* Cooper, E. L., Ed., Vol. 4, Plenum Press, New York, 1974, 179.

38. **Schwemmler, W.,** Endosymbionts: factors of egg pattern formation, *J. Ins. Physiol.,* 20, 1467, 1974.

39. **Schwemmler, W.,** Zikadenendosymbiose: Ein Modell für die Evolution höherer Zellen? Zur Verifikation der Endosymbiontentheorie der Eukaryontenzelle, *Acta Biotheoretika,* 23, 132, 1974.

40. **Schwemmler, W., Hobom, G., and Egel-Mitani, M.,** Isolation and characterization of leafhopper endosymbiont DNA, *Cytobiologie,* 10 (2), 249, 1975.

41. **Schwemmler, W.,** Allgemeiner Mechanismus der Zellevolution, *Naturwiss. Rdschau.,* 28 (10), 351, 1975.

41a. **Schwemmler, W.,** Evolution der Urzeller: Rekonstruktion des Lebensursprunges, *Natur und Museum,* 108 (2), 49, 1978.

41b. **Schwemmler, W.,** *Mechanismen der Zellevolution. Grundriss einer modernen Zelltheorie,* De Gruyter Verlag, New York, 1978.

42. **Schwemmler, W.,** Die Zelle: Elementarorganismus oder Endosymbiose? *Biologie in unserer Zeit,* 7 (1), 7, 1977.

43. **Schwemmler, W.,** Endocytobiose und Zellforschung, *Naturwissenschaften,* 66, 366, 1979.

44. **Schwemmler, W. and Herrmann, M.,** Oszillationen im Energie-stoffwechsel von Wirt und Symbiont eines Zikadeneies. I. Analyse möglicher stoffwechselphysiologischer Korrelationen beider Systeme, *Cytobios,* 25, 45, 1979.

45. **Schwemmler, W. and Herrmann, M.,** Oszillationen im Energie-stoffwechsel von Wirt und Symbiont eines Zikadeneies. II. Analyse möglicher endogener Rhythmen beider Systeme, *Cytobios,* 27, 193, 1980.

46. **Schwemmler, W.,** Endocytobiose: Modell zur molekularen Analyse von Circadianrhythmik, Eimusterbildung und Krebs, *Naturwiss. Rdschaus,* 33 (2), 53, 1980.

47. **Schwemmler, W.,** Endocytobiosis: General principles, *Biosystems,* 12, 111, 1980.

48. **Schwemmler, W. and Schenk, H.,** Endocytobiology. Endosymbiosis and Cell Biology. Synthesis of Recent Research, *Int. Coll. Proc.,* Vol. I, Walter de Gruyter, New York, 1980.

49. **Schwemmler, W. and Luuring, B.,** Methodical analogies between endocytobiont and DNA-organelle research. in Origins and Evolution of eukaryotic intracellular organelles. Fredrick, J., Ed., *Ann. N.Y. Acad. Sci.,* 361, 284, 1981.

50. **Schwemmler, W.,** Experimental consequences of endocytobiotic studies for eukaryotic cell research, *Ber. Deutsch. Bot. Ges.,* 94, 591, 1981.

51. **Schwemmler, W.,** The periodic system of cells and the endocytobiotic cell theory, *Acta Biotheoretica,* 31, 45, 1982.

52. **Schwemmler, W.,** Analysis of possible gene transfer between an insect host and its bacteria-like endocytobionts, *Int. Rev. Cytol.,* 14, 247, 1983.

53. **Schwemmler, W.,** Endocytobiosis as intracellular ecosystem, in *Endocytobiology II. Intracellular Space as Oligogenetic Ecosystem, Proc.,* Vol. II, Schenk, H. and Schwemmler, W., Eds., Walter de Gruyter, Berlin, New York, 1983, 363.

54. **Sutcliffe, D. W.,** The chemical composition of hemolymph in insects and some other arthropods in relation to their phylogeny, *J. Comp. Biochem. Physiol.,* 9, 121, 1963.

55. **Ziegler, H.,** La sêve des tubes criblés, in *Traité de Biologie de l' abeille,* Vol. III, Mason et Cie, Paris, 1968.

Chapter 9

GENERAL MECHANISMS OF EVOLUTION

The basic information on cell evolution presently available through induction and deduction has been presented in the preceding chapters. The mechanisms of the separate phases of evolution have been thoroughly discussed. Now we should like to derive general evolutionary principles from these individual mechanisms, such as the principles of modular construction, phase, periodicity, and triality.

I. MODULAR PRINCIPLE

We have already mentioned the modular principle in the introductory chapter on the goals and problems of evolution research[26,28,29] (see Chapter 1, Figure 3). According to this principle, each higher level of evolution is reached by the integration of different representatives (mutations, variants) of the next lower level. The result would be a *hierarchy* of component systems of ever increasing *complexity*, something like a child's set of nesting boxes. We can differentiate at least eleven such systems, as summarized in Figure 1. The modular principle is generally accepted for the systems of atoms, molecules, biomers, preorganelles (abioids), multicellular organisms (humans), and cultures. It is also assumed that elementary particles are composed of sub-particles (quarks*, antiquarks*), although this is not universally accepted.[8,18,25] The development of eucytes from various procytes is the subject of the endocytobiotic cell theory. It is in principle no more than the application of the modular principle to the cellular level. The endocytobiotic cell theory, at least in its modest form (only mitochondria and chloroplasts originated from formerly free-living procytes), is becoming widely accepted. It seems very unlikely that precyte evolution should be the only form of evolution not to have proceeded according to this mechanism. Rather, the modular principle should be regarded as one of the generally valid principles of evolution. For this reason, the mechanism of precyte evolution suggested in this book has also been based on the modular principle, or the endocytobiotic cell theory taken in its broadest sense (see Chapter 5, III).

II. PHASE PRINCIPLE

The evolution of ever more complex systems is not a continuous process. Rather, it is driven by a constant alternation between *divergent* and *convergent* phases.[13,14] This general phase principle has already been presented (Chapters 3 and 5). According to this principle, each newly developed evolutionary system soon gives rise to many variants or mutations with similar chances for survival, since raw materials or sources of nutrition are present in excess. No major changes occur; this is a divergent phase. However, the explosive growth of the organisms gradually exhausts their resources. As soon as new variants appear which serve a new purpose, namely the utilization of new sources of energy or materials, a convergent phase begins. The convergent phase is strongly selective, and the selection is directed. Only those representatives of the new variant survive which are best adapted for the new purpose. The transition from the divergent to the convergent phase is, of course, initiated by chance, but if the population is large enough and the time periods involved are long enough, this change must occur. It is not the event itself, but only the time at which it occurs which is a matter of chance. This effect is typical of evolving systems. The convergent and divergent phases of evolution are shown in Figure 2.

---

* The subunits of elementary particles, whose existence is yet unproved.

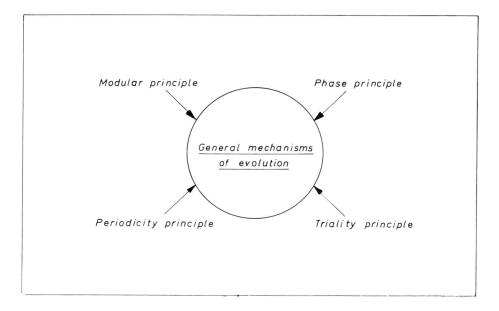

FIGURE 1.   Schematic representation of the hierarchical system of a modular principle of evolution. This system can be called the *modular construction principle*, because each higher level is built upon the next lower level, and there are manifold combinations between the individual levels.[28,29]

It must be mentioned that not all scientists accept the above phase principle. To be sure, it is generally agreed that chance and necessity have channeled all evolution. However, there are differences in opinion as to whether the decisive factors are chance events or inner forces within the system.[1] Neither of the opposing points of view is presently provable. We find only evidence which tends to support one or the other alternative standpoint. However, the observed periodicity in the divergence of evolutionary systems seems to be strong evidence in favor of the causality of the evolutionary process.

III. PERIODICITY PRINCIPLE

The periodicity of material systems has been particularly well studied for atoms. Atoms

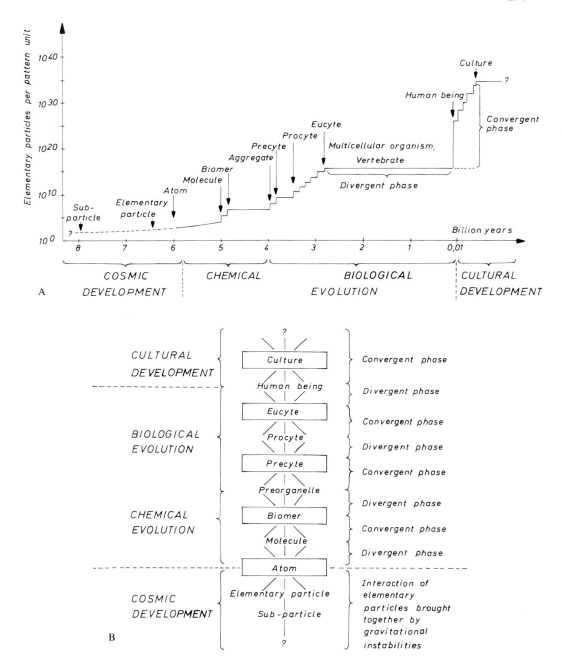

FIGURE 2. Model of the phase principle of evolution. Chemical, biologial, and cultural evolution develop through the repetition of convergent and divergent phases. Selection, in the biological sense, does not occur in cosmic evolution. The condensation of sub-particles to elementary particles to atoms occurs whenever the physical conditions are appropriate.

(A) In the divergent phases, starting with self-replicating systems, the number of elementary particles per pattern unit (unit taken from Bresch[1]) does not significantly increase, until a variation/mutation serves a new purpose. In the subsequent convergent phase, the "evolutionary carousel" turns until an optimum form for this purpose is found. The result is an increase in the number of elementary particles per pattern unit, and thus further development or an evolutionary leap.

(B) The evolutionary leap occurs when different variants of one evolutionary level (divergent phase) unite to form a new, more complex integrational unit or evolutionary stage (convergent phase). See also Figure 1, modular principle.

are the smallest elements of the system of chemical evolution. They can be arranged in a periodic system, whose underlying principle is understood today. This system will be briefly reviewed in the following.

A. Periodic System of the Atoms

As early as the begining of the 19th century, when only a few chemical elements (kinds of atoms) were known, Döbereiner saw "in one chemical system the essence of all chemical experience". He arranged the first triade of

Calcium (Ca) — Strontium (Sr) — Barium (Ba).

Then in 1869, Dimitri Mendelejeff (1834—1907) and Lothar Meyer (1830—1895) independently arranged the atoms according to increasing atomic weight. They found that atoms with similar properties were found at certain intervals, and Mendelejeff arranged the elements in a table in which those elements with similar chemical properties stood under each other in columns. They called the horizontal rows in this representation *periods*, and made predictions about the remaining empty spaces, that is, the elements missing in these periods. The chemical and physical properties of the elements which were found later were in accordance with their predictions. It was later possible to iron out difficulties in the order of the elements, when it was discovered that the deviations were due to the presence of mixtures of isotopes. This is the reason, for example, that the element argon, (Ar, no. 18) with an atomic weight of 39.948 comes before potassium (K, no. 19) which has an atomic weight of only 39.102.

There are now 105 known chemical elements.[12] They are arranged in seven periods (see the complete periodic table of the elements, Figure 3). The first period contains two elements, the second and third, 8 elements each, the fourth and fifth, 18 elements each, the sixth, 32, and the seventh,18. Read vertically, these seven periods are arranged in eight *groups*, in which the atoms have similar properties. This is expressed by the group names, such as alkali metals (group I), earth alkali metals (group II), boron group (group III), chain formers (group IV), nitrogen group (group V), ore formers or chalcogens (group VI), salt formers or halogens (group VII), and noble gases (group VIII). The horizontal (periods) and vertical (groups) relations are not the only ones, however. If one draws a diagonal from the upper left to the lower right corner of the periodic table, one divides the *metallic elements* (below the diagonal) from the *non-metals* (above it). The *amphoteric elements*, which have both metallic and non-metallic characteristics, are arranged along the diagonal.

The arrangement of the elements in periods and groups was at first purely empirical, but its physical basis was later rationally founded in the quantum theory developed by Max Planck (1858—1947) and Albert Einstein (1879—1955), and in the *atomic theory* developed by Johannes Schrödinger (1897—1961) and Wolfgang Pauli (1900—1958). Rutherford (1911) and his pupil Bohr (1913) had proposed an intuitive model of atomic structure (Figure 4). According to this *atomic theory*, every atom consists of a *nucleus* surrounded by an *electron shell*. The atomic nucleus is made up of positively charged elementary particles, the *protons* (p^+), and neutral particles, the *neutrons* ($n^{\pm}$). The position of an atom in the periodic system depends on the number of protons in its nucleus, which determines the nuclear charge or *atomic number*. The outer shell of the atom is made up of negatively charged elementary particles, the *electrons* (e^-), which can occupy various energy levels, or *orbitals*, making up the *atomic shells*. There are seven shells which are occupied in the known elements, and these have been designated K, L, M, N, O, P, and Q. Each period represents a group of elements whose outermost electrons are in the same shell, the lower-energy shells all being already occupied. There are thus exactly as many periods as shells. Going from left to right in each period, the number of electrons in the outer shell increases by one for each element.

PERIODIC SYSTEM OF ELEMENTS *

Group / Period (Shell)	0 (VIIIB)	I A	I B	II A	II B	III A	III B	IV A	IV B	V A	V B	VI A	VI B	VII A	VII B	VIII A	VIII B (0)
1 K		1H 1.008_0												1H 1.008_0			2He 4.00260
2 L	2He 4.00260	3Li 6.94_1		4Be 9.01218		5B 10.18		6C 12.011		7N 14.0067		8O 15.999_4		9F 18.9984			10Ne 20.17_9
3 M	10Ne 20.17_9	11Na 22.9898		12Mg 24.305		13Al 26.9815		14Si 28.08_6		15P 30.9738		16S 32.06		17Cl 35.453			18Ar 39.94_8
4 N	18Ar 39.94_8	19K 39.10_2	29Cu 63.54_6	20Ca 40.08	30Zn 65.3_7	21Sc 44.9559	31Ga 69.72	22Ti 47.9_0	32Ge 72.5_9	23V 50.941_4	33As 74.9216	24Cr 51.996	34Se 78.9_6	25Mn 54.9380	35Br 79.904	26Fe 55.84_7 27Co 58.9332 28Ni 58.7_1	36Kr 83.80
5 O	36Kr 83.80 / 68	37Rb 85.467_8	47Ag 107.868	38Sr 87.62	48Cd 112.40	39Y 88.9059	49In 114.82	40Zr 91.22	50Sn 118.6_9	41Nb 92.9064	51Sb 121.7_5	42Mo 95.9_4	52Te 127.6_0	43(Tc)	53J 126.9045	44Ru 101.0_7 45Rh 102.9055 46Pd 1064	54Xe 131.30
6 P	54Xe 131.30	55Cs 132.9055	79Au 196.9665	56Ba 137.3_4	80Hg 200.5_9	57/71 ΣLa¹⁾	81Tl 204.3_7	72Hf 178.4_9	82Pb 207.2	73Ta 180.947_9	83Bi 208.9806	74W 183.8_5	84Po 210	75Re 186.2	85(At)	76Os 190.2 77Ir 192.2_2 78Pt 195.0_9	86Rn 222
7 Q	86Rn 222	87(Fr) –		88Ra 226.0254		89/103 ΣAc²⁾		104Ku –		105Ha –							

1) ΣLa =Lanthanides: 57 to 71   La $138,905_5$

Ce	Pr	Nd	(Pm)	Sm	Eu	Gd
140.12	140.9077	144.2_4	–	150.4	151.96	157.2_5
Tb	Dy	Ho	Er	Tm	Yb	Lu
158.9254	162.5_0	164.9303	167.2_6	168.9342	173.0_4	174.97

2) ΣAc = Actinides: 89 to 103   Ac –

Th	Pa	U	(Np)	(Pu)	(Am)	(Cm)
232.0381	231.0359	238.029	237.0482	–		
(Bk)	(Cf)	(Es)	(Fm)	(Md)	(No)	(Lv)
–	–	–	–	–	–	–

* Elements which do not occur naturally are shown in parentheses.

FIGURE 3.   Periodic table of the elements (from Klemm;[12] details, see text). The bioelements are emphasized by gray shading (see Rahmann[21]).

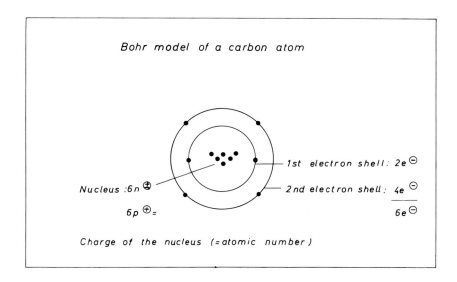

FIGURE 4.   The Bohr model of atomic structure, with carbon as the representative example (details, see text). This model is presented here instead of the generally accepted Schrödinger model, because it is easier to comprehend and shows the atomic shells more clearly.

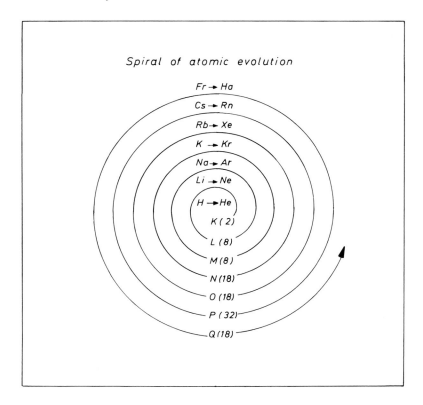

FIGURE 5.   Schematic representation of successive development of the seven atomic shells in the form of an evolutionary spiral (symbol of elements see Figure 3).

Therefore, all the atoms of the first group have just one electron in the outer shell, all the atoms in the second group have two electrons in the outer shell, and so on. For quantum mechanical reasons, the first shell has "room" for only 2 electrons, the next two for only 8 electrons each, and so on. Also, the larger outer shells are subdivided into subshells, which produces elements (metals) with rather similar chemical properties. This is why they can be divided into just 8 groups instead of 18 or even 32. The number of electrons in the highest shell is the same as the group number of the element; furthermore, it is the major factor determining the chemical behavior of an element; this accounts for the similarity of the elements in each group.

As we have seen in Chapter 2 on the formation of the atoms, the light elements were formed first, followed by progressively heavier ones. In a sense, the progressve increase in the number of electrons occupying the atomic shells (atomic number) from period 1 to period 7 is also the order in which they arose, an "atomic phylogeny" (Figure 5). The process of "atomic evolution" can be described by a spiral which increases in diameter as it rises. The importance of the periodic system cannot be overemphasized. It summarizes so much information about the relatonships between the atoms and their compounds, and about their evolution, that it was this system which first made a deeper understanding of chemistry possible. The atomic theory and the periodic table of the elements, which can be logically deduced from it, is thus the central theory for the explanation of the structure, function, and "evolution" of the chemical elements. The question now arises, whether the phenomenon of periodicity must be limited to atoms, or whether a similar ordering principle might be discovered for cells, the elementary "particles" of biological evolution.

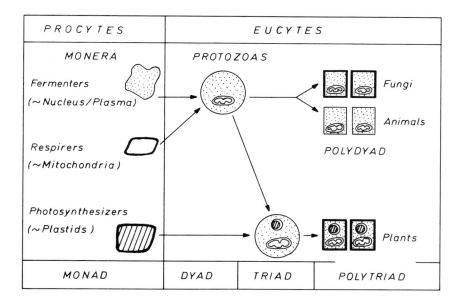

FIGURE 6. Classification of procytes and eucytes on the basis of endocytobiotic cell theory with reference to the number of genetic systems and protein biosynthesis machineries, according to a suggestion of Taylor[32] and Margulis.[16]

B. Hypothetical Periodic System of Cells

According to general scientific procedure, each discipline first describes the phenomena of its research area, then analyzes these phenomena, and finally categorizes them into a system. Chemistry has followed this classical pattern, having developed its central system in the form of the periodic system of the elements. To date, biology has lacked such a system for its smallest building blocks, the cells. The cell is the smallest element of life, analogous to the atom which is the smallest element of matter. The theory of evolution, like the quantum theory for atoms, explains certain central evolutionary mechanisms. However, a generally accepted theory of the organization of the cell, comparable to the atomic theory, has yet to be proposed. Considering the facts and conclusions described (Chapter 7), the endocytobiotic cell theory (in its short form, without the flagella symbiosis) should provide a satisfactory explanation for structure, function, information, and evolution of the various cell types. It is therefore reasonable to attempt to arrange the confusing variety of cellular forms on this basis.

There has been no lack of attempts to arrange the rich variety of cell types in some sort of system. One of the most recent attempts was made by Harald Riedl.[22] Using plants as examples, he sketches a model of evolution which could apply indirectly to cells. Riedl differentiates between cellular *organizational types*, which have arisen in the course of evolution (evolutionary types) and cellular *functional types*, which recur at every new level of evolution. He describes the evolutionary process as a spiral, which with each turn (evolutionary type) passes through the same angular locations (functional types). Seen in this light, the organizational or evolutionary types of cells correspond to the periods of the elements, and the cellular functional types correspond to the groups of elements. However, Riedl has not to date specified the individual cellular organizational or functional types of this system.

According to a suggestion made by F. J. R. Taylor[32] and Margulis,[16] the procytes or moneras are designated as monads, since they have only one genetic system with a single protein biosynthesis machinery (Figure 6). Fungus-like and animal unicellular organisms (protozoas) are thus dyad, i.e., they have not only the genetic system and protein biosynthesis machinery of the plasma-nucleo-system but also the systems of the mitochondria. Multiple cellular fungi and animals (metazoas, metaphytes) can thus be designated as polydyads. The unicellular plants are, in contrast, triadic because of the additional plastidal protein biosynthesis apparatus and the multicellular plants are polytriadic (metaphytes). Finally, animal, fungal and plant cells can integrate one or several kinds of endocytobiotic protein biosynthesis machinery and become thus correspondingly tetrads*, pentads, hexads, heptads*, and so on.

On the basis of the endocytobiotic cell theory (in its limited form), it is now possible to conceive a system for classification of cells within the evolutionary framework envisaged by Riedl and using the terminology of Taylor and Margulis. This system can be called the *periodic system of cells* (Table 1). It is, of course, at best a preliminary model, since a final periodic system for cells can only be proposed on the basis of a systematic comparison of DNA and RNA sequences of each cell type and of each DNA organelle and endocytobiont. We are a long way from this goal. The preliminary periodic system of cells has seven horizontal periods and eight vertical groups, corresponding to the periodic system of the elements.

The single monad of the procytes (bacteria, blue-green algae), as well as the two dyads of fungi and animals and one triad of plants represent different evolutionary types; these can be designated as the first, second, third, and fourth periods of cell evolution (Table 1.A). The animal, fungal, and plant period is divided each into two subperiods, one containing unicellular or monodyad (monotriad) protozoas and one containing multicellular or polydyad (polytriad) metazoas or metaphytes. The representatives of each period are homologous to one another. In a comparison of the cell types of all periods, similar, i.e., analogous groups of cellular functional types emerge, according to the special milieu and substrate conditions. As in the basic metabolic types presented earlier (Chapter 6), we can distinguish generally fermentative, respirative, photergic, and photoassimilative cell function types. The biotype for the fermentative cell types is dark, organotrophic, chemotrophic, C-heterotrophic; for the respiring types dark, organotrophic/anorganotrophic, chemotrophic, C-heterotrophic; for the photergic cell types light, organotrophic, phototrophic, C-heterotrophic; and for the photosynthesizing cells light, organo-inorganotrophic (anaerobic) or inorganotrophic (aerobic), phototrophic, C-autotrophic (compare Chapter 6, Figures 3 and 4). Each of these functional types can again be subdivided into an anaerobic and aerobic group, resulting in eight different groups. In the first and second group we find mostly fermentative cell types with resorption and storage functions such as mycel, intestine, fat, root, tuber cells, and zooflagellates or ciliates. Groups 3 and 4 are composed mostly of respiring, growth-active, and supportive cell types such as spore, gamete, embryo, muscle, wood cells, and amoebas or sporozoas. These groups also include the prokaryotic anaerobic and aerobic chemoautotrophs (biotype: dark, organo-anorganotrophic, chemotrophic, C-hetero-/C-autotrophic). Groups 5 and 6 consist in general of photergic, pigmented, and photo-moving cells, e.g., certain cells of the retina (rods) and flower cells. Finally, 7 and 8 include mostly photoassimilative cell types such as leaf cells and phytoflagellates.

Fungus-like and animal dyads as well as plant triads have continued, in the further course of evolution, the up-take and incorporation of fermentative, respirative, photergic, and photoassimilative endocytobionts, which then leads to the formation of the following post-cytic periods of cell evolution (Table 1B). In the case of endocytobiotic integration of only one monad, there emerged from the dyad of the fungi the 5th period in the form of a postcytic triad. From the dyad of the animals there evolved by endocytobiosis the postcytic triad of the 6th cell period, and, finally, from the triad of the plant the postcytic tetrad of the 7th

---

* At these points Taylor and Margulis used Latin terms (quadrad, septad) which do not agree with the otherwise Greek terminology.

cell period. By taking up dyad or triad systems (animal or plant cells) or several monads (bacteria, cyanobacteria) at once, even pentads, hexads, and multiple monads (oligads) in the form of monooligadic and polyoligadic postcytes emerged in individual periods.

Among the postcytes one can also distinguish eight groups of analogous functional types in a comparison of the various representatives of the cell periods (see Chapter 8, III and IV). The postcytic fermentative type belongs to group one and two and is characterized by endocytobiosis with mostly fermentative procytes (accessory symbiosis), e.g., rickettsia symbioses of the cicadas and cockroaches, and gall bacteria symbioses of the succulent plants. Groups 3 and 4 of the postcytic respiring type are usually composed of endocytobiosis with respiring procytes (companion symbiosis), e.g., the bacterial symbiosis of amoebas and flagellates (*Pelomyxa, Mixotricha*) and the actinomycete symbiosis of alders and willows. The representatives of group 5 and 6 are of the photergic type (primary symbiosis). They form endocytobioses with generally photoactive, pigmented bacteria such as the luminescent bacteria symbioses of squid. Finally, the cell functional types of the 7th and 8th groups belong to the postcytic photosynthesis type (auxiliary symbiosis). They are made up of endocytobioses with mainly photoassimilative bacteria and blue-green algae, such as the syncyanoses of fungi and plastid-free algae. The three postcytic periods underline the importance of endocytobiosis formation in eucyte evolution. They indicate that the process of cell evolution is by no means finished but is rather in full swing, whereby further endocytobionts are continuously taken up and integrated as a kind of additional DNA-containing cell organelle.

As can be seen in detail in Table 1, the periodic system of cells not only displays many horizontal and vertical relationships, but also diagonal relationships. If one draws a diagonal from the upper left to the lower right, one finds the primarily photoactive, typical *plant cell* types above the line, and below it primarily respiring-fermenting, typical *animal cells*. The ambivalent *fungal* or *fungal-like* cell types lie in the region close to the diagonal. The latter have cellulose walls like plant cells, but, on the other hand, they do not contain plastids and are in some cases surrounded by chitin, like some animal cells (Table 2).

It is thus apparently possible to organize cells, as well as atoms, in a rationally deducible periodic system. According to this system, cell evolution appears as a spiral of constantly increasing numbers of protein synthesis apparatuses (Figure 7). This much, however, is already certain — the endocytobiotic cell theory could play a role in the creation of a central classification system for biology that is similar to that once played by the atomic theory in the postulation of such a system for chemistry.

With the help of this rationally deducible system it is not only possible to classify the various cell forms according to homologous evolutionary types and analogous functional types. This classification system also promotes a deeper understanding of the complex interactions between the nucleo-cytoplasm system and the DNA-containing cell organelles and endocytobionts on a functional-causal level. Furthermore, precise statements on the nature and existence of as yet unidentified endocytobiotic cell systems can be made and such systems can also be produced artificially, as in the case of photoassimilating animal cells obtained by infecting mouse fibroblasts with chloroplasts.[17] This last-mentioned type of system occurs physiologically in the gut of water snails.[33,34]

However, it must be stressed that the hypothetical model of a periodic system of cells is valid independent of the verification of the endocytobiotic cell theory. The construction of such a system on the basis of today's knowledge would be about the same, for all practical purposes, if this system were based on the classic compartmental cell theory. The various physiological niches and protein biosynthesis machineries which define the categories of the system are present whether we consider them to be exogenically or endogenically derived. However, the system appears to be logical and more satisfying on the basis of the endocytobiotic cell theory; this can possibly be seen as an indication of the validity of the endocytobiotic cell theory.

Table 1A

EVOLUTIONARY TYPES (homology series) →

FUNCTIONAL TYPES (analogy series) →

PERIOD \ GROUP	subgroup	FERMENTATION 1. Anaerobic	2. Microaerobic	RESPIRATION 3. Anaerobic	4. Aerobic	PHOTERGY 5. Anaerobic	6. Aerobic	PHOTOSYNTHESIS 7. Anaerobic	8. Aerobic
1. MONAD (Monera)		Fermenters (eg. Fermentative bacteria)		Respirers (eg. Eubacteria)		Photergers (eg. Halophilic bacteria)		Photosynthesizers (eg. Photobacteria, Blue-green algae)	
2. DYAD (Fungus cells)	Mono-dyad	Myxoflagellates (eg. Myxomycophyta)		Myxamoeba (eg. Myxomycophyta)		?		?	
	Poly-dyad	Resting cells (eg. Mycel, Gemma)		Growth cells (eg. Gametes, Spores)				Lichen (eg. Green-a. blue-green algae)	
3. DYAD (Animal cells)	Mono-dyad	Trichoprotozoas (eg. Flagellates, Ciliates)		Sporoprotozoas (eg. Amoebas, Sporozoas)		?	Rod cells (eg. Eye)	?	
	Poly-dyad	Storage cells (eg. Intestine, Fatty tissue)		Building cells (eg. Gametes, Blastema, Muscle)					
4. TRIAD (Plant cells)	Mono-triad	Plastid-free algae cells (eg. Glaucocystis)				Chromoflagellates (eg. Yellow and brown algae)		Phytoflagellates (eg. Green algae)	
	Poly-triad	Amyloplast Leucoplast cells (eg. Root, Tuber)		Proplastid cells Etioplast cells (eg. Gametes, Stem, Wood)			Chromoplast cells (eg. Flower)		Chloroplast cells (eg. Leaf)
	TYPE	RESORPTION-STORAGE-TYPE		GROWTH-SUPPORT-TYPE		LIGHT-MOVEMENT-TYPE		PHOTOASSIMILATION-TYPE	

Classification of procytes, eucytes, and postcytes on the basis of the endocytobiotic cell theory, with reference to homologous evolutionary types and analogous functional types, in the form of a hypothetical periodic system of cells.

There are eight groups (columns) of metabolic ecological niches and seven periods (rows) depending on the number of independent protein-synthesis apparatuses present. The cells are also grouped according to plant, animal, and fungal cells (details see text).

(A) Periodic system of procytes and eucytes with selected examples. (B) Periodic system of postcytes with selected examples.

Table 1B

EVOLUTIONARY TYPES (homology series)

PROCYTES — POSTCYTES

FUNCTIONAL TYPES (analogy series)

PERIOD	GROUP	FERMENTERS 1. Anaerobic	FERMENTERS 2. Microaerobic	RESPIRERS 3. Anaerobic	RESPIRERS 4. Aerobic	PHOTERGERS 5. Anaerobic	PHOTERGERS 6. Aerobic	PHOTOSYNTHESIZERS 7. Anaerobic	PHOTOSYNTHESIZERS 8. Aerobic
5. TRIAD-OLIGAD (Fungus symbiosis)	Mono-triad								
	Poly-triad	Mycorrhiza: endogenous (eg. Orchis)			?		?		Blue-green algae symbiosis (eg. Geosiphon)
6. TRIAD-OLIGAD (Animal symbiosis)	Mono-triad								
	Poly-triad	Rickettsia symbiosis (eg. Ciliates, Cockroaches, Bugs)		Bacteria symbiosis (eg. Cicadas: some, Flagellates, Amoebas)		Luminescent bacteria symbiosis (eg. Squid, Tunicates, Fish)		Chloroplast symbiosis (eg. Amoebas, Ciliates, Snails, Mouse artificial)	
7. TETRAD-OLIGAD (Plant symbiosis)	Mono-tetrad								
	Poly-tetrad	Plantgall bacteria symbiosis (eg. Succulents)		Actinomycete symbiosis (eg. Alder, Willow)		Chromobacteria symbiosis (eg. Some cicadas)		Blue-green algae symbiosis (eg. Glaucocystis, Cycadina)	
		ACCESSORY-TYPE		AUXILIARY-TYPE		PRIMARY-TYPE		COMPANION-TYPE	

Table 2

EVOLUTIONARY TYPES (homology series) →
FUNCTIONAL TYPES (analogy series) ↓

GROUP	I	II	III	IV	V	VI	VII	VIII
	FERMENTATION		RESPIRATION		PHOTERGY		PHOTOSYNTHESIS	
	Anaerobic	Microaerobic	Anaerobic	Aerobic	Anaerobic	Aerobic	Anaerobic	Aerobic
	Dark / Organotrophic / Chemotrophic / C-Heterotrophic		Dark / Organotrophic /Inorganotrophic / Chemotrophic / C-Heterotrophic (C-Autotrophic)		Light / Organotrophic / Phototrophic / C-Heterotrophic		Light / Organotrophic /Inorganotrophic / Phototrophic / C-Autotrophic	
	RESORPTION-STORAGE-TYPE		REPRODUCTION-SKLERO-TYPE		LIGHT-MOVEMENT-TYPE		PHOTOASSIMILATION-TYPE	
PROCYTE — Cytoplasm (Period 1, Monad) — Precyte / Eobiont	Fermenters (eg Fermentative bacteria)		Respirers (eg Eubacteria) (Chemoautotrophs)		Photergers (eg Halophilic bacteria)		Photosynthesizers (eg Photobacteria, Blue-green algae)	
FUNGAL EUCYTE (Period 2, Dyad) — Mono-dyad: Cytoplasm; Poly-dyad: Mitochondrion	Myxoflagellates (eg Myxomycophyta: some) / Resting cells (eg Mycel, Gemma)		Myxamoeba (eg Myxomycophyta: some) / Growth cells (eg Gametes, Spores)		?		Lichen (eg Green- and blue-green algae) / ?	
ANIMAL EUCYTE (Period 3, Dyad) — Mono-dyad: Cytoplasm; Poly-dyad: Mitochondrion	Trichoprotozoas (eg Flagellates, Ciliates) / Storage cells (eg Intestine, Fatty tissue)		Sporoprotozoas (eg Amoebas, Sporozoas) / Building cells (eg Gametes, Blastema, Muscle)		Rod cells (eg Eye)		?	
PLANT EUCYTE (Period 4, Triad) — Mono-triad: Cytoplasm; Poly-triad: Mitochondrion, Plastid	Plastid-free algae cells (eg. Glaucocystis) / Amyloplast Leucoplast cells (eg Root, Tuber)		Proplastid Etioplast cells (eg Gametes, Stem, Wood)		Chromoflagellates (eg Yellow and brown algae) / Chromoplast cells (eg Flower)		Phytoflagellates (eg Green algae) / Chloroplast cells (eg Leaf)	
FUNGAL POSTCYTE (Period 5, Triad-oligad) — Mono-triad: Cytoplasm; Poly-triad: Mitochondrion, x Endocytobionts	Mycorrhiza: endogenous (eg Orchis) / ?		?		?		Blue-green algae symbiosis (eg Geosiphon) / ?	
ANIMAL POSTCYTE (Period 6, Triad-oligad) — Mono-triad: Cytoplasm; Poly-triad: Mitochondrion, x Endocytobionts	Rickettsia symbiosis (eg Ciliates) / (eg Cockroaches, Bugs)		Bacteria symbiosis (eg Flagellates, Amoebas) / (eg Cicadas: some)		Luminescent bacteria symbiosis / (eg Squid, Tunicates, Fish) / ?		Chloroplast symbiosis (eg Amoebas, Ciliates) / (eg Snails, Mouse artificial) / ?	
PLANT POSTCYTE (Period 7, Tetrad-oligad) — Mono-tetrad: Cytoplasm; Poly-tetrad: Mitochondrion, Plastid, x Endocytobionts	Plantgall bacteria symbiosis (eg Succulents) / ?		Actinomycete symbiosis (eg Alder, Willow) / ?		Chromobacteria symbiosis (eg Cicadas: some) / ?		Blue-green algae symbiosis (eg Glaucocystis, Cyanophora) / (eg Cycadina, Water fern) / ?	
	ACCESSORY-SYMBIOSIS-TYPE		AUXILIARY-SYMBIOSIS-TYPE		PRIMARY-SYMBIOSIS-TYPE		COMPANION-SYMBIOSIS-TYPE	

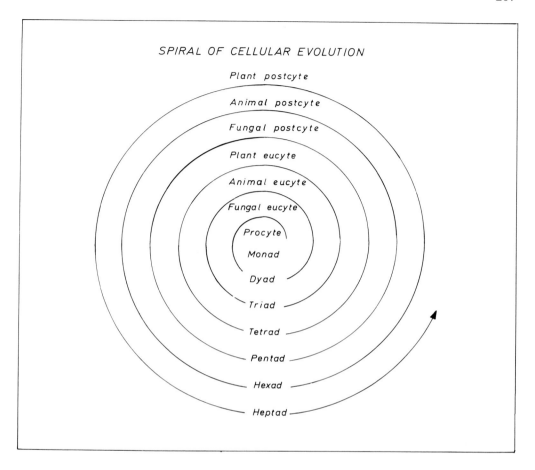

FIGURE 7. Schematic representation of the successive evolution of the polygenomic systems of eucytes in the form of the spiral of evolution (see also Figure 5).

IV. TRIALITY PRINCIPLE

The question now arises, how the *analogies* between the periodic system of the atomic elements and that of cells[30] (see Chapter 3) can be derived. It is clear that the basis for any such explanation must lie in the laws of physics, or be in principle traceable to them. The constant alternation between convergent and divergent phases appears to be the basis for the dialectic which is the prerequisite of a system in which periodicity evolves (phase principle; see Section II, Figure 2). It provides a basically trialistic mechanism of evolution which generates a trialistic structure of the evolving systems. The principles of the periodic system appear to be derived from this triality principle which we find in various fields of scholarship.[30] *

A *general philosophical triality* law was formulated by G. Hegel (1770—1831), who suggested that the development of a concept, or of reality itself, is based on contradiction *(dialectic)*. He wanted this law to be understood not only as the general process of logical procedure (logical technique), but also as "the true nature of all being itself". As soon as a particular condition in thought or in reality has been attained *(thesis)*, it is transformed

---

* Translator's note: the author uses here the German word "Wissenschaft", which encompasses all fields of scholarship. I have used the word "scholarship" in this sense, and the term "natural science" for the more limited range of fields expressed by the German word "Naturwissenschaft".

into its opposite *(antithesis)*, from which a third condition arises, which includes the previous two but also goes beyond them *(synthesis)*. The synthesis then becomes the thesis, so that the ''game' can begin again. The most recent natural philosophical derivations of triality principles stem from Karl Popper[20] Konrad Lorenz,[15] and Rupert Riedl.[23] They postulate a division of the world into three parts. The real world *(World 1)* of physical objects and conditions stands opposed to the subjective world *(World 2)* of our conscious conditions. The arts and sciences *(World 3)* attempt to objectivize the objects of World 1 and the subjects of World 2 in materially fixed culture. That is, World 3 represents the response to the questions raised by our ego, the subjective World 2 of the questioner, about the objects of World 1. John Eccles[5] has even attempted with some success to locate the structures in the brain associated with the three worlds. The *law of mass action* developed by C. Guldberg (1836—1902) and P. Waage (1833—1900) is an example of a dynamic triality principle in chemistry. It says, in part, that the forward reaction of two substances is in dynamic equilibrium with the back reaction of the product. Finally, one of the most important dynamic triality principles of biology was developed by Darwin (1809—1882): *biological evolution* of the various organisms occurs through the interaction of *mutation* and *selection*. Evolution occurs as the best adapted organisms are selected from the palette of possibilities offered by mutation. The list of triality phenomena described in the natural sciences might be continued indefinitely. However, the few examples mentioned here may indicate that there is a comparable principle or mechanism on which all of them may be based, namely the general principle of thesis, antithesis, and synthesis as formulated by Hegel, taken in the widest sense. Closer consideration of a concrete example can even reveal possible causal relationships between this abstract triality principle and the periodicity phenomenon.

The study of the evolutionary dynamic of the various evolutionary systems is well suited to this purpose. The evolutionary dynamics of individual atoms or molecules, one-celled or muticellular organisms, or whole populations, are primarily determined by the ratios of their rates of formation and decomposition, or of birth and death, or, more generally of *increase and decrease*. These relationships were recently described by Eigen.[7] He distinguishes three possible forms of reaction, or basic strategy, for the rates of increase and decrease. The rates of increase and decrease may, for example, be dependent on the size of the population, i.e., changes in the rate of increase have the same sign (positive or negative) as changes in the population level. This *conforming strategy,* which is symbolized S + in game theory, can be taken as thesis in the triality principle. It is opposed by the *contrary strategy*, S − , in which the changes in the population level call forth the opposite tendency in rate of increase. In this context, it can be regarded as the antithesis. In the *indifferent strategy*, S 0, the rates of increase and decrease are not affected by changes in the population level. It corresponds to the synthesis* of this triality. The combination diagram (Figure 8) shows how the various strategies of increase and decrease qualitatively affect the population level. There are four basic conditions: *stable, unstable, indifferent* and *variable*. Under *stable conditions*, the rates of increase and decrease are in dynamic equilibrium, so that a given population level is not changed on the average. *Unstable conditions* are characterized by the fact that slight changes in the population level affect the rates of increase and/or decrease in such a way that the change is amplified. The result is a population explosion or extinction. Under *indifferent conditions*, the population can have any level, without a consequent regulatory effect. The *variable conditions*, finally, are composites of the other three basic types. These may, depending on the quantitative circumstances, appear alternately, thus driving the population from one condition to the other. The variety of atomic and molecular populations is so great that all strategies and conditions are represented among them. In biological populations of unicellular and multicellular organisms, however, the variable strategies or

---

* A general use of the term synthesis.

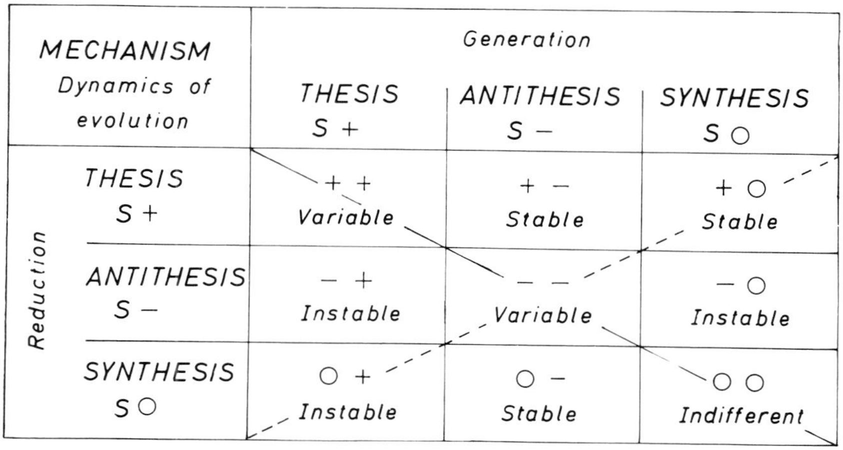

MECHANISM Dynamics of evolution	Generation		
	THESIS S +	ANTITHESIS S −	SYNTHESIS S ○
THESIS S +	+ + Variable	+ − Stable	+ ○ Stable
ANTITHESIS S −	− + Instable	− − Variable	− ○ Instable
SYNTHESIS S ○	○ + Instable	○ − Stable	○ ○ Indifferent

(left vertical label: Reduction)

FIGURE 8. Combination diagram of the conformist (S +), contrary (S −) and indifferent (So) strategies of rates of increase and decrease of populations of atoms, molecules, unicellular and multicellular organisms. The diagram is formulated according to the triality principle as general mechanism of evolutionary dynamics of these populations. There are interesting horizontal, vertical, and diagonal trialistic relationships (details, see text; Schwemmler[30]).

conditions occur most often. In the diagram, these are represented on the diagonals indicated by dotted lines. They are chracterized by regulation and self organization, and are found on the level of self-organizing macromolecules, differentiating or reproducing cells, and in the morphogenesis of organisms. However, the combination diagram suggests horizontal, vertical, and diagonal functional pattern which is caused by the triality principle.

The triality mechanism seems to be also reflected in the structure of evolving systems. This appears to apply to the central evolutionary systems of physics, chemistry and biology (Figure 9). According to atomic theory, the atom is the essential synthesis of positively charged protons (p^+) and negative electrons (e^-), which can be considered thesis and antithesis with respect to their charges (see Chapter 2). The *precyte* arose, according to the modular construction principle, from the fusion of the anabolic acid nucleo abioids and katabolic basic plasma abioids[28,29] (see Chapter 5, Section III), whose internal milieus are supposed to have been related to each other, with respect to charge, degree of acidity and metabolic activity, as thesis and antithesis. The *eucyte* evolved, according to the endocyto-biotic cell theory, primarily by combination of an anaerobic, fermenting procyte with an aerobically respiring one (see Chapter 7). Anaerobic fermenters and aerobic respirers are to a certain degree related to one another as thesis and antithesis, namely with respect to their ecological niches. The individual systems of each evolutionary level may then have been formed by systematic variation of these trialistic basic structures according to the triality mechanism. At the level of physical evolution, the structural pattern of matter organized itself (see Chapter 2, Section III), at the chemical level, the aggregation pattern of the molecules (see Chapter 3), and at the biological level, the reproducing pattern of the cells (see Chapters 5, 6, 7). The large trialistic structural patterns which we now attempt to arrange in the periodic systems of atoms and cells may thus have arisen in the course of evolution.

If we summarize the facts and conclusions discussed so far to a general abstract trialistic combination diagram, the result is a hypothetical *triality scheme* which may contain the essential characteristics of a periodic system in condensed and short form (Figure 10). The *horizontal relationships* of this scheme are expressed in three horizontal sequences. The thesis, antithesis, and synthesis of each horizontal sequence are characterized by possessing

STRUCTURAL PATTERN Evolutionary systems		THESIS	ANTITHESIS	SYNTHESIS
PHYSICS (matter organization)	THESIS	+ ○ Proton (p ⊕)	○ − Electron (e ⊖)	+ − Atom (⊕)
CHEMISTRY (molecule self aggregation)	ANTITHESIS	○ + Plasma abioid (pH > 7, catabolic)	− ○ Nucleo abioid (pH < 7, anabolic)	− + Hypercycle (pH ± 7, metabolic)
BIOLOGY (cell autono- mous reproduction)	SYNTHESIS	+ + Plasmatic fermentation (Dark, anaerobic)	− − Mitochondrial respiration (Light, aerobic)	○ ○ Eucyte (Dark / light, anaerobic/aerobic)

FIGURE 9.   Hypothetical, trialistic structural pattern of the most important evolutionary systems: atoms, molecules, and cells (details, see Chapters 3, 5, and 7; Schwemmler[30]).

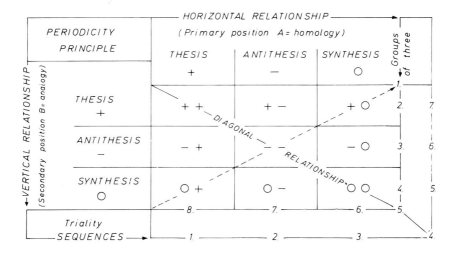

FIGURE 10.   Generally formulated, hypothetical trialistic combination diagram, which appears to include the most essential characteristics of the periodic system in condensed form. The homologous horizontal relationships appear to correspond to the groups, the analogous vertical relationships to the periods (details, see text; Schwemmler[30]).

the same pattern element in the primary position A. The thesis, antithesis, and synthesis of each position are, with reference to the A position, *homologous* to each other. They are representatives of the same evolutionary level. The secondary position B is trialistically varied, and analogous with respect to the A position. The *vertical relationships* are thus characterized by three analogous vertical sequences, whose thesis, antithesis, and synthesis are representatives of different evolutionary levels. There is also one sequence of *diagonal relationships*. In addition to the total of seven trialistic sequences, we can also observe 8 groups of three in the diagram: 3 horizontal, 3 vertical, and 2 diagonal. Parallels to the periodic system of elements seem obvious. The homologous horizontal relationships would correspond to the groups, the analogous vertical relationships to the periods, and finally, the diagonal relationships to those of the periodic system.

In summary, the phenomenon of periodicity is at present not understood, and certainly cannot be deduced. In the phase and triality principles, however, we can see the direction in which the explanation of this phenomenon is to be possibly sought and found. The triality principle can be defined as the most general mechanism of evolution, working in the widest sense according to the Hegelian principle of thesis, antithesis, and synthesis, both in the individual subsystems and in the whole system. It is also reflected in the structure of evolving systems, and finds its clearest expression in the periodic systems of the individual levels of evolution. In comparing the 3 components of the triality principle, one sees that only the thesis and antithesis have equal stature: the synthesis is on a higher level. As a mechanism and its structural reflection, however, the thesis, antithesis, and synthesis compose a uniform, trialistic principle. The term trialistic is thus more encompassing than the term dialectic, and seems fully justified for the phenomena of the general structure and function of evolution discussed here.

It might be suspected that the smallest building blocks of cosmic evolution, the *elementary particles*, can also be organized in a periodic system (e.g., Ne'emann,[18] Schopper,[25] Georgi[8]). If this were so, then the periodicity would prove to be a general phenomenon of evolution. It might even be possible to develop a scientifically derivable system or model for cultural or sociopolitical phenomena among men. It would seem that the need for such a classification system becomes greater, as the flood of contradictory cultural and political ideologies becomes greater. In such a derivation, however, it must be kept in mind that not necessarily all the mechanisms of evolution discussed can be directly applied to the cultural phase of evolution.

V. GENERAL MECHANISMS OF CULTURAL EVOLUTION

Human beings, like all living things, are the product of natural evolution. Their phylogenetic development from hominoids is well documented. Life in hunting bands transformed primitive humans at the beginning of the modern period into *social organisms* (see Chapter 1, Figure 3, and Chapter 2, Figure 4). This promoted the development of a highly developed system of communication, *symbolic speech*, which is possessed only by human beings. It is not inborn, but must be learned. The development of speech, associated with a high capacity for learning, was the prerequisite for *culture*. Human cultures are now evolving in their own right (Eccles[6]). Humans are thus subject both to biological and to cultural evolution. The latter is analogous to the former in several respects, but differs in several essential points, according to Osche:[19]

1. There is no inheritance of acquired properties in biological evolution; new and advantageous mutations and gene combinations must permeate the population slowly through selection and heredity, since they can be passed on only to the possessor's own offspring. Cultural evolution rests exclusively on the transfer of acquired properties. Through speech, and later writing and modern communication technologies, inventions of the individual can rapidly become the common possession of a large part of humanity. Cultural evolution therefore proceeds much more rapidly than biological evolution, and thus today there are both primitive peoples living at a stone age level and a highly developed technology which makes space travel possible.*

---

* Here the term "inheritance" is taken in somewhat too narrow a sense. The "transmission of acquired information", on the other hand, would apply both to biological and to cultural evolution. Furthermore, "traditions" among animals develop in a way similar to that described here for cultural evolution.

2. In biological evolution, properties are adapted to the conditions of the environment. In cultural evolution, man adapts the environment to his own traits (needs).*

3. In biological evolution, a differentiated exploitation of the environment is achieved by adaptive radiation and the formation of different species with their own ecological niches. In cultural evolution, the adaptation to niches (professions) occurs by differentiation of ''tools'', without speciation. Although man has created numerous ecological niches, there is only one species *Homo sapiens*.**

4. Biological evolution works opportunistically and uses only its successes. It cannot ''learn'' from failure. The same disadvantageous mutations and gene combinations can occur again and again, and species have repeatedly specialized in a way which leads to their extinction when the environmental conditions changed. Cultural evolution also leads to specialization, but man as a biologial organism remains ''unspecialized'' and can learn from his mistakes and avoid them in the future.

At the beginning of the central (biological) evolutionary process, as well as following this process, we observe phases which are based on comparable mechanisms, such as periodicity, but also those which are founded on completely different principles. The principles of mutation and selection do not apply to the formation of atoms, which occurs as a side reaction of cosmic development. Rather, they are the product of the interaction of all the elementary particles in a galaxy, caused by gravitational instabilities. Cultural development does not depend on the inheritance of chance mutations which are best adapted to the environment, but rather on the directed inheritance of acquired properties. However, our rapidly expanding knowledge will soon enable us to derive the many special mechanisms of evolution from a few principles. *These central mechanisms of the real world which we hope to derive, such as those of evolutionary processes in the widest sense, will not replace the elementary laws of physics, but will show that these laws are special cases of a higher physical regularity.*[8,25]

On the other hand, our initial optimism with regard to an exhaustive analysis of the phenomena of single and multicellular systems using exclusively the methods and categories of physics and chemistry has, in the meantime, become subdued. This optimism has given way to the realization that, between the molecular and phenomenological levels, there are independent transition stages in which quantity is transformed suddenly into a new quality. An adequate analysis of these transitions requires special methods and special, more complex categories of description.

REFERENCES

1. **Bresch, C.**, *Zwischenstufe Leben; Evolution ohne Ziel?* Piper Verlag, Munich, 1977.
2. **deChardin, T.**, *Der Mensch im Kosmos,* Verlag C. H. Beck, Munich, 1959.
3. **Clarke, B.**, Darwinian evolution of proteins, *Science,* 168, 1009, 1970.
4. **Czihak, G., Langer, H., and Ziegler, H.**, *Biologie. Ein Lehrbuch für Studenten der Biologie,* Springer Verlag, New York, 1976.
5. **Eccles, J. C.** *Das Gehirn des Menschen,* R. Piper Verlag, Munich, 1975.

---

* There are also examples in biological evolution in which the individual species adapts the environment to its needs, e.g. beavers.

** There are, however, various races and subraces. Since these are still relatively young, phylogenetically speaking, it may be that there simply has not been enough time for speciation to take place.

6. **Eccles, J. C.,** Die menschliche Persönlichkeit: ein wissenschaftliches und ein philosophisches Problem, *Naturw. Rdschau.,* 34 (6), 227, 1981.

7. **Eigen, M. and Winkler, R.,** *Das Spiel,* Piper Verlag, Munich, 1975.

8. **Georgi, H.,** Vereinheitlichung der Kräfte zwischen den Elementarteilchen, *Spektrum der Wissenschaft,* 6, 71, 1981.

9. **Halbach, U.,** Modelle in der Biologie, *Naturwiss. Rdschau,* 27 (8), 293, 1974.

10. **Haskell, E.,** *Full Circle,* Gordon and Breach, New York, 1972.

11. **Kimura, M. and Ohta, T.,** Protein polymorphism as a phase of molecular evolution, *Nature,* 229, 467, 1971.

12. **Klemm, W.,** Anorganische Chemie, de Gruyter Verlag (Sammlung Göschen), Berlin, 1971.

13. **Kuhn, H.,** Model consideration for the origin of life, *Naturwissenschaften,* 63, 68, 1976.

14. **Kuhn, H.,** Evolution biologischer Information, *Ber. Bunsen-Gesellsch. phys. Chem.,* 80 (11), 1209, 1976.

15. **Lorenz, K.,** *Die Rückseite des Siegels,* R. Piper Verlag, Munich, 1973.

16. **Margulis, L.,** Genetic and evolutionary consequences of symbiosis (a review), *Exper. Parasitol.,* 39, 277, 1976.

17. **Nass, M. M. K.,** Uptake of isolated chloroplasts by mammalian cells, *Science,* 165, 1128, 1969.

18. **Ne'emann, Y.,** Das Atom, *Bild der Wissenschaft* 10 (9), 1056, 1972.

19. **Osche, G.,** *Evolution. Grundlagen — Erkenntnisse, Entwicklungen der Abstammungslehre,* Herder Verlag (studio visuell), Freiburg, 1972.

20. **Popper, K.,** *Objektive Erkenntnis. Ein evolutionärer Entwurf,* Verlag Hoffmann und Campe, Hamburg, 1971.

21. **Rahmann, H.,** Die *Entstehung des Lebendigen.* Gustav Fischer Verlag, Stuttgart, 1972.

22. **Riedel, H.,** A model proposed for the progress of evolution with special reference to plants, *Acta Biotheoretica,* 21, 63, 1972.

23. **Riedel, R.,** *Die Ordnung des Lebendigen: Systembedingungen der Evolution,* Paul Parey Verlag, Hamburg, 1975.

24. **Riedel, R.,** *Biologie der Erkenntnis. Die stammesgeschichtlichen Grundlagen der Vernunft.,* 2 Aufl., Paul Parey Verlag, Hamburg, 1980.

25. **Schopper, H.,** Die jüngste Entwicklung des Bildes von der Materie, *Naturwiss.,* 68, 307, 1981.

26. **Schwemmler, W.,** Allgemeiner Mechanismus der Zellevolution, *Naturwiss. Rdschau,* 28 (10), 351, 1975.

27. **Schwemmler, W.,** Die Zelle: Elementarorganismus oder Endosymbiose? *Biologie in unserer Zeit,* 7 (1), 7, 1977.

28. **Schwemmler, W.,** Evolution der Urzeller: Rekonstruktion des Lebensursprunges, *Natur und Museum,* 108 (2), 49, 1978.

29. **Schwemmler, W.,** *Mechanismen der Zellevolution. Grundriss einer modernen Zelltheorie,* De Gruyter Verlag, New York, 1978.

30. **Schwemmler, W.,** The triality principle as possible cause of the periodicity of evolving systems, *Acta Biotheoretica,* 29, 35, 1980.

31. **Schwemmler, W.,** The endocytobiotic cell theory and the periodic system of cells, *Acta Biotheoretica,* 31, 45, 1981.

32. **Taylor, F. J. R.,** Implications and extensions of the serial endosymbiosis theory of the origin of eukaryotes, *Taxon,* 23, 229, 1974.

33. **Trench, R. K.,** Chloroplasts as functional endosymbionts in the mollusc *Tridachia crispata* B. (Opistobranchia, Sacoglossa), *Nature,* 222, 1071, 1969.

34. **Trench, R. K., Greene, R. W., and Bystrome, B. G.,** Chloroplasts as functional organelles in animal tissues, *J. Cell Biology,* 42, 404, 1969.

35. **van Valan, L.,** Two modes of evolution, *Nature,* 252, 298, 1974.

APPENDIX I. ABBREVIATIONS

A	adenine		MW	molecular weight
Å	Ångström (10^{-10} m)		Mg	magnesium
AA	amino acid (s)		mg	milligram (10^{-3} g)
Ala	alanine		μg	microgram (10^{-6} g)
Arg	arginine		min	minute
Asn	asparagine		mℓ	milliliter (10^{-3} liter)
Asp	aspartic acid		μm	micrometer (10^{-6} meter)
Cys	cysteine		μmol	micromole 10^{-6} mole)
Gln	glutamine		molecules:	
Glu	glutamic acid		CH$_4$	methane
Gly	glycine		CO$_2$	carbon dioxide
His	histidine		CO	carbon monoxide
Ile	isoleucine		H$_2$	hydrogen
Leu	leucine		H$_2$O	water
Lys	lysine		HCN	hydrogen cyanide
Met	methionine		H$_3$PO$_4$	phosphoric acid
Phe	phenylalanine		H$_2$S	hydrogen sulfide
Pro	proline		N$_2$	nitrogen
Ser	serine		NH$_3$	ammonia
Thr	threonine		O$_2$	oxygen
Tyr	tyrosine		mosmol	milliosmol
Val	valine		mRNA	messenger RNA
C	carbon		n	haploid chromosome number
C	cytosine		n$^\pm$	neutron
°C	degrees Celsius		ν	frequency (oscillations per second; nu,
^{12}C	carbon nucleide with mass number 12;			Greek)
	analogously, ^{13}C, ^{14}C		Na	sodium (natrium, Latin)
Ca	calcium		NAD(P)	nicotinamide adenine dinucleotide
cal	calorie (calor, Latin)			(phosphate)
CoA	coenzyme A		NAD(P)H	reduced nicotinamide adenine
cpm	counts per minute			dinucleotide (phosphate)
DNA	deoxyribonucleic acid		NC	nucleotide(s)
Δt°C	freezing point depression in degrees		ADP	adenosine diphosphate
	Celsius		AMP	adenosine monophosphate
E	energy		ATP	adenosine triphosphate
e$^-$	electron		nm	nanometer (10^{-9} meter)
E$_{kin}$	kinetic energy		O	oxygen
ECT	endocytobiont theory		osmol	unit of osmotic pressure
ESH	endosymbiont hypothesis		P	phosphorus
FAD	flavin adenine dinucleotide		P$_i$	inorganic phosphorus
FMN	flavin mononucleotide		p$^+$	positron
Fe	iron (ferrum, Latin)		pH	negative logarithm of the hydrogen ion
G	guanine			concentration (measure of the acidity
G	free energy			or basicity)
g	gram		poly(A)	polyadenylic acid
H	hydrogen		P-P$_{(i)}$	pyrophosphate
h	hour		R	organic residue
He	helium		RNA	ribonucleic acid
J	joule		rRNA	ribosomal RNA
K	potassium (kalium, Latin)		S	sulphur
K	degrees Kelvin		S	Svedberg unit (measure of sedimentation)
kcal	kilocalorie		sec	second
ℓ	liter		T	thymine
λ	wavelength (lambda, Greek)		t	time
Ly	light year		tRNA	transfer RNA
log	logarithm		U	uracil
m	meter		UV	ultraviolet
μ	micro (mu, Greek)			

APPENDIX II. GLOSSARY

The following works have been consulted in the preparation of this glossary:

— *Brockhaus A B C Biologie*. VEBF F. A. Brockhaus Verlag, Leipzig, 1975.
— Dose, K. and Rauchfuss, H., *Chemische Evolution und der Ursprung lebender Systeme*, Wissenschaftliche Verlagsgesellschaft, Stuttgart, 1975.
— Kaplan, R., *Der Ursprung des Lebens*. Georg Thieme Verlag, Stuttgart, 2. Aufl., 1978.
— Rieger, R. et al., *Glossary of genetics and cytogenetics,* Springer Verlag, Berlin, 4th ed.,1978.
— Pschyrembel, W., *Klinisches Wörterbuch,* W. de Gruyter Verlag, Berlin, New York, 251st ed., 1972.
— Tischler, W., *Ökologie,* G. Fischer Verlag, Stuttgart, 1975.

Technical terms used in the text or pertinent to the subject are given short and easily understandable definitions. (↗) Means that the term is defined in the glossary.

abioenzyme abiogenically ↗ formed protein or peptide, usually with low catalytic activity. Abioenzymes are believed to have been capable of splitting sugar phosphates, thereby forming energy-rich diphosphates (Chapter 5, Figures 9, 10, 11).

abiogenic not formed by a living organism.

abioid cell-isomorphic structure or aggregate (system ↗) composed of prebiotic or artificial, abiogenically synthesized elements (abiomers ↗; Chapter 5, Figures 9, 12), produced experimentally or deduced to have existed. Some abioids are/were surrounded by abiogenic ↗ membranes ↗ (abiomembranes ↗ Chapter 5, Figure 8).

abiomembrane membrane ↗ composed of an abiogenic ↗ lipid ↗ double film with abiogenic globular proteins ↗ embedded on both sides, or abiogenic tunnel proteins ↗ penetrating it (Chapter 5, Figures 6, 7, 8). It is the limiting element deduced for the hypothetical plasma ↗ and nucleo abioids ↗.

abiomers abiogenically formed components of the precursors of the first cells (precytes ↗). They include essentially all the basic molecule types present among biomers ↗, i.e., sugars ↗, nucleobases ↗, amino acids ↗ and fatty acids ↗. However, in some cases, their bonds are unphysiological and they include the optical isomers ↗ not formed biologically. Like the biomers, the abiomers can be subdivided into abiomonomers, abiooligomers and abiopolymers (Chapter 3, Figure 4, Tables 3, 4).

abiomonomer see abiomers.

abiooligomer, abiopolymer see abiomers.

abiotic not occurring in organisms and not associated with them.

accessory bacteriocyte, -symbiosis, -symbiont see symbiosis (Chapter 8, Figures 15, 17).

activation the act of bringing a molecule into a reactive condition. For example, amino acids are activated for peptide bond formation in protein biosynthesis ↗ by coupling to transfer RNA molecules.

activation energy the amount of energy which must temporarily be supplied to reacting molecules to start the reaction. The less reactive the substances, the greater the activation energy. Catalysts ↗ can be used to reduce the activation energy considerably.

aerobic living in an oxygen-containing milieu; tolerating oxygen or using it in metabolism.

alpha (α) particles see radiation.

amoeba a member of a subgroup of protozoa ↗ which normally moves by plasma extension (pseudopodia).

amino acids substituent of proteins; organic molecules with at least two functional groups, an acid (-COOH) and an amino (-NH$_2$) group. The side chain (R) determines the character of the amino acid, of which 20 occur naturally (Chapter 5, Table 2).

General formula:

$$\text{General formula:} \quad R-\underset{\underset{\text{H}}{|}}{\overset{\overset{\text{NH}_2}{|}}{C}}-COOH$$

amyloplast an unpigmented plastid ↗ for storage of carbohydrates.

anabolite see metabolite.

anaerobic capable of life in an oxygen-free milieu. Facultatively anaerobic organisms can live in either oxygen-free or oxygen-containing milieus, while obligate anaerobes can survive only in the absence of oxygen, which is poisonous to them.

analogy similarity of organs or organelles with respect only to their function, but not with regard to their historical development, origin or location (opposite of homology ↗).

annihilation tranformation of mass into energy occurring when a particle collides with the corresponding antiparticle ↗.

anthropomorphism the use of human terms and behavior patterns to describe non-human systems or beings.

antibiotic a substance produced by living organisms which inhibits or kills other organisms. Examples are chloramphenicol, tetracycline, penicillin.

anticodon a series of three nucleotides on a transfer RNA ↗ molecule which interacts as a base triplet ↗ with the corresponding complementary base triplet on the mRNA ↗ (codon ↗) through the formation of hydrogen bonds. (See Chapter 4, Figure 7.) Example:

$$\begin{array}{ccc} \text{codon} & \text{U} \quad \text{C} \quad \text{G} \\ & \bullet \quad \bullet \quad \bullet \\ \text{anticodon} & \text{A} \quad \text{G} \quad \text{C} \end{array}$$

antiparticles (also called antimatter) elementary particles which, when they collide with the corresponding particles of ordinary matter, undergo mutual annihilation. The electric charge on an antiparticle is the opposite of the charge on the corresponding particle: the negatively charged electron corresponds to the positively charged positron, but their other properties are identical.

antiquarks see sub-particles.

apatite a group of minerals with the composition

$$Ca_5 \,[(F, Cl, OH, 1/2 \, CO_3) \, (PO_4)_3]$$

assimilation incorporation of a substance such as CO_2, N_2 or H_3PO_4 into the cell substance.

asymmetric carbon atom see (optical) isomer.

atmosphere the shell of gases around a celestial object. The primary atmosphere of the primitive earth was composed of H_2 and He. These were supplemented by volcanic outgassing to form the secondary atmosphere of H_2, H_2O, CH_4 and NH_3, which was changed by constant solar radiation to the tertiary atmosphere of N_2, CO, CO_2, H_2O. The activity of living organisms (photosynthesis) created the quaternary, present atmosphere of the earth: N_2, O_2, CO_2, H_2O. The first three atmospheres were reducing, whereas the quaternary is oxidizing (Chapter 3, Figure 3).

ATP (adenosine triphosphate) energy-rich compound (nucleotide ↗) consisting of adenine, ribose, and three phosphate groups. It is the most common source of chemical energy for cellular metabolism (Chapter 4, Figure 2).

autocatalysis a chemical reaction which produces its own catalyst ↗. A self-reproducing system is also always autocatalytic.

autonomous morphogenesis the creation of form from within; the most important criterion of life, which includes the capacity for self-regulating metabolism ↗, identical reproduction ↗ and inheritable mutability ↗.

autotrophy self-feeding: the state of living only from inorganic substances (inorganic substrate ↗), in particular with CO_2 as the sole source of carbon and light or energy-rich organic or inorganic compounds as energy source. Examples are all photosynthetic ↗ or chemoautotrophic ↗ organisms (Chapter 6, Figure 3).

auxiliary bacteriocyte see symbiosis.

auxiliary symbiosis, auxiliary symbiont see symbiosis.

bacteriocyte eukaryotic host cell with bacteria-like endocytobionts (see also postcyte ↗; Chapter 8, Figure 15).

bacteriome a special organ in a host (symbiont organ) containing its bacteriocytes ↗ (Chapter 8, Figure 2).

bacteriophage a virus which specifically infects bacteria and replicates within them, which finally destroys the bacteria cell.

bacteriorhodopsin protein with bound carotenoid ↗ (chromoproteid) in the membranes of the halobacteria ↗.

basal body a structure at the base of flagella ↗ and cilia ↗ which serves to anchor them in the cytoplasm ↗. The structure of the basal body is similar to that of the centriole ↗ (axial skeletal pattern 9 + 0; Chapter 7, Figure 3).

base pairing see bases.

bases (nucleic acid bases, nucleobases) components of the nucleotides and thus of nucleic acids, which store genetic information in the sequence of their bases. RNA ↗ contains the four bases adenine (A), guanine (G), cytosine (C), and uracil (U); while DNA ↗ contains A, G and C, but thymine (T) instead of uracil (see Chapter 4, Figure 4). Complementary bases can pair by forming hydrogen bonds, the partners determining each other by spatial fit. G normally pairs with C by forming three H-bonds (G≡C), and A with T or U by forming two H-bonds (A=T, A=U; Chapter 4, Figure 7).

base triplet sequence of three nucleic acid bases (codon ↗ — anticodon ↗), which determine one amino acid or the beginning or end of protein biosynthesis ↗ (Chapter 4, Figure 9).

Bdellovibrio bacteria a group of bacteria parasitic on other bacteria.

big bang the generally accepted theory that the expansion of the universe began at a definite time from a state of enormous density and pressure (Chapter 2, Figure 3).

biocenosis a community of plants and/or animals which interact through mutual dependence and influence.

bioelements those chemical elements found in living matter: carbon (C), hydrogen (H), oxygen (O), nitrogen (N), sulfur (S), phosphorus (P), and several trace elements (Chapter 9, Figure 3).

bioevolution phylogenetic development from primitive cells to human beings. The main factors in bioevolution are hereditary variation (mutation ↗) and natural selection ↗ of hereditary variants on the basis of their more successful reproduction in the given milieu (Chapter 1, Figure 3, Chapter 9, Figures 1, 2).

biogenic produced by a living system.

biogenesis generation of life from non-living material under certain physicochemical conditions on the primitive earth = evolution of the first cells (precytes ↗).

biogenetics science of the origin of life or primitive cells (biogenesis ↗).

biogenetic principle a rule first formulated by Haeckel in 1903, according to which ontogenesis (development of the individual from the germ cells) recapitulates, to a certain extent, phylogenesis (development of species).

bioids cell-isomorphic structure not capable of independent life; system ↗ or aggregate of biogenic materials (in contrast to abioids ↗).

biomer biogenically ↗ formed element of organisms or cells. Biomonomers include sugars, amino acids, nucleobases, and fatty acids; biooligomers and biopolymers include the polysaccharides, proteins, nucleic acids, and lipids (biomer contrasts with abiomer ↗).

biotic related to life.

biotope the natural, limited territory occupied by a living community adapted to it.

black hole a region of space containing so much mass that its gravitational attraction prevents even light from escaping. It is believed that some very massive stars collapse into black holes at the end of their existence.

blepharoblast basal body ↗ of a flagellum ↗.

blue-green algae (Cyanophyceae) photosynthesizing ↗ group of prokaryotes ↗.

Calvin cycle the photosynthetic ↗ formation of carbohydrate by fixation of a CO_2 to a pentose bisphosphate, which is then regenerated in a cyclic process. Carbohydrates for other purposes are also withdrawn from this cycle (Chapter 6, Figure 13).

carbon source compound used by an organism to generate biomass (compare with energy substrate ↗).

carbohydrate general term for all sugars and related substances. Low molecular weight forms serve as energy sources, high molecular weight (insoluble) forms as reserve material.

carotenoid orange-colored pigments located in the inner membranes of plastids ↗ and bacteria, etc. Their structures are based on the carotin skeleton, which is composed of eight isoprene units ↗.

carrier a protein which transports a substance across a membrane.

catabolite see metabolite.

catalysis the process of increasing the rate of a chemical reaction by a catalyst ↗. In living organisms, essentially all reactions are catalyzed by enzymes (proteins). They effect catalysis by binding the molecules to be changed (substrate ↗) to a particular area on their surface (active site) in such a way that the reaction occurs quickly and easily.

catalyst a substance which increases the rate of a chemical reaction without being permanently changed itself, and without changing the equilibrium of the reaction.

cell (cyte) the smallest unit of life capable of independent reproduction, function and evolution.

cell wall a rigid layer outside the plasma membrane ↗, surrounding plant cells and bacteria.

centrifugal directed away from the center.

centrioles a pair of organelles lying in the neighborhood of the nucleus ↗ of some animal and flagellated plant eucytes ↗; their structure resembles that of a basal body ↗, from which they can be generated by division, and vice versa. They consist of a membraneless hollow cylinder of nine filaments (triplets of microtubuli, axial skeletal pattern 9 + 0) and are involved in spindle formation (Chapter 7, Figure 3).

centripetal directed toward the center.

centromere (kinetochor) the point of attachment of the spindle fiber to a chromatid, and of the sister chromatids to one another during mitosis ↗ and meiosis ↗. Its structure is not clear; it may have developed from the basal body or the centriole ↗ (see endocytobiotic cell theory ↗, Chapter 7, Figure 3).

chain form spatial arrangement of chain-like macromolecules stabilized by ionic interactions, disulfide bridges, hydrophobic interactions, and H-bond formation (Chapter 3, Table 5).

chelate molecular complexes which arise through interaction of unbound electron pairs with bound H atoms (chelate bonding = "scissors" binding).

chemical bonds: Primary bonds (electron pair formation, ionic bonds, Chapter 3, Table 5) are very strong at room temperature. *Secondary bonds* (hydrogen bonds, hydrophobic interactions, Chapter 5, Table 3) are much weaker. Hydrogen bonds are responsible for the pairing of nucleobases ↗. Hydrophobic interactions account for the tendency of fats or hydrophobic peptide chains to separate out of aqueous solution.

chemoautotroph one of a group of C-autotrophic ↗ bacteria which subsist on completely inorganic sources. Their energy is obtained fom chemical processes, their only carbon source for the synthesis of biomass is CO_2 (Calvin cycle ↗; Chapter 6, Figure 13A).

chemoevolution abiogenic formation of the most important organic compounds, the so-called abiomers ↗, and their self-organization to prebiotic aggregates, or abioids ↗ (Chapter 1, Figure 3; Chapter 3, Figures 2, 4).

chemofossils see fossils, paleobiochemistry (Chapter 6, Table 1).

chemotrophy that form of metabolism possessed by fermenters ↗, respirers ↗, and chemoautotrophs ↗, which obtain their energy (ATP) exclusively through chemical processes (Chapter 6, Figures 5, 9, 16, 19) in contrast to phototrophy ↗.

chlorophyll the green pigment of chloroplasts ↗ and photosynthesizing bacteria.

chloroplasts photoassimilating eucyte ↗ organelles ↗, the most important type of plastid in typical plant cells. They are surrounded by a double membrane and contain DNA in the form of circular chromosomes ↗. Their inner membranes fold inward to form thylakoids, on which the photosynthesis pigments (chlorophyll ↗) are localized (Chapter 7, Figure 5)

chromatids (sister chromatids) the identically replicated halves of a chromosome ↗, which are joined by a common centromere ↗ They separate during mitosis ↗ or meiosis ↗ by division of the centromere to become independent chromosomes.

chromatin DNA-protein complex found in a cell nucleus ↗, intensively colored by basic dyes.

chromoplast a pigment-containing (carotenoids), but non-green plastid.

chromosome in *prokaryotes* ↗: a nearly protein-free circular DNA molecule (double helix ↗), usually associated with the membrane, but lying free in the cytoplasm, i.e., not surrounded by a nuclear membrane. In *eukaryotes* ↗: a thread-like structure, separated from the cytoplasm by the nuclear membrane (nucleus) which contains various RNAs and proteins in addition to the chain-like giant DNA molecules. Each nucleus contains several chromosomes, the number and structure of the chromosomes being species specific.

cilia hair-like motile structures on a cell surface. They resemble flagella in structure, but are shorter and are usually connected under the cell membrane into groups capable of coordinated motion.

citrate cycle (Krebs cycle, tricarboxylic acid cycle) the cyclic degradation of activated acetic acid to CO_2 and H_2O under aerobic conditions. At the same time, redox equivalents, which are respired to form ATP, are generated.

coacervate (biotic aggregate/abioid/preorganelle) liquid precipitate of various kinds of macromolecules ↗ (e.g., gum arabic, histones) in the form of droplets. The oldest precyte model, introduced by Jong and Oparin (1924), in connection with simulation experiments ↗.

coccus round to oval bacterium.

codases (aminoacyl tRNA synthetases, amino acid activases) enzymes which couple the 20 different biotic ↗ amino acids to their respective specific tRNAs, thereby activating them for protein synthesis ↗. They are thus important for the translation of the genetic code ↗.

code, genetic ''dictionary'' for the translation ↗ of genetic information; the scheme which determines how the instructions contained in the genes for the structure of proteins ↗ (transcription ↗) are to be carried out. One code word (codon) consisting of three nucleotides ↗ (base triplet) corresponds to one amino acid. The anticodon is the base triplet on the transfer RNA molecule ↗ which pairs with the corresponding code on the messenger RNA ↗ by means of hydrogen-bond formation (Chapter 4, Figure 7; Chapter 5, Figure 14).

code to determine the amino acid sequence ↗ of a protein in terms of the nucleotide sequence. A gene which codes a particular protein contains the genetic information on the structure of the protein.

codon sequence of three nucleotides (base triplet ↗) on a messenger RNA molecule which corresponds in the genetic code to a particular amino acid.

colloid a substance whose particles or molecules are small enough to remain in suspension indefinitely, but too large to form a clear solution.

companion bacteriocyte see symbiosis.

companion symbiosis, companion symbiont see symbiosis.

compartmentation membranous subdivision of the eucyte into different reaction spaces, e.g., those for heredity (nucleus↗), fermentation (cytoplasm), respiration (mitochondria ↗) and photosynthesis (chloroplasts ↗).

compartment hypothesis proposition that the eucyte ↗ developed directly from one procyte ↗ by successive differentiation ↗ and compartmentation ↗ (Chapter 7, Figure 2).

conjugation parasexual form of transfer of genetic information through cell contact between a donor and a recipient cell. Recombination ↗ of the transferred piece of chromosome with the homologous section of the recipient's chromosome can then occur.

consumer an organism which uses the organic materials produced by the producers ↗ (plants) as nutrients, either directly (herbivores) or indirectly (carnivores).

conversion hypothesis according to this hypothesis, procytes ↗ developed first as fermenters ↗, from which the photosynthesizers ↗ arose, and from these, by loss of the photosynthetic apparatus, the chemoautotrophs ↗ and respirers ↗. The respirers, finally, gave rise to the photergers ↗ (Chapter 6, Figure 1.A, in contrast to the splitting hypothesis ↗).

corpuscular radiation see radiation.

cosmic background radiation macrowave radiation with a temperature of 2.7 K (Kelvin) which arrives uniformly from all directions of space; thought to be weak residual radiation from the big bang (Chapter 2, Figure 4).

cosmic development the development of the universe, including the development of atoms (Chapter 1, Figure 3; Chapter 2, Figures 1-4).

cosmic radiation see radiation.

cosmology the science of the development of the universe.

cosmos universe (Chapter 2, Table 1, Figures 1-3).

crista see mitochondrion.

crossing over exchange of chromatin between two homologous chromosomes ↗ during meiosis ↗. It is significant as a mechanism for recombination ↗ of genetic material.

cultural evolution the development of hominids through the rise of speech and culture (Chapter 1, Figure 3; Chapter 2, Figure 4).

cyanelle endocytobiotic ↗ blue-green alga ↗ with the character of an organelle.

cytochrome an iron-containing organic compound, involved in electron transport during respiration ↗ and photosynthesis ↗ (occurs in mitochondria and chloroplasts).

cytoplasm the protoplasm, including the cell organelles and membrane systems, but not the nucleus.

dalton unit of molecular mass; water, for example, has a molecular mass of 18 daltons.

decomposer an organism which completely degrades organic remains to simple inorganic compounds. Along with producers ↗ (of biomass: plants) and consumers ↗ (of biomass: animals), decomposers (mostly bacteria) are essential members of ecosytems ↗.

deoxyribonucleic acid see nucleic acids.

dictyosome structural and functional unit of the golgi apparatus ↗.

differentiation internal differentiation is the formation of different parts, organs, or organelles in or on an organism or cell.

diffusion the distribution of the molecules of a substance throughout a space, caused by thermal molecular motion. Diffusion depends on the concentration gradient, molecular weight, and temperature. It occurs whether or not the space in question is concurrently occupied by another substance, although its rate can be very much affected by the other substance(s) present.

dinoflagellate representative of a class of algae (dinophyceae) characterized by two flagella, usually different from each other and a clearly visible cell nucleus with permanent chromosomes of a primitive type.

diploid the conditions of a nucleus ↗ (nuclear equivalent ↗) or cell with two complete sets of chromosomes.

dissipation the transformation of more readily usable energy into a less usable form, such as electric energy into heat.

dissociation breakdown of molecules or molecular complexes. For example, the DNA double helix dissociates into single strands on being heated (see Chapter 5, Figure 13).

DNA see deoxyribonucleic acid.

double helix the double twist form of two DNA strands, in which the complementary base pairs lie opposite each other on the inside of the helix. This is the usual state of DNA, but two complementary RNA strands can also form a double helix (see Chapter 4, Figure 6).

duplet code assumed primitive code of the eobionts ↗, in which the codon ↗ and anticodon ↗ consisted of three bases (triplets ↗), of which only the first two bases of each were significant (Chapter 5, Figure 24).

duplicase see polymerase.

duplication structural mutation ↗ of the chromosome ↗, involving a doubling of a gene or section of chromosome.

ecological niche sum of the interactions between species ↗ and environment in an ecosystem ↗.

ecosystem the relationships of organisms to one another and to their environments (biotope ↗, biocenosis ↗).

electrical discharges, (gas discharges) passage of an electric current through a gas, producing ions. In the simulation experiments in which abiomers ↗ were synthesized, various intensities of discharge were used.

electron see elementary particles.

elementary particles term for the simplest known physical system ↗. They cannot be disintegrated by the energies available for experiment; rather they are transformed into each other or disintegrate into other stable elementary particles. The stable elementary particles, the electron (e^-), proton (p^+) and neutron ($n^\pm$), are the substituents of atoms. Other particles have been discovered in cosmic radiation and in nuclear processes (photon γ^0, neutrino ν^0).

endocytobiology a new synthetic field of research between endosymbiosis and cell biology, which analyzes the question if on one side DNA containing cell organelles originate from free-living endocytobionts, or if on the other hand certain endocytobionts have characteristics of DNA-organelles.

endocytobiosis, endocytobiont see symbiosis.

endocytobiotic cell theory the theory that eucytes ↗ evolved by the successive incorporation of various procytes ↗. A fermenting bacterium, as the original host, is thought to have taken up aerobic bacteria (which became the mitochondria ↗) and photosynthesizing blue-green algae (later chloroplasts ↗) as endocytobionts and to have incorporated them. A more comprehensive version suggests that the first endocytobiont was a flagellar symbiont (later locomotive organelles, including the flagellum ↗ basal body ↗, centriole ↗, spindle, centromere ↗), (Chapter 7, Figure 2) in contrast to the compartment hypothesis ↗.

endocytosis uptake of solid (phagocytosis ↗) or liquid (pinocytosis) matter into the cell through invagination of the membrane followed by pinching off of the membrane section to form a vacuole.

endoplasmic reticulum (ER) a system of double membranes ↗ penetrating the entire cytoplasm ↗ of the eucyte ↗, and communicating with the nuclear and plasma membranes, so that the hollow space between the reticular membranes opens to the outside.

endorhythms rhythmically occurring endogenous oscillations in metabolism ↗ (for example, energy metabolism: the ratios of ATP ↗ to ADP and to AMP) of uni- and multicellular organisms (Chapter 8, Figure 10).

energy is the ability of a system ↗ to do work.

energy flow is the production of energy in chemical form and its transfer through an ecosystem ↗, starting with the energy-fixing producers ↗, passage through energy-consuming consumers ↗ to the decomposers ↗, which release the last remaining energy.

energy substrate substrate for the energy metabolism (see H-donors ↗, H-acceptors ↗).

entropy degree of disorder of a system; measure of probability of the state of a system. Random states are more probable than ordered states. Negative entropy is a measure of order.

environmental factors ecological factors which affect the ability of individuals of populations ↗ or biocenoses to survive, and which also affect the expression of genetically determined characteristics (phenotype ↗): *abiotic environmental factors* physical and chemical factors; *biotic environmental factors* interactions of other organisms with the organism in question.

enzyme (biocatalyst ↗, reaction accelerator) a substrate-specific protein ↗ which accelerates a particular metabolic reaction and emerges unchanged. Its particular catalytic ability is based on the special folding of the amino acid chain, which produces a cleft on the surface of the molecule (active center) into which the specific substrates fit in such a way that the reaction occurs much faster than in the absence of the enzyme. The folding of the chain, and thus the specificity of substrate and reaction, is determined by its amino acid sequence.

eobiont some authors use this term for protobionts ↗; here it denotes a representative of primitive cells (precytes ↗) which have developed past the stage of protobionts but have not reached the level of the procytes ↗ (prokaryotes). It would have had about 50 genes, a duplet code ↗ and RNA double helices ↗ for the storage of the genetic code ↗ (Chapter 5, Figure 24).

epigenetic in contrast to syngenetic, not originally present but introduced later; the concept applies to the identification of chemical fossils ↗.

episome additional genetic material (circular DNA ↗) in bacteria which lies free in the cytoplasm or is integrated into the bacterial chromosome ↗ and which determines supplementary characteristics such as resistance to antibiotics or conjugation behavior.

episymbiosis symbiosis ↗ on the surface of an organism or cell.

equilibruim, dynamic see system.

erythrocyte red blood cell, which has in mammals no nucleus.

etioplast plastids ↗ in the leaf cells of plants which have grown in complete darkness, characterized by the presence of a prolamellar body.

eukaryote an organism whose cell(s) contain(s) true nuclei. The eukaryotes include unicellular organisms (protists), fungi, plants and animals. They are separate from the group of (mostly unicellular) prokaryotes ↗ which consists of procytes ↗ lacking true cell nuclei.

eucyte cell of eukaryote ↗ containing a nucleus ↗ (surrounded by a double membrane and containing chromosomes ↗) mitochondria ↗, endoplasmic reticulum ↗ and, in some cases, plastids ↗. The cell nucleus divides by mitosis, thus differing from the procytes ↗, which simply split in two (see Chapter 7, Figure 3).

evolution in the biological sense, the development of organic species through natural selection ↗ (Darwin); in the widest sense, the process which has led from the origin of the universe to humans. It is divided into a chemical, a biological, and a cultural phase. The essential products of these phases are cells, the human organism, and civilizations.

evolution research deductive reconstruction of the historical process of evolution from fossils.

evolution, general mechanisms of those mechanisms which apply to all phases of evolution (chemical, biological, and cultural). We regard the modular construction principle ↗, the phase principle ↗, the periodicity principle ↗, and the triality principle ↗ as general mechanisms (Chapter 1, Figure 3, Chapter 9, Figures 1,2 Table 1).

evolution, theory of derived by Charles Darwin and published in 1859: of the representatives of a system, those which are best adapted to the internal and external conditions will be the most likely to reproduce themselves, i.e., to survive in the "struggle for life".

evolve, to develop through the process of Darwinian evolution.

exocytose excretion of cellular material by extension and pinching off of a segment of membrane.

facultative not essential to existence.

fatty acids see lipids.

fermenters cells capable of fermentation ↗: some bacteria and all eucytes (Chapter 6, Figure 5).

fermentation anaerobic (microaerobic) degradation of sugars to alcohol and carbon dioxide to produce energy (ATP ↗). Fermentation occurs in the cytoplasm (Chapter 6, Figure 5).

flagellates flagellated group of protozoa, divided into plastid-containing phytoflagellates and plastid-free zooflagellates.

flagellum a thread-like extension of many bacteria, flagellates, and motile cells of higher organisms (e.g., sperm cells), which provide locomotion. A bacterial (procyte ↗) flagellum consists of a single filament, while the flagellum of an eucyte ↗ consists of two single filaments surrounded by nine peripheral double filaments (microtubuli ↗; Chapter 7, Figure 3).

fluid mosaic model model for the structure of the membrane ↗ (unit membrane) which predicts that the lipid layer of a membrane behaves as a fluid medium in which the embedded proteins can move rapidly (Chapter 4, Figure 1).

fossils usually petrified impressions or remains of organisms in sediments. There are no fossils in volcanic or metamorphic rocks, or in those which predate life. In addition to biological fossils, there are chemical fossils (chemofossils) which are chemical substances of abiogenic or biogenic origin in geological formations ↗ (Chapter 5, Figure 2; Chapter 6, Table 1).

galaxy from the Greek word for the Milky Way ↗, one of the large groups of stars and gas held together by gravity (Chapter 2, Figure 1).

gametes the male and female reproductive cells, containing the haploid chromosome number.

gamma (γ) rays see radiation.

gas chromatography a method for separating mixtures of substances in which a mixture of fluid and gas is carried in a flow of inert gas through a stationary phase. The technique is used for qualitative and quantitative analysis and for purification of some substances.

gel a colloidal solution in a non-flowing state.

gene (hereditary factor) a section of DNA ↗, often about 1000 nucleotides ↗ long, usually containing the instructions for the synthesis of one protein (structural gene). The base sequence of the gene, translated according to the genetic code ↗ determines the amino acid sequence of the protein, and thus its function, for example as an enzyme ↗. The transfer RNAs ↗ and ribosomal RNAs ↗ are also coded by specific genes.

gene duplication see mutation.

gene mutation see mutation.

gene transfer exchange of genetic material between two separate genetic systems ↗. Gene transfer between the nucleus and mitochondria or plastids of the eucyte are assumed in the endocytobiotic cell theory ↗.

genobiosis life based only on nucleic acids, without protein biosynthesis ↗, or with genotype but no phenotype.

genome the sum of all the genes ↗ in the haploid chromosome set of the nucleus (eukaryote ↗) or the nuclear equivalent (prokaryote ↗).

geological formation layer of rock which was formed during a particular period.

germ cells (gametes ↗) eucytes ↗ involved in reproduction, i.e., egg and sperm cells and their precursors (in contrast to somatic ↗ eucytes).

geyser systems which periodically eject hot water and steam; associated with vulcanism.

globular proteins more or less spherical proteins, which serve in some cases as structural elements of membranes ↗.

glycocalix surface layer on the plasma membrane ↗ consisting of polysaccharide-lipids (glycolipids) and polysaccharide-proteins (glycoproteins) synthesized under genetic control. The glycocalix is species- and immuno-specific and is responsible for cell motion, exchange of materials and cell recognition in tissues ↗.

glycolysis the degradation of glucose to pyruvate (Chapter 6, Figure 6).

Golgi apparatus an eucyte ↗ organelle ↗ with membrane-limited spaces which are stacked like ''plates'' and are associated with secretion and other processes. The functional unit of the Golgi apparatus is the dictyosome ↗.

gram staining method of staining bacteria with taxonomic relevance. Whether or not a cell wall stains is related to its structure.

gravitation property of masses, which are mutually attractive.

gravitational energy: the energy arising from the gravitational force.

H-acceptor see H-donor.

halobacteria a group of bacteria capable of photergy ↗.

haploid cell or nucleus with a single set of chromosomes.

H-donor Oxidation is a process of removal of an electron from an atom or molecule; in organic molecules, the loss of an electron is often accompanied by the loss of an H^+. A substance which is easily oxidized can serve to reduce another compound, in which case the former is an electron donor, and the latter is the electron acceptor. If an H^+ accompanies the electron, as it often does in biological reactions, then the reducing agent is the H-donor, and the compound reduced is the H-acceptor.

helix coiled secondary structure of some macromolecules, such as proteins and nucleic acids. The helix is stabilized by hydrogen bonds or the stacking energy of the nucleobases.

heterotrophy requiring organic substances, in particular an organic carbon source, as food. All animals are heterotrophic (Chapter 6, Figure 3).

heterocyclic bases organic ring compounds containing nitrogen atoms as members of the ring; the N-atoms give the compound basic properties.

histones proteins associated with chromosomes ↗, containing a large fraction of basic amino acids ↗.

holobiosis complete living system with genotype ↗ and phenotype ↗, i.e., with genes ↗ and enzymes coded by them (protein biosynthesis ↗).

homology similarity of organs or organelles due to a common evolutionary history (in contrast to analogy ↗).

host see symbiosis.

hybridization experiments tests in which the ability of RNA or DNA segments to form double helices with nucleic acids of different origin is measured for the purpose of determining the degree of relatedness of the two systems.

hydrolysis splitting of molecules accompanied by uptake of H_2O (e.g., hydrolysis of sugars in glycolysis ↗).

hydrophilic ''water-loving'', soluble in water.

hydrophobic ''water-fearing'', not soluble in water.

hydrosphere the layer of water lying on a celestial body, in a more or less continuous layer, above the lithosphere ↗ (Chapter 3, Figure 3).

hyperbolic growth increasing by the same factor in ever shorter time. If uninhibited, such a system would become infinite within a finite time:

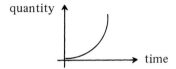

hypercycle the first coupling of working molecules (enzymes ↗) with information molecules (nucleotides ↗) in a reproductive system as a prerequisite for the beginning of life (Eigen). The information molecules (nucleotides) of the cycle reproduce not only themselves, but each one also produces an enzyme as well, which affects the reproduction of the neighbor nucleotide. Only in this way could a self-reproducing system have become stable under the conditions of the primitive earth, thus producing life. Under appropriate conditions, a hypercycle is capable of hyperbolic growth ↗, which means that such a system increases explosively. For this reason, the best-adapted variant of a hypercycle would soon drive all competition from the field. This explains the unity of the genetic code ↗ in all organisms on the earth (Chapter 5, Figure 15, Table 4).

imago adult animal.

infectious form see protoplastoids.

information can be considered to be a form of energy. In cybernetics, information is defined as a quantitative, and thus mathematically definable, property of signs within a prescribed code. Information is the ability of a particle not to interact with another particle in a random way.

intergalactic space the space in the universe between the galaxies ↗

in vitro occurrence of biotic ↗ processes in a test tube, for example the replication ↗ of DNA or protein synthesis ↗ in vitro.

in vivo occurrence of a biotic ↗ process in the intact organism.

ionizing radiation see radiation.

isomers molecules containing the same number and kinds of atoms, but arranged differently (e.g., leucine, isoleucine): *optical isomers* (stereoisomers, antipodes) are mirror images of each other (like the right and left hands), formed by an asymmetric carbon atom. Optical isomers rotate the plane of a beam of polarized light ↗ to equal degrees but in opposite directions, i.e., to the left and right. Examples are D- and L-amino acids, D- and L-ribose.

isoprene a structural unit of natural rubber and other natural compounds, such as the carotenoids:

$$CH_2 = C - CH = CH -$$
$$| $$
$$CH_3$$

isotope dating determination of the age of an object by calculating the ratio of radioactive and radiogenic (arising by radioactive decay) substances.

leucoplast an unpigmented plastid ↗.

L-forms (L stands for the Lisfer Institute, where these forms were discovered by Klienberger) mycoplasma-like ↗ forms of bacteria which lack most of their cell walls, induced by certain antibiotics ↗ (e.g., penicillin).

lichen symbiosis ↗ between a fungus and an alga.

light year the distance travelled by light in one year; at a velocity of about 300,000 km/s, this is about 9.5×10^{12} km (Chapter 1, Figure 2, Chapter 2, Table 1).

light systems I and II (photosystems I and II) closely related pigment systems in the plant cell which trap light energy and convert it to chemical energy (ATP ↗; redox systems ↗). Both systems contain chlorophyll ↗ a and b as main components, located in the thylakoids of the chloroplasts.

lipids collective term for fats and fat-like substances; amphipathic molecules with a water-soluble (hydrophilic ↗) head portion and a water-insoluble (hydrophobic ↗) tail section (Chapter 5, Figure 7). Lipids form monomolecular layers on the surface of water, whereby the hydrophobic part of the molecules point away from the water and the hydrophilic parts toward the water.

lithosphere the sum of the solid layer of a celestial body (Chapter 3, Figure 3).

lithotrophy (inorganotrophy) using only inorganic substrates as nutrients (photosynthesizers ↗, chemoautotrophs ↗; Chapter 6, Figures 3, 15, 19).

luminescence, difference in the difference between the brightness of a star in two different regions of the spectrum, e.g., the blue intensity minus the yellow intensity. The intensities are expressed in logarithmic classes.

lysozyme an enzyme, discovered by Flemming in 1922 in hen's eggwhite and body fluids which dissolves bacteria by hydrolyzing their cell walls (muropeptides).

macromolecules molecules with high (above about 5000) molecular weights, usually consisting of polymerized chains of identical or similar atomic groups (monomers ↗) (Chapter 1, Figure 2).

meiosis two-stage division of a diploid ↗ eucyte ↗ nucleus; in the first stage, the number of chromosomes is halved in a regular fashion (reduction division), and, in the second, the previously formed daughter chromosomes are mitotically distributed among the daughter cells (equation division). Thus four haploid ↗ cells arise from the diploid cell.

membrane (unit membrane) structure surrounding the cell (plasma lemma) and most of its organelles (organelle membranes, endoplasmic reticulum). It is about 10 nm ↗ thick, and appears in the electron microscope (when stained) as consisting of three layers. It is made up of lipids ↗ and proteins ↗. The lipids are almost certainly arranged in a bilayer; the proteins appear to be embedded in this bilayer in most cases (fluid mosaic model: Chapter 4, Figure 1). *artificial membranes* experimentally produced monolayers or bilayers of lipids, in some cases with embedded proteins. When the membranes are dispersed through an aqueous medium in the form of small droplets, they are called liposomes (Chapter 5, Figures 6, 7).

membrane model see artificial membranes.

mesosomes folds in the plasma membrane ↗ of the bacterial cell, morphologically and functionally similar to mitochondria ↗.

messenger RNA (mRNA) single-stranded nucleic acid molecule formed by transcription ↗ of the DNA ↗. It serves as a ''messenger'', bringing the information from the gene ↗ to the ribosome ↗, where it is translated ↗ into a protein molecule. The base sequence determines the amino acid sequence of the protein molecule to be synthesized, via genetic code ↗. Since the genomes of some viruses are single-stranded RNA, these can serve as their own messengers without previous transcription.

metabolism the sum of chemical reactions in a cell; in anabolism ↗, cell components are generated from the nutrients; in catabolism ↗, energy substrate or cell components are broken down.

metabolite individual compound generated and used in metabolism. Catabolites are generated by degradation of other molecules, while anabolites are the products built by the cell.

metaphyte multicellular plant.

metazoa multicellular animals.

meteorites solid fragments of celestial bodies which land on the earth's surface.

microaerophilic (microaerobic) capable of optimal growth at reduced O_2 and high CO_2 concentrations.

micromolecule molecules with molecular weights of less than about 5000; the expression is occasionally used to distinguish smaller molecules (e.g., abiomonomers ↗) from the macromolecules ↗ (e.g., abiopolymers ↗) (Chapter 1, Figure 2).

microsphere (abiotic aggregate ↗, abioid ↗, pre-organelle ↗) cell-like microparticle formed spontaneously when a hot proteinoid ↗ solution is cooled. Introduced by Fox (1965) as ''protocells'' models in simulation experiments ↗.

microtubules tube-like filaments in cells, composed of identical protein subunits arranged in lengthwise fibrils. The centriol ↗, basal body ↗, flagellum ↗, and spindle apparatus ↗ are largely composed of microtubuli (Chapter 7, Figure 3).

Milky Way a traditional name for the band of stars stretching across the night sky, made up of the stars in the equatorial plane of our galaxy.

minimal diet an artificial diet containing the minimum composition and quantities of nutrients.

mitochondrion important eucyte organelle in which respiration occurs, containing a circular DNA chromosome ↗ and surrounded by a double membrane. The inner membrane folds toward the inside to form tubules or cristae, on which the respiratory enzymes responsible for respiratory synthesis of ATP ↗ are located (Chapter 7, Figure 4).

mitosis nuclear division in somatic eucytes ↗ (in contrast to the division of germ cells, meiosis ↗) in which the chromosomes first replicate and the daughter chromatids ↗ are then distributed between the two daughter nuclei by means of the spindle fibrils ↗ (Chapter 7, Figure 3). In this way each new nucleus ↗ contains a copy of every chromosome and thus of every gene.

modification a non-genetic change in the phenotype ↗ of an individual caused by environmental factors.

modular construction principle general principle of evolution, according to which each level of evolution is attained by combination of several representatives of the next lower level (Chapter 1, Figure 3; Chapter 9, Figure 1).

molecular weight the weight of a molecule expressed as a multiple of the weight of a hydrogen atom (dalton ↗).

monomer a small molecule capable of polymerizing to a macromolecule ↗.

monophyletic having arisen from a common ancestor; in contrast to polyphyletic: having arisen from different ancestral forms.

montmorillonite a clay-containing mineral, silicate with a banded structure, with the general composition Al_2 (Si_4O_{10}) $(OH)_2 \cdot n \cdot H_2O$ Al^{3+} can be partly replaced by Mg^{2+}.

multienzyme cycle a complex of two or more enzymes catalyzing successive steps of a biochemical reaction sequence, for example the splitting of sugar bisphosphates and simultaneous ATP ↗ synthesis.

multi-list hypothesis according to this hypothesis, the first cells (primitive cells ↗) arose by combination of several different aggregates (abioids ↗); (Chapter 5, Figure 1.B) in contrast to the multi-step hypothesis ↗.

multi-step hypothesis hypothesis that the first living cells developed linearly and successively from differentiating and complex-forming abiogenically ↗ formed aggregations (abioids ↗); (Chapter 5, Figure 1.A) in contrast to the multi-list hypothesis ↗.

mutability degree to which a gene is subject to mutation, frequently measured as the rate of mutation per gene or cell and unit of time (e.g., generation).

mutant see mutation.

mutation a permanent change in the genetic information. A mutation can involve a change in the number of chromosomes or copies (ploidy) of the genome per nucleus (genome mutation), a structural change in a chromosome (chromosome mutation) or a substitution of one base for another in the DNA sequence (gene mutation, Chapter 4, Figure 13). A mutant is an organism whose genome carries a mutation.

mycoplasms a group of primitive bacteria lacking a cell wall and usually living in the cells of another organism. Most representatives are pathogens of plants or animals.

myxobacteria mobile slime bacteria, not flagellated.

mycetome a special organ in a host ↗ (symbiont organ) containing its mycetocytes ↗.

mycetocyte usually eucytic ↗ cells (postcyte ↗) harboring yeast-like symbionts (endocytobionts ↗). The mycetocytes of a host ↗ can be collected into a special symbiont organ, the mycetome ↗.

neutrino see elementary particles.

neutron elementary particle with zero charge; together with protons, a subunit of the atomic nucleus.

neutron star the remains of an exploded star (supernova ↗); its matter is so dense that it consists of tightly packed neutrons ↗.

nm (nanometer) unit of length: 1 nm $= 10^{-9}$ m $= 10$ Å; 1000 nm $= 1$ μm.

non-histones proteins associated with chromosomes and containing a relatively large fraction of acid amino acids ↗.

nova a star which suddenly becomes, for a short time, one hundred to one million times brighter.

nucleic acid macromolecule built up of nucleotides ↗. There are two kinds, deoxyribonucleic acid (DNA) containing the sugar deoxyribose and the bases adenine (A), guanine (G), cytosine (C), and thymine (T); and ribonucleic acid (RNA) containing the sugar ribose and the bases A, G, C, and uracil (U) instead of thymine.

nucleic-acids-first hypothesis a thesis developed mainly by H. Kuhn (1972) according to which biogenesis began with the nucleic acids (Chapter 5, Figure 1.A), in contrast to the proteins-first-hypothesis ↗).

nucleo abioid hypothetical, abiogenically ↗ formed aggregate of the complexity of a preorganelle ↗, surrounded by an abiomembrane ↗ and possessing the capacity for autocatalytic ↗ simultaneous synthesis ↗ of polypeptides ↗ on polynucleotides ↗, and for RNA replication ↗ (Chapter 5, Figures 12, 13).

nucleobases (nucleic acid bases) see bases.

nucleoid the nuclear ↗ equivalent of bacteria, or the nucleic acid portion of viruses ↗.

nucleoside compound consisting of a nucleobase ↗ bound to a sugar, normally ribose or deoxyribose. Nucleosides are equivalent to nucleotides ↗ minus the phosphate groups (Chapter 4, Figures 2, 3).

nucleoside triphosphate (NTP) nucleoside with three phosphate groups in a chain (Chapter 4, Figure 2). The NTP, with ribose as sugar moiety (ATP, GTP, CTP, UTP) are the energy-rich substrates for RNA synthesis, while the dNTP (dATP, dGTP, dCTP, dTTP) with deoxyribose moieties are polymerized to DNA. They are joined together by excluding a pyrophosphate unit in the course of replication or transcription to form nucleic acids ↗.

nucleotide subunit of nucleic acids ↗, consisting of one of the four nucleobases ↗ (Chapter 4, Figure 3), a sugar (ribose or deoxyribose), and a phosphate group.

nucleus approximately spherical organelle of eucytes ↗, separated from the cytoplasm by a double membrane and containing the chromosomes ↗, i.e., the genome ↗ of the cell. Division normally occurs by mitosis. The nucleus is the "regulator" of the cell, because the genes in it control protein biosynthesis via the messenger RNAs (mRNA). Procytes ↗, in contrast, contain only a nuclear equivalent (nucleoid ↗), a region of the cell plasma not bound by a special membrane, in which the DNA is located. The nucleoid does not divide by mitosis.

nuclear equivalent (nucleoid) see nucleus.

nuclear membrane a system consisting of two unit membranes ↗ which surrounds the nucleus. It contains pores and is connected to the outer membrane via the endoplasmic reticulum ↗.

nutritional endocytobiosis hypothesis according to this hypothesis, the contribution of the endocytobiont ↗ to its host consists of enriching the host's one-sided diet, that is, of supplementing it. The host ↗ and symbiont ↗ are thought to exchange only metabolites and gene products in the form of enzymes, but not the genes themselves (Chapter 8, Figure 1.A), in contrast to the organelle endocytobiosis hypothesis ↗.

nymph (larva) particular stage of animal development between embryo and imago ↗.

obligatory existentially, necessary.

oligomer molecule with a limited number of polycondensed monomers ↗.

ontogenesis development of an organism from the egg to death.

optical isomers see isomers.

organelle part of a cell with a characteristic structure, comprising several types of macromolecules, and a special function in life processes (analogous to the multicellular organ in a higher organism). The most important cell organelles are the membrane ↗, nucleus ↗ (nucleoid ↗), ribosomes ↗, endoplasmic reticulum ↗, mesosomes ↗, Golgi-apparatus ↗ (dictyosomes ↗), cell wall ↗, flagellum ↗, centriole ↗, mitotic spindle ↗, mitochondria ↗, plastids ↗.

organelle endocytobiosis hypothesis assumption that the endocytobionts serve not only to compensate for the one-sided nutrition of their host (nutritional endocytobiosis hypothesis ↗), but also assume functions of the DNA-containing cell organelles, mitochondria, and plastids (e.g., regulation of pH, osmotic pressure, and endorhythms). According to this hypothesis, this degree of integration presupposes a degree of genetic transfer ↗ between host and symbiont in the course of their mutual development (Chapter 8, Figure 1.B).

organotrophy feeding on organic substrate (fermenters ↗, photergers ↗, respirers ↗; Chapter 6, Figures 3, 5, 9, 10, 16).

osmosis diffusion ↗ through a semipermeable membrane ↗.

oxidative phosphorylation formation of ATP ↗ in the course of oxidative respiration ↗.

paleobiochemistry field of biochemistry or paleontology concerned with the collection and determination of bio-chemically important compounds in ancient sediments.

panspermia hypothesis the hypothesis suggested by S. Arrhenius, stating that the germs of life are transferrable from one celestial body to another, possibly through the radiation pressure of light.

parasexuality see conjugation.

parasitism a form of interaction or living together of different organisms to the one-sided benefit of one of them, and the detriment of the other (in contrast to symbiosis ↗).

Pasteur effect effect discovered by L. Pasteur (1822—1895) in which anaerobic fermentation ↗ is repressed in the presence of oxygen. This may be caused by pre-empting of inorganic phosphate by respiration ↗, so that it is no longer available for fermentation.

peptide, peptide bonding bond between two or more amino acids ↗ in which the COOH group of one bonds to the NH_2 group of the next, excluding H_2O. Proteins are chains of amino acids linked by peptide bonds (Chapter 3, Table 4).

peptidase protein ↗ component of the ribosome ↗ which catalyzes ↗ the bonding of the activated amino acids ↗ delivered by the tRNAs ↗ (Chapter 4, Figure 11).

periodic system see periodicity principle.

periodicity principle hypothesis, according to which evolution proceeds periodically, as if in an ever-widening spiral. With every turn of the spiral the number of species increases, and in addition, the same constructions are used repeatedly, at ever higher levels (for example the periodic systems of the elements or of cells: Chapter 9, Figures 3, 5, 7, Table 1).

phagocytosis uptake of solid substances by endocytosis ↗.

phase principle the hypothesis that evolution always proceeds in alternating divergent and convergent phases. In the divergent phases, many variants with similar chances of survival arise, until one variant suddenly fulfills a new function. This brings the beginning of a convergent phase, which is strongly selective. Those variants which best perform the new function survive (Chapter 3, Figure 5; Chapter 5, Figure 29; Chapter 9, Figure 2).

phenotype the sum of all characteristics of an organism, generated through the interaction of the genotype ↗ and the environment.

phosphorylation addition of a phosphate residue to a compound, for example, ADP + phosphate → ATP.

photerger cell capable of photergy ↗. The only known modern representatives are the halobacteria (Chapter 6, Figure 10).

photergy a type of metabolism using a light-driven proton pump. Membrane-bound chromoproteins (bacteriorhodopsin ↗) absorb light, which leads to the formation of an electro-chemical protein gradient. The energy stored in the gradient is used for ATP ↗ synthesis (Chapter 6, Figure 10).

photophosphorylation a reaction in which light is used as the energy source for phosphorylation ↗.

photosynthesis (photoassimilation) formation of sugars from water (H_2O) and carbon dioxide (CO_2) with light as a source of energy, and releasing O_2. Some bacteria, all blue-green algae, and green plants photosynthesize (chloroplasts ↗ as photosynthesis organelles; Chapter 6, Figure 15).

photosynthesizers (photoassimilators) cells capable of photosynthesis ↗ (photoassimilation): purple bacteria, blue-green algae, and plant cells with special photosynthesis organelles, the chloroplasts ↗ (Chapter 6, Figure 15).

phototrophy extraction of energy (ATP ↗) through photergic processes (Chapter 6, Figures 3, 10, 15) by photergers ↗ and photosynthesizers (in contrast to chemotrophy ↗).

phylogenesis, phylogeny historical development of species, in particular the branching of the hereditary lines through species differentiation, or accumulated genetic changes.

physiochemical types see physiochemistry.

physiochemistry (physicochemistry) the chemical (ions, molecules) and physiological (including pH, osmotic pressure) properties of the intra- and extracellular fluids of organisms and cells. There are certain phylogenetic regularities in the ratios of inorganic to organic components, and of ions to molecules, on which the three basic physiochemical types are based (Chapter 6, Figure 2; Chapter 8, Figure 11, Table 2).

plane polarized light light which oscillates in only one plane. It is composed of equal portions of left and right circularly polarized light.

plasma abioid hypothetical abiogenically formed aggregate of the complexity of a preorganelle ↗. It is assumed to have been enclosed by an abiomembrane and to have been capable of primitive sugar hydrolysis coupled to the formation of high-energy phosphates, catalyzed by abioenzymes ↗ (Chapter 5, Figures 9, 10, 11).

plasma lemma (plasma membrane) see membrane.

plasmid see episome.

plastids plant organelles, including green chloroplasts (location of photosynthesis ↗) and their derivatives, the colorless chlorophyll-free leucoplasts or amyloplasts (for carbohydrate storage), and the red and yellow chromoplasts (ATP synthesis through photergy ↗ ?).

polarized light light which oscillates in only one plane (isomers ↗, optical).

polymer macromolecule ↗ composed of many identical or similar groups of atoms connected as in a chain (monomers ↗).

polymerase enzyme catalyzing polymerization or its reverse. RNA and DNA polymerases, for example, polymerize nucleotides to RNA and DNA, respectively. Duplicases (replicases ↗) are polymerases which synthesize a nucleic acid strand complementary to the originally present, template strand (Chapter 4, Figure 8).

polymerization chemical reaction through which monomers ↗ are joined to form polymers ↗.

polynucleotide polymer ↗ of nucleotides ↗. A polynucleotide containing only one kind of base, such as adenine, would be abbreviated poly(A) (Chapter 4, Figures 5, 6).

polypeptide polymer of amino acids joined by peptide bonds ↗.

polyphyletic see monophyletic.

polyploid state of possessing more than two sets of chromosomes.

population the sum of the individuals of a species ↗ living in a particular continuous biotope ↗, in general displaying genetic continuity over several generations.

porphyrins ring-shaped compounds, composed of four pyrrole rings; important natural compounds such as heme, chlorophyll ↗ and the cytochromes ↗ are porphyrins.

positron see elementary particles.

postcyte cellular system that is the host for intracellular symbionts ↗ like prokaryotes ↗ (e.g., bacteriocyte ↗) or eukaryotes as yeast (e.g., mycetocyte ↗). The formation of postcytes can be considered to be a continuation of the eukaryotic cellular evolution.

prebiotic occurring before the development of life.

Precambrium the oldest period in the earth's history, from the origin of the earth about 4.5×10^9 years ago to about 6×10^8 years ago.

precyte (abioid, preorganelle, precell, primitive cell, protocell) all cell-like structures arising through self organization ↗ of completely or partly abiotically ↗ formed constituents. The term applies to any phylogenetic precursor of living cells, such as plasma abioids ↗, nucleo abioids ↗, protobionts ↗ and eobionts ↗.

preorganelles organelle-like structures formed abiotically ↗ by self-organization ↗ of macromolecules ↗. They were not phylogenetic precursors of modern organelles. Examples: abiomembrane ↗, nucleo abioid ↗, plasma abioid ↗ (Chapter 5, Figures 8, 9, 12). Protoorganelles, in contrast, were the precursors of organelles (for example, primitive ribosomes, Chapter 5, Figure 25).

primal soup the primitive seas, lakes, and ponds, whose waters were solutions (probably not very concentrated) of abiomers ↗ (Chapter 3, Figure 4).

primary bond see chemical bonds.

primary bacteriocyte see symbiosis.

primary structure, protein see sequence.

primary symbiosis, primary symbiont see symbiosis.

primitive cells see precyte.

primitive gene (protogene) abiogenic polynucleotide ↗ of a protobiont ↗ whose base sequence was translated into a polypeptide ↗ in the translation apparatus ↗ of the protobiont.

primitive glycolysis hypothetical, primitive breakdown of sugar bisphosphates by ur-enzymes ↗ with the simultaneous formation of energy-rich diphosphate (Chapter 5, Figure 21).

primitive procytes the first procytes ↗ with fermentation ↗, according to one hypothesis, derived from the eobionts ↗ (Chapter 5, Figure 27).

prokaryote an organism composed of one, or rarely, several procytes ↗ without true cell nuclei. Instead the genome consists of a circular, protein-free DNA molecule in a membrane-less nucleoid. Procytes ↗ also lack mitochondria, endoplasmic reticulum, plastids, and other organelles found in the eukaryotes ↗; prokaryotes do not divide by mitosis ↗ or meiosis ↗. Examples are bacteria and blue-green algae (Chapter 7, Figure 1).

procyte cell of a prokaryote ↗, generally without internal compartmentation ↗, with a membraneless nuclear equivalent (nucleoid ↗) instead of a true nucleus, no mitosis ↗ or meiosis ↗, but simple crosswise cell division.

producers all plants which form organic material (biomass) photosynthetically or chemosynthetically from inorganic substances.

protobiont hypothetical primitive cell (precyte ↗), the first capable of reproduction, formed by integration of various preorganelles ↗ (aggregates, abioids ↗) which had arisen from abiotic materials. The protobiont is hypothesized to have had a simplet code, about 15 genes and RNA double helices for storage of information (Chapter 5, Figure 15).

protoenzyme see ur-enzyme.

protons see elementary particles.

protoplastoids primitive, primarily intracellular prokaryotes, newly recognized as a group, whose representatives have two structural forms (infectious and vegetative) as do the mycoplasms ↗ and rickettsiae ↗ (Chapter 7, Figure 8).

protoprosthetic group (pre-coenzyme) phylogenetic ↗ precursor of the prosthetic group ↗.

protozoa eucytic one-celled organisms.

proplastid a plastid ↗ which has not yet fully developed.

prosthetic group a metal ion or a small organic compound bound to an enzyme which is active in catalysis.

protein long chain of various amino acids joined by peptide bonds (more than 100). Along with the nucleic acids, proteins are the most important functional molecules (biopolymers) of living cells. They are synthesized through translation ↗ of the genetic information of the genes ↗ into amino acid sequence.

protein biosynthesis the genetically controlled synthesis of the proteins of an organism. Protein biosynthesis consists of transcription ↗ and translation ↗.

proteinoid protein-like material created by Fox under simulated primitive-earth conditions by heating amino acid mixtures on hot lava. Material from which microspheres ↗ were formed.

proteins-first hypothesis suggestion by S. W. Fox (1965) that biogenesis started with proteins (Chapter 5, Figure 1.B).

protists (unicellular organisms, protozoa) eukaryotic ↗ one-celled organisms and other little-organized eukaryotes.

pulsar (pulsing star) probably a very rapidly rotating neutron star ↗ from which we receive very regular pulses of radio waves.

pulsation theory a hypothesis formulated in 1963 by Hughes, according to which the cosmos periodically contracts and, after the next big bang ↗, expands.

purine member of a class of compounds based on two fused rings, one with six atoms and one with five atoms, each containing both C and N atoms (heterocyclic compounds ↗). The nucleobases adenine and guanine are purines, as are coffeine and uric acid (Chapter 4, Figure 4).

purple bacteria photosynthetic group of bacteria.

pyrimidine one of a class of compounds based on a six-membered heterocyclic ↗ ring containing four C and two N atoms. The nucleobases ↗ cytosine, uracil, and thymine are pyrimidines (Chapter 4, Figure 4).

quarks see sub-particles.

quasar (*quasi*stellar radio source) objects with puzzlingly high radiation, believed to be the most distant objects in the universe.

quaternary structure the three-dimensional structure of a protein made up of subunits.

radiation:

    (1)*corpuscular radiation* all radiation consisting of particles in motion, such as atoms, molecules, electrons, ions, neutrons and mesons.

    (2)*cosmic radiation* high-energy corpuscular radiation from space, consisting of 90% protons, 9% alpha (α) particles and 1% heavy ions.

(3)*ionizing radiation* forms of radiation which impart so much energy to the matter with which they interact that electrons are stripped from the atoms, producing ions.

(4)*radioactive radiation* radiation arising from radioactive decay (α and β particles, γ rays). α particles are equivalent to $^4He^{2+}$, β particles to e^-, and γ rays to high-energy electromagnetic waves.

(5)*ultraviolet radiation* electromagnetic radiation $\nearrow$ with wavelengths shorter than the visible range. UV includes the interval from about 5 to 400 nm.

radical, chemical very reactive molecule or atom with one or more unpaired electron(s).

radioactivity see radiation.

recent modern, designating the present geological epoch, i.e., since the end of the pleistocene.

recombination, genetic formation of new combinations of genetic information, for example through exchange of sections of homologous chromosomes during meiosis $\nearrow$, a process called crossing-over $\nearrow$. The new combination is then tested by selection $\nearrow$.

red giant an enormously expanded star at the end of its development, which radiates mostly red light.

red shift of spectral frequencies the shift in the wavelengths of known spectral lines toward the red end of the spectrum, caused by the Doppler effect as stars move away from us.

redox system a system in which one partner donates, and the other accepts an electron. In the cell, a series of cytochromes aligned on the membrane can transfer an electron from one cytochrome to the next, each one being first reduced (accepting the electron) and then oxidized as it passes the electron on to the next.

redundant DNA segments segments of DNA with identical nucleotide $\nearrow$ sequence arising by duplication $\nearrow$.

relativity, general theory of a (field) theory of the cosmos developed by Einstein, based on the postulate of equivalent distribution of matter in space and the geometry of space, and predicting certain properties of gravitation (Chapter 2, Figures 2, 3).

repetitive sections of DNA nearly identical sections of DNA, of which several copies are present.

replicase enzyme which catalyzes the replication of DNA (in precytic systems, of RNA) (Chapter 4, Figure 8).

replication (reduplication) the synthesis of a "daughter" strand of nucleic acid, complementary to the "parent" strand on which it is synthesized by base pairing between the parent nucleotide and the newly added daughter nucleotide. DNA is replicated semiconservatively, meaning that when the two strands separate, a new daughter strand is synthesized for each (Chapter 4, Figure 8).

reproduction formation of new organisms by old ones (parents), in which the characteristics of the parents are inherited by the offspring.

respiration (aerobic) respiration in the presence of oxygen leading to energy production (ATP $\nearrow$) through oxidation of organic compounds (redox systems $\nearrow$; Chapter 6, Figure 16).

respiratory chain biological redox system $\nearrow$ for the production of ATP $\nearrow$ from ADP and inorganic phosphate. In eucytes, it is located on the inner mitochondrial membrane (Chapter 6, Figure 17A).

respirer cells which are capable of respiration $\nearrow$: bacteria, all eucytes $\nearrow$ with special respiratory organelles, the mitochondria $\nearrow$ (Chapter 6, Figure 16).

ribosome cell organelle, present in large quantities, consisting of two subunits. These are composed of RNA and globular proteins. The ribosome serves as a universal "printing press" in protein biosynthesis $\nearrow$ (Chapter 4, Figure 10).

ribosomal RNA (rRNA) see ribonucleic acid.

ribonucleic acid nucleic acid $\nearrow$ consisting of nucleotides $\nearrow$ with ribose as the sugar moiety. RNA has a number of cellular functions. Ribosomal RNA (rRNA) is a constituent of the ribosomes $\nearrow$, transfer RNA (tRNA) mediates between the messenger and activated amino acids $\nearrow$ (Chapter 4, Figures 11, 12). Messenger RNA (mRNA) carries the genetic information from the DNA to the ribosomes, where protein synthesis occurs.

rickettsiae primitive, usually pathogenic group of bacteria, which exist intracellularly and are of varying forms (pleiomorphy).

RNA see ribonucleic acid.

saccharide (carbohydrate) compound made up of one or more simple sugar $\nearrow$ units (mono-, di-, oligo-, and polysaccharides).

secondary bonding (coordinative bonding) see chemical bonds.

secondary structure, protein the folding of polypeptide chains into pleated sheets or their coiling into α-helices.

sedimentary rock rock which was formed as a sediment under the water surface and later compressed and hardened.

selection one of the two main forces in evolution; the other is mutation. Selection denotes the survival and successful reproduction of those variants of a species which are best adapted to their environment.

self-organization of matter the ability of some macromolecular systems to aggregate spontaneously under the appropriate conditions. The 50 different subunits of ribosomes, for example, have been shown to aggregate spontaneously. The formation of preorganelles $\nearrow$ is thought to have occurred similarly.

self-reproduction ability to direct the formation of an identical object; for example, the reproduction of a nucleic acid molecule, catalyzed by an enzyme.

semiconservative replication see replication.

semipermeable membrane membrane which can be penetrated by some molecules, such as water, but not by others, such as proteins ↗.

sequence the order in which different monomers ↗ are arranged in a macromolecule ↗. In nucleic acids, it is the order of the four kinds of nuleotides (genetic information), and in proteins, the specific order in which the 20 different kinds of amino acids are arranged (which determines the form of the protein, and thus its function).

sexuality all morphological and functional properties associated with sexual reproduction.

simplet code hypothesized primitive code of the protobionts ↗ in which the codon and anticodon consisted of three bases each (triplet ↗), whereby only the middle bases had to be complementary (Chapter 5, Figure 15).

simulation experiment experiment carried out under conditions intended to reproduce those existing on the primitive earth. The first such experiment was carried out by S. Miller in 1953 (Chapter 3, Figure 1).

sol a colloidal ↗ solution with the properties of a viscous liquid.

somatic cells all the eucytes in an organism with the exception of the egg and sperm cells (gametes ↗) and their precursors.

species a unit of classification defined as a natural, continuous breeding community, or as the sum of those individuals which share all essential characteristics with each other and with their offspring.

spectrum the electromagnetic spectrum includes the entire range of electromagnetic waves from the longest radio waves to the shortest-wavelength photons in cosmic rays. ''Light'' is conventionally divided into infrared, including heat waves with wavelengths from 0.75 to 1.4 μm, visible light from 400 to 750 nm, and ultraviolet from 5 nm to 400 nm.

spindle the structure formed by the microtubuli attached to the chromosomes during mitosis or meiosis. The spindle is usually shaped like two opposing cones meeting at their bases (the equator); the microtubuli draw the chromatids apart toward the apexes of the cones, where the two daughter nuclei form.

spirillum spiral-shaped, rather inflexible bacterium.

spirochetes a group of long, spiral-shaped, flexible bacteria with characteristic axis filaments and coat.

splitting hypothesis according to this hypothesis, the first procytes ↗ to evolve were fermenters ↗, from which the anaerobic respirers arose. These then diverged into four branches, the anaerobic/aerobic photergers ↗, the anaerobic/aerobic photosynthesizers ↗, the anaerobic/aerobic chemoautotrophs ↗, and finally the anaerobic/aerobic respirers ↗ (Chapter 6, Figure 1.B) in contrast to the conversion hypothesis ↗.

spontaneous generation the idea, held by Aristotle (384-322 B.C.) that life is continuously arising from dirt and mud.

spore structure produced by procytes ↗ (e.g., bacteria) and eucytes ↗ (e.g., fungi, protists), surrounded by a tough, multilayered wall, for the purpose of reproduction and survival under adverse conditions.

sporozoa a subgroup of the protozoa ↗ which forms spores ↗.

steady-state hypothesis a proposal, now outdated, that the expanding universe continuously creates new matter, so that the average density of matter in the universe remains constant.

stromatoliths calcium-rich fossils ↗ of algae-like organisms.

substrate the substance caused to react by an enzyme ↗. In nutritional physiology, the substrate is food.

substrate specificity the property of many enzymes, which catalyze the reaction only of particular compounds; there is a specific enzyme for each compound.

substrate-level phosphorylation formation of ATP ↗ in the course of anaerobic fermentation ↗.

sub-particle subunits of elementary particles, predicted by physicists but not yet proved to exist: quarks and antiquarks.

sugar a carbohydrate; a sugar contains an aldehyde or ketone and one or more hydroxyl groups. The names of sugars end in -ose; ribose and deoxyribose are the most important 5-carbon sugars (pentoses); glucose and fructose are the major 6-carbon sugars (hexoses).

sugar splitting, primitive see plasma-abioid.

supernova the explosion of a massive star, in which hundreds or millions of times the energy of a normal nova ↗ is released.

Svedberg (S) a unit in which sedimentation constants are expressed. The sedimentation constant is related to the molecular weight of a macromolecule, but not linearly.

symbiont see symbiosis.

symbiosis a mutually beneficial partnership of organisms of different species. The concept was introduced in 1879 by De Bary. In analogy to parasitic relationships, the larger partner is called the host, and the smaller partner the symbiont. A special form is intracellular symbiosis, or endocytobiosis, in which the symbiont, usually a procyte, lives in the cytoplasm or nucleus of the host cell, which is usually a eucyte (postcyte ↗) (see also endocytobiotic cell theory ↗). A mutually beneficial partnership of postcytes ↗ with symbionts ↗ of the photergic ↗ type could be characterized as primary symbiosis, those with photosynthesizing symbionts as auxiliary symbiosis, those with respiring symbionts as companion symbiosis, and those with fermenting symbionts as accessory symbiosis.

syncyanosis partnership between blue-green algae ↗ and zooflagellates ↗, fungi, or plastid-free phytoflagellates ↗.

syngenetic see epigenetic.

synthetase an enzyme catalyzing a synthesis reaction.

system in thermodynamics, an open system is one which exchanges matter and energy with the surroundings. It never reaches a true thermodynamic equilibrium, but only a steady state (expression for a dynamic steady state: $\Delta G \neq 0$). A closed system, by contrast, is one which exchanges neither matter nor energy with the surroundings.

tertiary structure the three-dimensional folding of a polypeptide chain, which may also include the relationships of coiled regions with the rest of the molecule.

thermodynamics a discipline concerned with the various forms and transformations of energy.

thylakoid see chloroplast.

tissue functional association of cells differentiated in the same way (eucytes ↗).

transfer RNA (tRNA) see ribonucleic acid.

transcriptase enzyme ↗ which catalyzes transcription ↗; that is, the synthesis of RNA from a DNA template (Chapter 4, Figure 11).

transcription the enzymatic (transcriptase ↗) process in which the information on a DNA strand is "rewritten" in the complementary base sequence of an RNA strand (mRNA ↗, rRNA ↗) (Chapter 4, Figure 11).

translation enzymatically controlled protein biosynthesis on the ribosomes ↗ in which the base sequence of the RNA bound to the ribosome is translated in terms of the genetic code into the amino acid sequence of the protein (Chapter 4, Figure 11).

transport, active movement of a substance through a membrane ↗ against the concentration gradient. The process requires energy and is accomplished by carriers ↗ (including tunnel proteins ↗); in contrast to passive transport, which is driven only by the concentration gradient (diffusion ↗).

triality principle the hypothetical, most general mechanism of evolution, working in the widest sense according to the Hegelian principle of thesis, antithesis, and synthesis, both in the individual subsystems and in the whole system. It is also reflected in the structure of evolving systems, which finds its clearest expression in the periodic systems of the individual levels of evolution (Chapter 9, Figures 8-10).

tubulus see mitochondrion.

tunnel protein proteins extending all the way across the biological membrane ↗ which provide hydrophilic pores, through which some substances may be transported (compare Chapter 5, Figure 8).

ultraviolet radiation (UV) see radiation.

unit membrane see membrane.

unphysiological bonding chemical bonds between abiomers ↗, especially in oligomers or polymers, which are not found in biologically generated biomers ↗.

ur-enzyme the first proteins with enzymatic function produced under genetic control by primitive cells (precytes ↗). The protobiont ↗ is thought to have had ur-enzymes for the splitting of sugars with substrate phosphorylation (primitive glycolysis ↗; Chapter 5, Figure 21).

vegetative form see protoplastoids.

virus submicroscopic nucleic acid strands without metabolism, but usually with a protein coat in the infective stage and requiring a host cell in which to reproduce. When the virus infects a cell, it injects its own genetic material, which then takes over the control of cell functions.

white dwarf a very small star with a high temperature. The white dwarf condition is thought to be the last stage in the development of a normal star.

AUTHOR INDEX

A

Allsopp, A., 130, 162 (ref. 1)
Altmann, R., 128, 162 (ref. 2)
Arnold, C. G., 145, 150, 162 (refs. 3—7)
Avery, O. T., 43 (ref. 1)
Ayala, F. J., 4 (ref. 1)

B

Baars, S., 88 (ref. 28)
Baldwin, R. L., 88 (ref. 51)
Barckhausen, R., 29 (ref. 1), 36, 39, 43 (ref. 2),
162 (ref. 7a)
Barrow, J. D., 17 (ref. 1)
Barton, I., 192 (ref. 1)
Berckhemer, H., 29 (ref. 2)
Bergamini, D., 17 (ref. 2)
Berger, J., 88 (ref. 47)
Berthold, M., 162 (ref. 8)
Bogorad, L., 130, 162 (ref. 9)
Bonen, L., 147—149, 162 (refs. 10—12)
Bonik, K., 4 (ref. 2)
Brandt, S., 17 (ref. 3)
Bresch, C., 4 (ref. 3), 15, 17 (ref. 4), 38, 43 (ref.
3), 87 (ref. 1), 212 (ref. 1)
Breuer, R., 17 (ref. 5)
Broda, E., 91, 98, 126 (refs. 1, 2)
Brown, R. M., 128, 143, 144, 164 (ref. 62)
Buchanan, R. E., 126 (ref. 3)
Buchner, P., 165, 168, 192 (ref. 2)
Buetow, D. E., 162 (ref. 13), 164 (ref. 91)
Buvet, R., 29 (ref. 3)
Bystrome, B. G., 213 (ref. 34)

C

Calvin, M., 29 (refs. 4, 5), 52, 53, 87 (refs. 2—4,
22)
Capra, F., 4 (ref. 4)
Carr, N. G., 163 (ref. 55)
Cavalier-Smith, T., 130, 149, 162 (ref. 14)
Chambon, P., 162 (ref. 15)
Chardin, T., 212 (ref. 2)
Chase, D., 163 (ref. 47)
Clarke, B., 212 (ref. 3)
Cohen, S. S., 162 (ref. 16)
Cook, C. B., 162 (ref. 17), 192 (ref. 3)
Cramer, W. A., 126 (ref. 4)
Crick, F. H. C., 35, 43 (ref. 11)
Cunningham, R. S., 162 (ref. 12)
Czihak, G., 212 (ref. 4)

D

Darwin, C., 2, 4 (ref. 5), 31, 73, 208
Dayhoff, M. O., 105, 147, 149, 164 (ref. 64)
Decker, W. A., 87 (ref. 23)
Dehnen, H., 10, 16, 17 (ref. 6)
DeJong, H. G. B., 47, 87 (ref. 23)
Dickerson, R. E., 105, 118, 122, 126 (ref. 5), 162
(ref. 18)
Dill, B. C., 163 (ref. 32)
Dobereiner, I. W., 198
Dobzhansky, T., 4 (ref. 7)
Dodson, E. O., 134, 162 (ref. 19)
Doolittle, W. F., 147—149, 162 (refs. 10—12)
Dose, K., 21, 29 (refs. 6, 7), 45, 51, 87 (refs. 5,
19), 95, 126 (ref. 6)
Drews, 162 (ref. 21a)
Duchateau, G., 180, 192 (ref. 4)
Duthoit, J. L., 193 (ref. 34)
Duysens, L. N. M., 126 (ref. 7)

E

Eccles, J. C., 208, 211, 212 (ref. 5), 213 (ref. 6)
Egel-Mitani, M., 164 (ref. 73), 194 (ref. 40)
Ehrhardt, P., 165, 193 (refs. 5, 6)
Eigen, M., 45, 46, 56, 57, 70, 72—74, 87 (refs.
6—13), 162 (ref. 20), 208, 213 (ref. 7)
Einstein, A., 12, 13, 17 (ref. 7), 198
Elson, E. L., 88 (ref. 51)
Epstein, W., 126 (ref. 8)

F

Famintzin, A. S., 128
Florkin, M., 192 (ref. 4)
Folsome, C. E., 87 (ref. 16)
Forsterling, H. D., 87 (refs. 14, 15)
Fox, C. F., 89 (ref. 64)
Fox, G. E., 147, 148, 164 (ref. 88)
Fox, S. W., 29 (ref. 7), 47, 49—51, 87 (refs. 17—
19)
Franzen, J. L., 5 (ref. 14)
Fredrick, J., 193 (ref. 7)
Fredrick, J. F., 105, 126 (ref. 9), 147, 148, 162
(ref. 21)
Freund, F., 20
Fuchs, G., 106, 126 (ref. 34)

G

Gaffal, K. P., 162 (ref. 7)
Gaffron, H., 126 (ref. 10)
Garret, R. A., 56, 87 (ref. 20)
Georgi, H., 211, 213 (ref. 8)

SUBJECT INDEX